CATI
OSCI
AND

THE CATHODE-RAY OSCILLOSCOPE AND ITS USE

G. N. PATCHETT
B.Sc.(Eng.), Ph.D., C.Eng., F.I.E.E., F.I.E.R.E., M.I.E.E.E.

Professor of Electrical Engineering
University of Bradford

LONDON
NORMAN PRICE PUBLISHERS LTD

NORMAN PRICE (PUBLISHERS) LTD
17 TOTTENHAM COURT ROAD, LONDON, W1P 9DP

ISBN 0 85380 054 5

Printed in Great Britain by
Biddles Ltd, Guildford, Surrey

ACKNOWLEDGEMENTS

The author wishes to thank the following for supplying literature on their equipment:

COSSOR ELECTRONICS LTD
DATA LABORATORIES
GRUNDIG ELECTRONICS
HEWLETT PACKARD
MEDELEC LTD
PYE UNICAM LTD (PHILIPS)
SCHLUMBERGER INSTRUMENTS
SHACKMAN INSTRUMENTS LTD

and particularly the following for more detailed information and demonstration equipment:

ADVANCE ELECTRONICS LTD (now GOULD ADVANCE LTD)
SCOPEX INSTRUMENTS LTD
TEKTRONIX INC (INCLUDING TELEQUIPMENT).

The author also wishes to thank the University of Bradford for their help and for the facilities provided to produce the oscilloscope trace photographs.

AUTHOR'S PREFACE

The purpose of this book is to explain in terms as simple as possible the theory, construction and use of the cathode-ray oscilloscope. The oscilloscope is a very powerful tool, both in fault-finding and design, and a reluctance to use it shown by some technicians is regrettable.

It is essential that one gets to know how to use the equipment and the author feels that one cannot competently do so without knowing something about its principles and possible limitations. In no way is this book intended to supplant the instruction book usually provided with the oscilloscope. One must first study the instructions with care and then use the instrument to become familiar with the controls. All too frequently this is not done and in consequence the user either finds the instrument difficult to use or obtains incorrect or misleading results.

The versatility of the cathode-ray oscilloscope and its use in so many fields imposed limitations on coverage and many special types are not included. Owing to the very extensive range of oscilloscopes, explanations in general terms presented difficulties as did quoting typical figures for sensitivity, etc. It should therefore be appreciated that some oscilloscopes may operate rather differently and that some may have figures quite different from those quoted.

CONTENTS

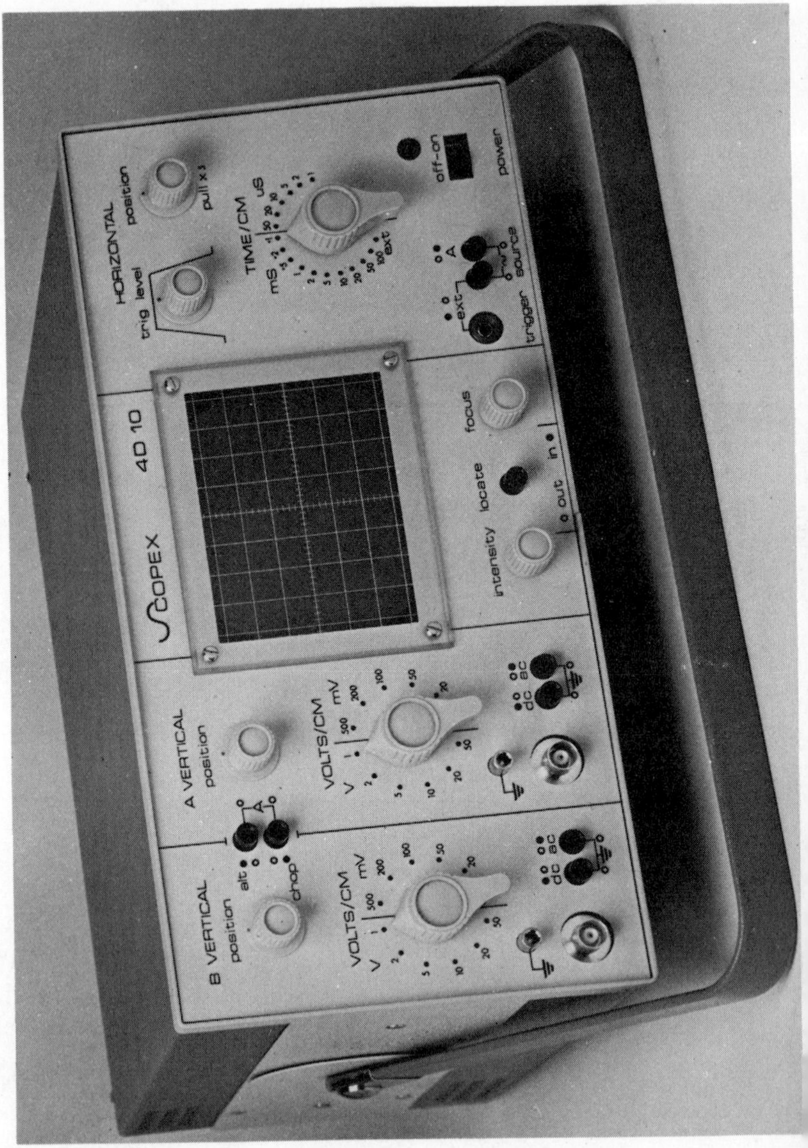

TYPICAL DUAL-TRACE, DUAL-TIMEBASE C.R.O.

1
Introduction

WHAT is an oscilloscope? One may consider it as being a graph plotter, which may operate at slow speeds, but more commonly at high speeds. The usual display is two dimensional like a normal graph, as in figure 1.1, where the X information is displayed horizontally and the Y information vertically. This is a type of graph in rectangular co-ordinates since the two axes are at right angles. Although a normal graph is a two-dimensional display it is possible to effectively have a three-dimensional display by variation of the brightness of the trace. This is called the Z-axis and is at right angles to the X and Y axes, shown diagramatically in figure 1.1.

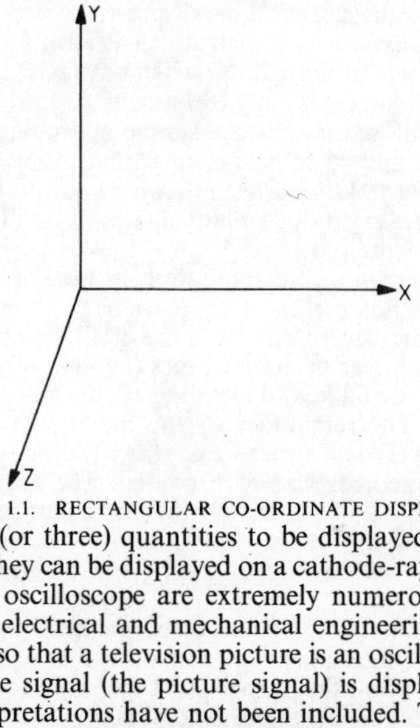

FIG. 1.1. RECTANGULAR CO-ORDINATE DISPLAY

Provided the two (or three) quantities to be displayed can be turned into equivalent voltages they can be displayed on a cathode-ray oscilloscope. Thus, the uses of such an oscilloscope are extremely numerous, covering a wide range of fields from electrical and mechanical engineering to medical work. One may consider also that a television picture is an oscilloscope used in such a way that a variable signal (the picture signal) is displayed on the Z-axis. However, such interpretations have not been included. The types of oscilloscope described are predominantly those for use in electrical engineering rather than those designed for particular applications such as computer displays, medical patient monitors – to give just two examples.

Some 20 years ago the cathode-ray oscilloscope was a comparatively simple instrument, but of great value at the time. Since then oscilloscopes have changed enormously and some have become extremely complex in order to cope with the design and testing of modern apparatus. The frequency range has been much expanded and the oscilloscope has changed from being more of a display equipment to a *measuring* and display equipment.

As stated the two axes can be made to represent any quantity provided a voltage can be obtained proportional to the quantity. However, in a large number of the applications the horizontal axis is proportional to time, and the display is the way in which some quantity, *e.g.* voltage or current, varies with time. This is usually referred to as a waveform. Much of the book is devoted to this type of display, but others have been included.

The applications of an oscilloscope, even if confined to electrical engineering, are so varied that it is difficult to quote typical figures for sensitivity, etc.

Where figures have been quoted it must be stressed that they are only examples and that there are oscilloscopes, particularly those for special purposes, having quite different figures. A simple oscilloscope would serve, say, a radio and television servicing technician where the requirements are easy to meet and cost is important. On the other hand a research engineer working on the frontiers of science would need a most sophisticated equipment, which may be difficult to design and be very expensive.

With their 4D10 oscilloscope, Scopex Instruments have taken very seriously the idea of making a simple oscilloscope which is easy to use and designed to prevent errors in taking readings. Simplification, and the omission of facilities that are never or hardly ever required, kept the cost down to about £130 (1975). Variable controls for Y-sensitivity and the timebase are excluded because the calibration of an instrument is no longer correct when they are used. Thus, at all times (assuming no faults) the calibration of the oscilloscope is correct on both axes. Automatic triggering is used and there are simple arrangements for a dual-trace display. Such an instrument is ideal for radio and television servicing but, of course, of limited use when advanced facilities are required. It is surprising how many servicing technicians appear to be alarmed at the idea of using an oscilloscope; given a simple instrument there should be no difficulties in using it and when proficient perhaps they will wonder how they ever managed without it. Of course, one can be put off by a complex oscilloscope. Telequipment also produce relatively simple oscilloscopes with variable controls for the servicing industry.

Tektronix produce some of the most sophisticated oscilloscopes which may cost £10,000 or more. Some of these are extremely versatile but difficult to use correctly and, in particular, to make full use of all their possibilities. The 7000 range consists of a number of main frames (*i.e.* the tube, power supplies and part of the X and Y amplifiers) depending on the type of tube required, *e.g.* normal or storage. The remainder of the facilities is provided by plug-in units such as amplifiers and timebases. Many plug-in units are available so that practically all facilities can be provided. The advantage of plug-ins is that one need buy only those necessary for immediate requirements and later add units as needed. Some of these plug-ins are described later. It would be impossible to include all facilities in a single instrument because of the size and cost.

Other manufacturers produce ranges of oscilloscopes between these two limits, both complete instruments and main frames with plug-ins.

As it is essential to know the properties of the tube before one can appreciate the equipment required to feed it the general principles and construction of the cathode-ray tube are first considered in Chapter 2. The basic principles of the cathode-ray oscilloscope are given in Chapter 3 so that one can understand the purposes of the various parts of the apparatus. Chapters 4 to 12 cover the various parts introduced in Chapter 3, together with added complications such as multitrace oscilloscopes and dual timebases. Chapter 13 is devoted to some uses, although examples of use are given in some of the other chapters. A record of the display is usually done by photographs, and Chapter 14 is devoted to this (which may be more involved than at first thought). Chapter 15 has brief descriptions of some of the arrangements used on complex oscilloscopes. Chapters 17, 18 and 19 are devoted to what might be termed special oscilloscopes but used extensively. These are very complicated, and only a short introduction has been given to them. Finally, there is a glossary of most of the terms and controls used.

2

The Cathode-Ray Tube

THE cathode-ray tube can be divided into three basic sections:

(a) The electron gun which produces the electrons and focuses them into a beam.

(b) The deflection part where the beam is deflected in two directions at right angles to each other.

(c) The screen where the beam of electrons is made visible due to the electrons falling on a fluorescent screen.

The simple basic oscilloscope tube will be considered under these three headings.

(a) Electron Gun

This consists of a cathode heated by a suitable heater; a grid to control the magnitude of the beam current; and two or more anodes which serve to accelerate the electrons and to form them into a beam.

A common arrangement is given in figure 2.1, where H is the heater and C is the cathode, the cathode being oxide coated. Close to the cathode is the

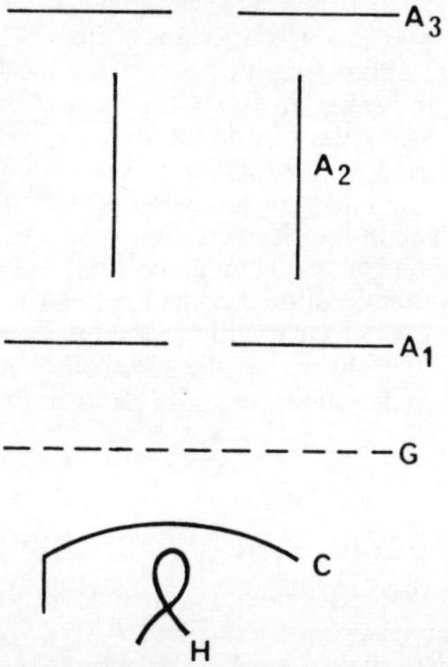

FIG. 2.1. BASIC CATHODE-RAY TUBE

grid G, consisting of a disc with a hole in it. The accelerating and focus systems consist of the three anodes A_1, A_2 and A_3. Anodes A_1 and A_3 are often in the forms of discs with a hole in the centre, while A_2 is a cylinder. The physical arrangement may differ considerably.

To explain the focusing action one must remember that an electron tends to travel along electrostatic lines going from the negative to the positive. This is the reverse of the correct definition of an electrostatic line, which is the direction in which a positive charge will move. In this book the positive direction of the electrostatic line will be taken as the direction in which an electron (a negative charge) moves. A simple diagram of the three-anode gun is shown in figure 2.2. The grid tends to concentrate the beam and controls the beam current; the more negative it is made in relation to the cathode the less the beam current and brightness of the trace. This forms the BRIGHTNESS

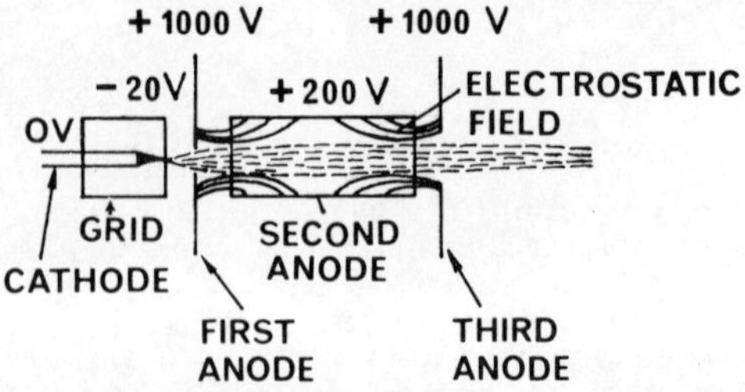

FIG. 2.2. ELECTROSTATIC LENS FORMED BY THREE ANODES

control. Although the beam passing through the hole of the first anode is of small cross-section it tends to spread out due to mutual repulsion between electrons. Thus a 'lens' is required to focus the electrons so that they arrive at the screen in a spot that is as small as possible. The electrostatic lens acts in a way very similar to an optical lens and forms an image of the cathode (or, more correctly, the crossover point near to the cathode) on the screen. The electrostatic lines between the anodes are shown in the figure, and it will be seen that they are in such a direction as to tend to bend the electrons round so that they are all focused to the same point. This obviously is a very simple explanation of a complex electrostatic lens system. The second anode commonly has a potential about one-fifth of that of the first and third anodes, which usually are at the same or about the same potential, as in the figure. By varying the potential of the second anode the effective focal length of the electrostatic lens is varied, hence this forms the FOCUS control.

This is known as a monoaccelerator tube because all the electron acceleration is done in one stage in the electron gun.

The velocity of the electrons leaving the final anode will depend on the voltage applied to the final anode, *i.e.* the greater the voltage the higher the velocity. The energy of an electron, due to the final anode voltage V_a, will be eV_a joules where e is the charge on the electron. The kinetic energy of the electron will be $\frac{1}{2}mv^2$ joules, where m is the electron mass and v is its velocity. Equating these:

$$\tfrac{1}{2}mv^2 = eV_a$$

or $v = \sqrt{2\frac{e}{m}V_a}$ metres/second where $\frac{e}{m} = 1{\cdot}759 \times 10^{11}$ coulomb/kg.

It will be seen later that this velocity is of some importance and it should be noted that the velocity is proportional to $\sqrt{V_a}$ *not* V_a. To give some indication, if $V_a = 5$ kV then the velocity is 42×10^6 metres/second, which is very high.

(b) Deflection Part

Electrostatic deflection is used almost universally for an oscilloscope. (Magnetic deflection is sometimes used for large screen demonstration oscilloscopes and is, of course, used in television). The deflection is done by two pairs of plates, one pair following the other at right angles to each other. The arrangement is shown in figure 2.3, where the beam passes through the final anode A_3 and between the two deflecting plates D_1 and D_2. If D_1 is made positive with respect to D_2, the field (as regards electron movement) is upwards, and during its travel between the plates the electron is accelerated in a vertical direction. After it leaves the field between the plates the electrons continue in a straight line. However, since the electrons now have a vertical

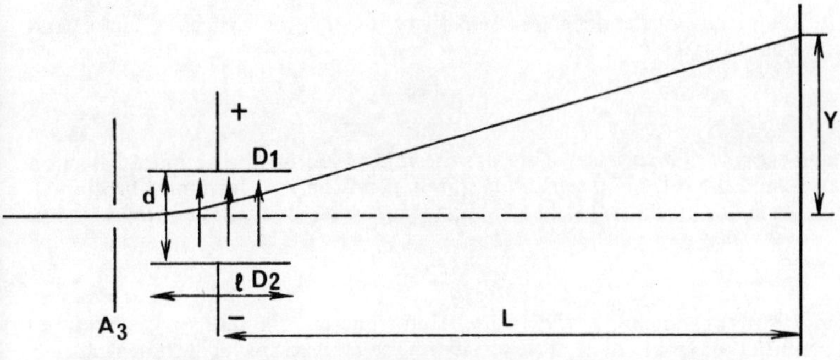

FIG. 2.3. BASIC ARRANGEMENT OF ELECTROSTATIC DEFLECTION

as well as a horizontal velocity, they travel in a line at an angle as shown and strike the screen at a distance Y from the centre. It can be shown that

$$Y = \frac{V_d l L}{2 d V_a}$$

where V_d = voltage applied between the deflecting plates
l = the length of the deflecting plates
L = distance from the centre of the plates to the screen
d = distance between deflecting plates
V_a = accelerating voltage, *i.e.* voltage on A_3.

It is seen that Y is proportional to V_d as it should be to obtain an undistorted trace. The ratio

$$\frac{Y}{V_d}$$

commonly expressed in cm/volt, is called the 'deflection sensitivity' of the tube and hence equals

$$\frac{l L}{2 d V_a}.$$

For a given tube, l, L and d are fixed, but V_a may be varied within limits. Hence, if the deflection sensitivity is quoted for a tube the value of V_a must also be quoted. Alternatively, it may be expressed in terms of V_a. Suppose

$$\frac{lL}{2d} = P,$$

then the sensitivity may be quoted as $\frac{P}{V_a}$ volts/cm/volt $\left(\text{or } \frac{P'}{V_a}\right.$ volts/cm/kilovolt where $\left.P' = \frac{P}{1000}\right)$. Thus, if P is 50 then the deflection sensitivity for a tube working on 2500 volts will be

$$\frac{50}{2500} \text{ cm/volt} = 0{\cdot}02 \text{ cm/volt.}$$

The inverse of the deflection sensitivity is called the 'deflection factor' and is equal to

$$\frac{2dV_a}{lL}$$

and expressed in volts/cm. Thus it is the voltage required to obtain a deflection of 1 cm. Like deflection sensitivity, this depends on V_a, which must be quoted, or the factor given as Q volts/cm volts (or Q' volts/cm kilovolts) where

$$Q = \frac{2d}{lL}$$

and is a constant for a given tube. Unfortunately, the terms are sometimes confused, the term 'deflection sensitivity' being used for 'deflection factor'.

In general, the larger the deflection sensitivity (and the less the deflection factor) the better because a smaller voltage is required from the X and Y amplifiers for a given size of display. The deflection sensitivity is

$$\frac{lL}{2dV_a}$$

and we will now consider how this may be increased.

(i) The distance from the deflecting plates to the screen, L, may be lengthened. This is generally limited for convenience as it increases the front-to-back dimensions of the oscilloscope. Also, increasing L makes the beam more likely to be deflected by stray fields, and it enlarges the spot size since it increases the magnification given by the electrostatic lens.

(ii) Increasing the length, l, of the plates will increase the sensitivity. This will increase the capacitance between the plates and will also limit the size of display for a given spacing. If very high frequencies are concerned, it will cause the high frequency response to fall at a lower frequency (see later in chapter).

(iii) Decreasing the spacing, d, will also increase the capacitance between the plates and again will reduce the maximum scan available. When parallel plates are used the effect of l and d on the maximum scan is shown in figure 2.4. If a bigger scan is attempted, the beam is intercepted by the plates. The scan can be increased without much change in sensitivity by using sloping plates, as in figure 2.5(a); by using bent plates as at (b); or by curved plates as at (c).

Although the normal trace may be limited to, say, a total of 6 cm vertically, a larger vertical deflection than this may be used (*e.g.*, in viewing a portion of the waveform in more detail). In this case the beam may be intercepted by the plates, and a current will flow in the plate circuit. It may be necessary to allow for this in the deflection amplifier. In some tubes, where maximum sensitivity has been aimed at, the available horizontal deflection may be limited and in this case there should be equal deflections available each side of the centre line. Owing to manufacturing tolerances this may not be so, and a small deflecting plate is added in the gun so that the beam can be centralised between the X-deflecting plates. A preset voltage is fed to this deflecting plate.

The first set of deflecting plates is followed by a second set at right angles, as in figure 2.6. The first set is normally designed to move the beam vertically, and the second set to move the beam horizontally. The reason for using this order is that the first set of plates (after A_3) gives the greater deflection sensitivity (for equal other factors) because

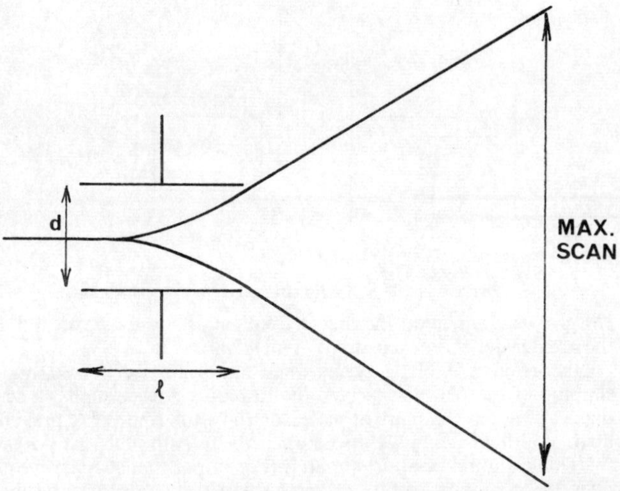

FIG. 2.4. MAXIMUM SCAN WITH PARALLEL PLATES

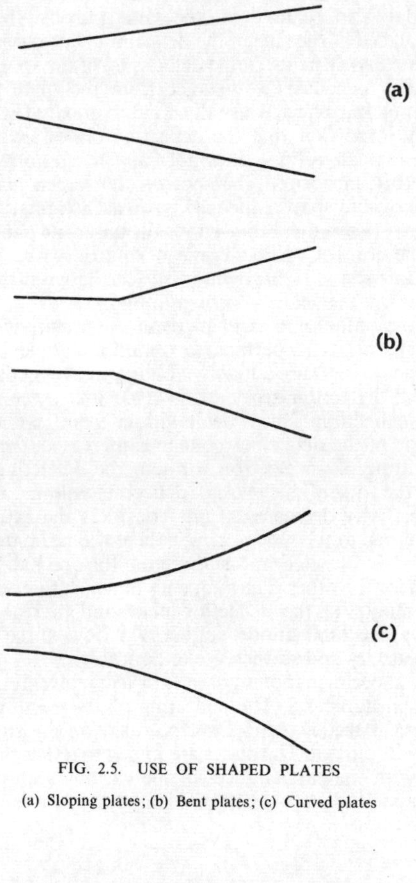

FIG. 2.5. USE OF SHAPED PLATES

(a) Sloping plates; (b) Bent plates; (c) Curved plates

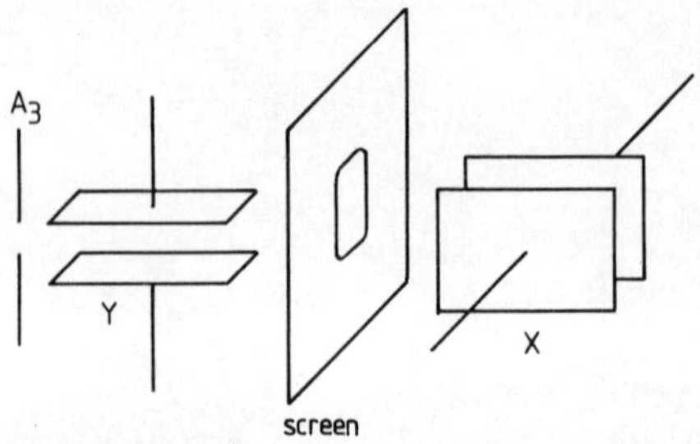

FIG. 2.6. THE X AND Y DEFLECTING PLATES

L is greater. In general the higher frequency is on the vertical axis, and the greater deflection sensitivity is an advantage.

Distortion is liable to occur when the beam is deflected, and some shaping of the plates may be done to reduce this. Usually, a screen is placed between the pairs of plates, as shown in figure 2.6, preventing a distorting field being set up between them. Although the potential of this shield should be approximately the same as that at A_3, some control over the geometry (barrel or pincushion distortion) can be made by varying its voltage slightly relative to A_3. Another electrode after the X-plates may also be used for the same purpose. Sometimes shield plates are placed at the side of the deflecting plates to reduce distortion and unwanted stray fields. Some defocusing of the spot will occur when it is deflected because the distance from the plate to the screen is increased, but this effect is usually small compared with others.

It is very important that the deflecting plates are fed in push-pull so that their mean voltage does not vary in relation to A_3, otherwise serious distortion occurs. The spot on the screen may not be round but may be oval in shape which is known as 'astigmatism'. It is usual to have a control to reduce this, either on the front panel or as a preset control. The control varies the mean voltage of the Y-plates relative to A_3 and forms a variable cylindrical lens between the plates and the final anode A_3. The effect is shown in figure 2.7.

(iv) Reducing the value of V_a will increase the sensitivity but, as will be seen later, V_a needs to be high to obtain adequate brightness of the trace. It is possible to use a low A_3 voltage and yet get adequate brightness by post deflection acceleration (PDA), *i.e.* accelerating the electrons after deflection. This is dealt with later in the chapter.

As the frequency of the deflecting voltage is increased, there comes a time when the transit time of an electron through the deflecting plate becomes comparable with the time of one cycle of deflecting voltage. When this occurs the deflection sensitivity decreases. If one considers the extreme case where the transit time is equal to the time of a cycle there will be no deflection because the electron will be accelerated in one direction for one half-cycle and in the opposite direction for the other. The frequency at which this occurs is increased by reducing the length of the deflecting plate and increasing the electron velocity by raising the final anode voltage V_a. Both these changes reduce the deflection sensitivity and so they are undesirable.

Greatly improved performance is possible by using DISTRIBUTED DEFLECTING PLATES, as shown in figure 2.8. The deflecting plates are now divided into a number of sections. Parallel plates have been shown for simplicity, but the overall 'plate' may be curved. The plates are joined together through inductors so that together with the capacitances of the various sections (and perhaps added external capacitances) a distributed or delay line is produced. If the

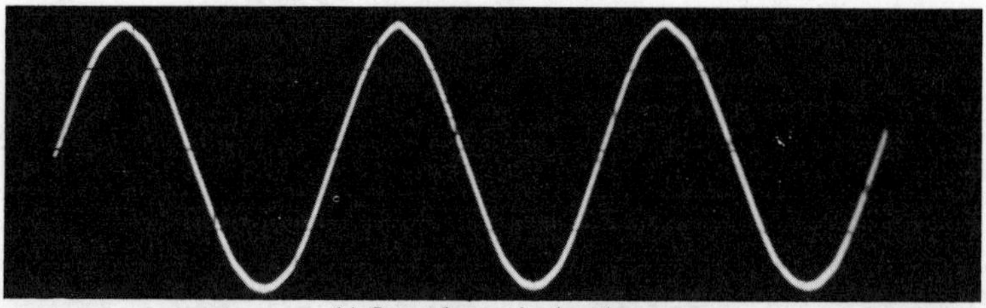

(a) Correct focus and astigmatism

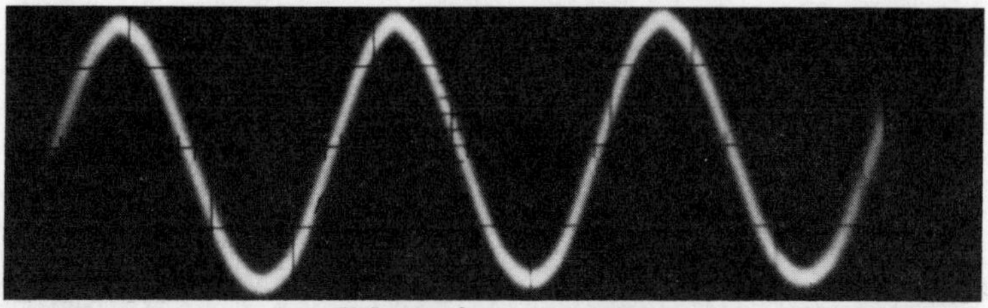

(b) Incorrect focus, no astigmatism

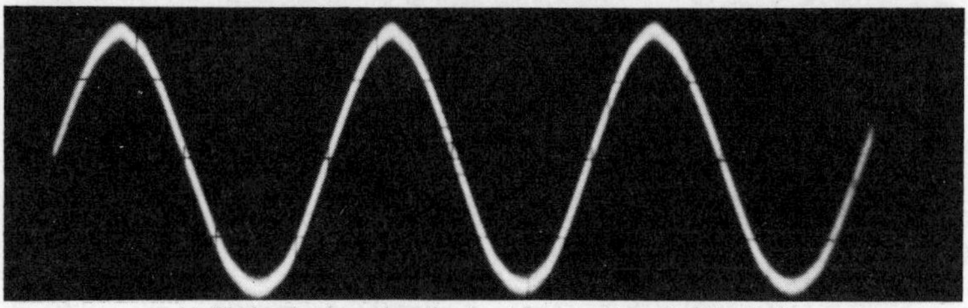

(c) Display showing astigmatism (spot drawn out in vertical direction)

FIG. 2.7. EFFECT OF FOCUS AND ASTIGMATISM CONTROLS

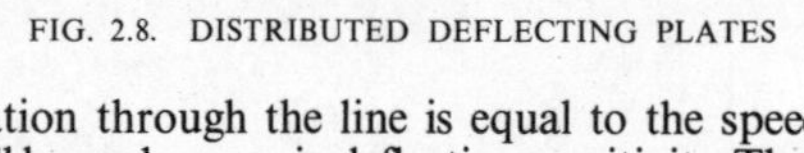

FIG. 2.8. DISTRIBUTED DEFLECTING PLATES

speed of propagation through the line is equal to the speed of travel of the electrons there will be no decrease in deflection sensitivity. The lines are matched by resistors R to prevent reflection; the line must be matched to the signal source. The speed of propagation through the line depends on frequency, whereas the electron speed is independent of frequency. Hence the operation can only be correct at one frequency. However, it does enable better performance to be obtained, and tubes operating up to 1000 MHz are possible. As 1000 MHz Y-amplifiers are not available at present, the tube can only be used at this frequency by using a direct connection to the plates. The signal to be investigated is not normally a push-pull one, so the signal is applied to only one plate, consequently asymmetrical deflection with its disadvantages has to be accepted. At present (1975) the limit of the Y-amplifier is about 500 MHz.

TABLE 2.1

Screen Type	Fluorescence Colour	Phosphorescence Colour	Afterglow		Approximate Relative Brightness	Approximate Relative Recording Speed
			To 10%	To 0·1%		
P_{31}	Green	Green	40 μs	32 ms	100%	50%
P_{11}	Blue	Blue	65 μs	20 ms	20%	100%
P_7	Blue	Yellow–Green	350 ms	1500 ms	40%	75%

For other constant factors the spot size with a P_7 screen is greater than that with a P_{31} or P_{11} screen.

(c) Screen

The purpose of the screen is to convert the energy of the electron beam into visible light energy. When the electrons strike the screen they produce light called 'fluorescence'. Some light is also produced when the beam of electrons has been removed; it is known as 'afterglow' and is due to phosphorescence. A number of phosphor materials are available, and the properties of the three most common are given in Table 2.1. Manufacturers have their own code numbers, those given being the JEDEC designations. Many other phosphors are available for special purposes. The figures for relative brightness are only approximate and will depend on operating conditions; the figure is highest for the P_{31} material because green corresponds approximately to the frequency of maximum sensitivity of the human eye. In a similar way the relative photographic recording speeds are approximate and the P_{11} phosphor gives maximum speed because the blue light corresponds to the frequency of maximum sensitivity of photographic materials. The P_{31} phosphor is used for normal purposes, while the P_{11} is mainly for photographic purposes where the highest writing speed is required. The P_7 is used where a long afterglow is required – mostly of advantage on low-frequency traces.

When the beam strikes the phosphor, light is emitted in all directions, hence only part of it is used when viewing the screen, much of the light being emitted backwards into the tube. This loss of light can be reduced by a thin coating of aluminium on the inside surface of the phosphor, known as 'aluminizing'. The light emitted into the tube is now reflected in the forward direction with an increase in brightness of the trace. There is a small loss of electron energy due to its having to penetrate the aluminium coating, hence the need for the coating to be thin. The aluminium coating lessens the possibility of screen burning.

It is possible to burn the phosphor screen, but not so easily with modern tubes as with early ones. The P_{31} phosphor is more resistant to burning than the P_7 and P_{11} by a factor of some ten times. When the electron beam hits the screen some of the electron energy is converted into light (about 10%) and the remainder into heat. If, therefore, the beam current is large and the beam is stationary the temperature will rise and the screen will be burnt. Thus this part of the screen will afterwards always show a dark mark. Modern tubes will not normally be burnt when a trace is being displayed. A burn is usually caused by a stationary spot of high intensity and is attributable to careless use. When a timebase is being used (since modern timebases are triggered and use unblanking techniques) a stationary spot is difficult to produce unless the brilliance control is turned up very excessively.

The contrast of the image may be improved by a filter in front of the screen. Of course, the brightness of the trace will be reduced but the contrast will be improved. This is because the emitted light has to pass through the filter only once, whereas the ambient light has to pass through the filter twice – from the outside to the screen and back again. A filter is therefore mainly of value under conditions of high ambient lighting. A grey or blue-grey filter may be used, or a green one with phosphors such as P_{31} or P_7. The effect is shown in figure 2.9. A blue filter should be used with P_{11} phosphor screens. If an amber filter is used with a P_7 phosphor the slow (long afterglow) component will be enhanced, and if a blue filter is used the short component will be enhanced.

TUBE SHAPE

Until fairly recently only circular-faced tubes were used, but now rectangular tubes are made with the idea that a larger proportion of the screen surface can be utilized. This certainly fits better in the display of multitrace oscilloscopes. If the horizontal trace on a circular tube is not truly horizontal it can easily be corrected by turning the tube round. This cannot be done with a rectangular tube, so a trace rotation coil must be fitted which will correct

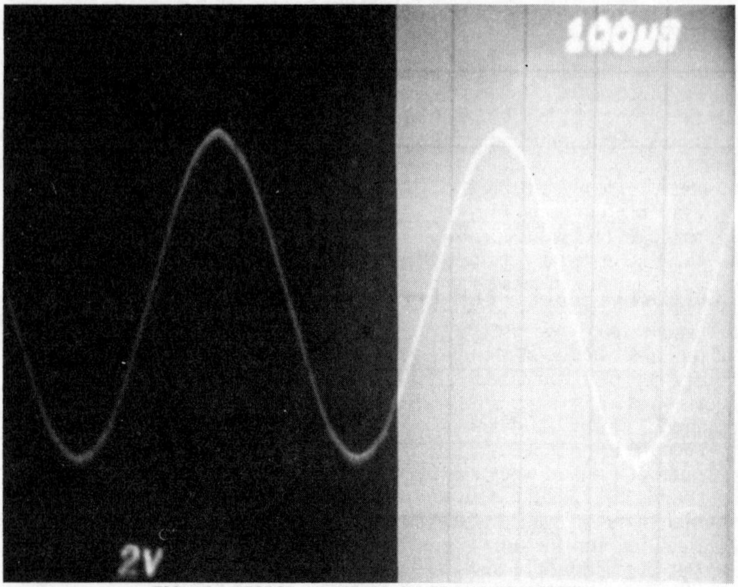

FIG. 2.9. EFFECT OF FILTER IN FRONT OF SCREEN

Green P31 screen with grey-blue filter on left-hand half; no filter on right-hand half, and bright ambient lighting

both the X and Y traces to be truly horizontal and vertical. Such coils are described later in connection with graticules.

POST DEFLECTION ACCELERATION (PDA)

As the frequency of operation is increased, especially where fast transients are to be seen or photographed, the trace has to be brighter. This can be done by increasing the beam current so that more electrons reach the phosphor. However, increasing the beam current means a larger spot size, therefore the beam current must be limited in value. The only other way of obtaining more brightness is to increase the velocity of each electron by raising the accelerating voltage. However, as has been shown the deflection sensitivity is inversely proportional to the accelerating voltage V_a applied before the deflecting plates. This is a serious disadvantage, as it is difficult to get a large deflecting voltage at high frequencies, particularly when transistors are used. The actual sensitivity will depend on the deflecting plate size and spacing; an older monoaccelerator tube with a final anode voltage of 4 kV had a deflection factor of 50 volts/cm. Thus, to get a 3 cm peak deflection (*i.e.* a total of 6 cm peak-to-peak) requires a deflecting voltage of $3 \times 50 = 150$ volts peak (or 300 V p-p). If the accelerating voltage becomes 8 kV then the required deflecting voltage is 300 volts peak or 212 volts r.m.s. The deflecting plates capacitance is, say, 10 pF and at 10 MHz the reactance is about 1600 Ω. The current into the deflecting plates with 212 volts is

$$\frac{212}{1600} = 0{\cdot}13 \text{ A (r.m.s.)}$$

which gives some indication of the problem. The volt-amperes to the deflecting plates $= 212 \times 0{\cdot}13 = 28$.

Post deflection acceleration (PDA), is now very commonly used in oscilloscopes. The general idea is that if the final anode of the gun is given a voltage of, say, only 1 kV, the electron beam will have a relatively low velocity when passing through the deflecting plates, and the deflection sensitivity will be high. If the electrons are accelerated after deflection it will have little or no effect on the deflection sensitivity. There are several methods of doing this. One is to use a high resistance helix (*e.g.* 500 MΩ) deposited on the inside of the tube, as in figure 2.10(a) overleaf. The screen end of the helix is connected to a high voltage, and the other end to a potential at or near that of the final anode A_3. An electrostatic field is produced as shown in figure 2.10(b), which does to some extent reduce the deflection sensitivity but by no means to the same extent that would result in applying the final PDA voltage to the final anode A_3.

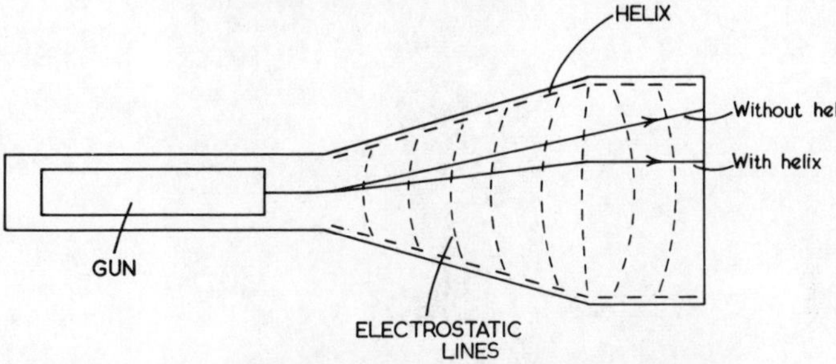

FIG. 2.10(b). PDA TUBE WITH SOME DECREASE IN DEFLECTION SENSITIVITY

Reduction in deflection sensitivity can be overcome by fitting a spherical mesh after the deflection plates, as in figure 2.11, and by suitable arrangement of the helix. In this case the beam travels at right angles to the equipotential surfaces, and is not, therefore, deflected by the electrostatic field. Conse-

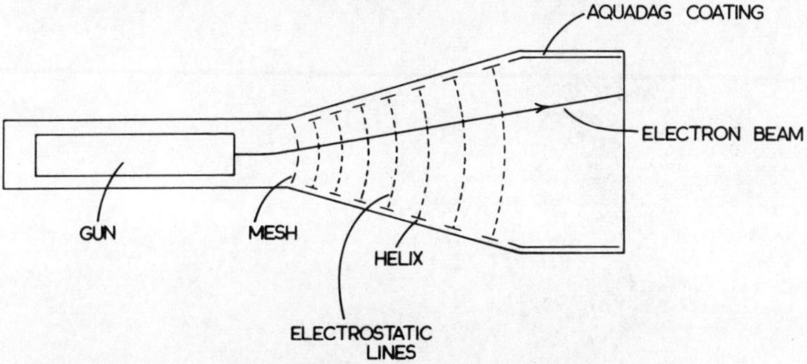

FIG. 2.11. PDA TUBE WHERE NO CHANGE OF SENSITIVITY OCCURS

quently, there is no reduction in deflection sensitivity due to the PDA voltage. By making the equipotential surfaces more convex one can get scan magnification, as in figure 2.12, which may be up to two times. The mesh structure, however, forms a pattern on the screen when the spot is defocused (but not visible when the tube is operated normally). The mesh collects some of the electrons, which reduces the effective beam current so far as the screen is concerned.

The voltage fed to the final anode A_3 is usually 1 to 2 kV, comparatively low so as to keep up the sensitivity. The final PDA accelerating voltage is commonly 5 to 10 kV, but may go to 20 kV.

Direct comparison with monoaccelerator tubes and PDA tubes is difficult because the deflection factor is also altered by changing the length of the

FIG. 2.10(a). HELIX TYPE PDA TUBE

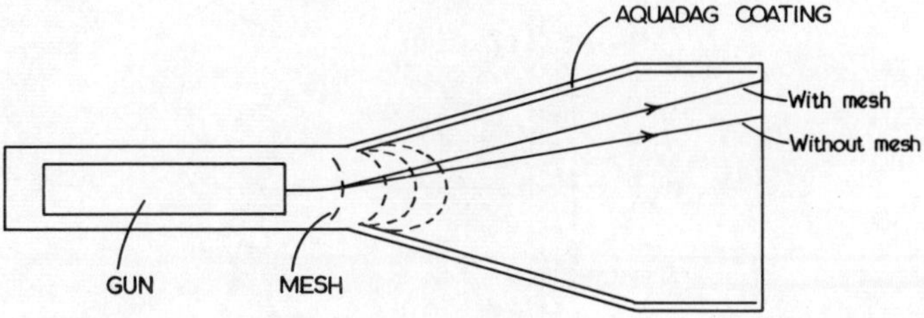

FIG. 2.12. PDA TUBE WHERE SCAN MAGNIFICATION OCCURS

deflecting plates and the spacing between them. Further, the effect of the PDA construction may reduce the sensitivity (increase the deflection factor) or increase it (decrease the deflection factor). Tubes are available with a deflection factor as low as about 4 volts/cm for the Y-plates for a PDA voltage of 10 kV, which is a great improvement on that given earlier of 50 volts/cm at 4 kV final anode voltage. The final anode voltage A_3 is about 1·5 kV for such tubes. The X-deflection factor is usually considerably more, say 10 to 15 volts/cm. By bringing out the deflecting plates leads to the side of the tube (rather than to the base) the capacitance is reduced to, say, 3 or 4 pF.

Thus, for a 3 cm peak deflection in the Y direction a peak voltage of only $3 \times 4 = 12$ volts peak or 8·5 volts r.m.s. is required. At 10 MHz, if the capacitance is assumed to be 4 pF, the reactance is 4000 Ω, hence the current is

$$\frac{8{\cdot}5}{4000} = 0{\cdot}002\ \text{A}.$$

The volt-amperes required for deflection now become only $8{\cdot}5 \times 0{\cdot}002 = 0{\cdot}017$. This should be compared with the 28 for the earlier calculation using a mono-accelerator tube having 8 kV final anode voltage.

Without the introduction of PDA tubes it would not have been possible to produce oscilloscopes with their present-day performance, both as regards high frequency operation and fast writing speeds.

Careful design of PDA tubes is essential or a distorted trace will be obtained which may be barrel or pincushion in shape.

GRATICULES

It will be explained later that practically all oscilloscopes have direct calibration of both voltage and time, therefore a graticule must be fitted to the screen so that the magnitude of the deflection can be read. One way of doing this is to fit a piece of transparent plastic suitably engraved or marked in front of the screen, as in figure 2.13. To make the engraving stand out, especially when using photographs, the graticule is illuminated from the edge by one or

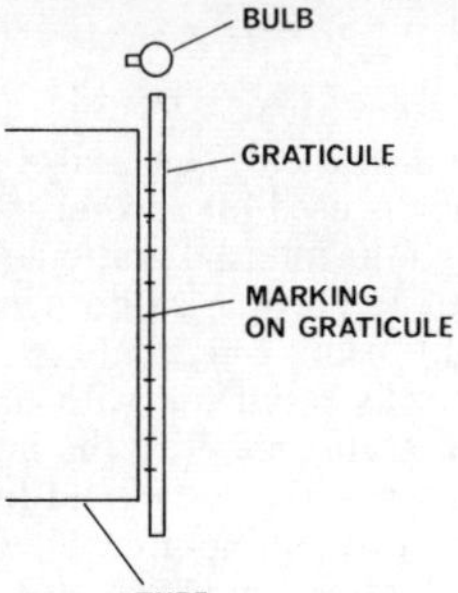

FIG. 2.13. SEPARATE GRATICULE

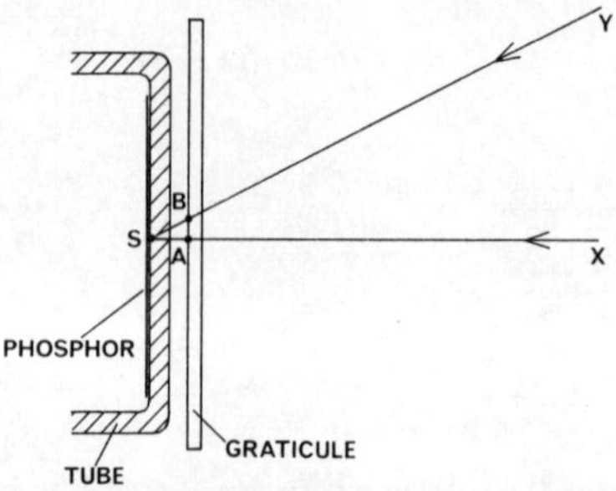

FIG. 2.14. PARALLAX ERRORS DUE TO SEPARATE GRATICULE

more bulbs. The brightness of the bulbs can be varied by a control knob, usually marked GRATICULE. The drawback to this type of graticule is that it suffers from parallax errors, as shown in figure 2.14. The glass of the cathode-ray tube is fairly thick, and the phosphor coating is on the inside. The graticule should be placed as close as possible with the front of the c.r.t., but as the face is never quite flat there will be a gap. Suppose that the screen spot corresponds to S. If this spot is now viewed from position X (at right angles to the screen) the reading on the graticule will correspond to A. If, however, the screen is viewed from position Y the reading on the graticule will correspond to point B. Hence the reading is going to vary according to the viewing angle. Always viewing from a position at right angles to the screen at the particular point will reduce the error but there is always likely to be some error. This can be largely remedied by marking a graticule on both sides of the plastic and a reading taken when both graticules appear on top of one another.

A parallax error also occurs when the trace is photographed because the camera looks at the screen from one fixed point.

Parallax errors can be avoided by making the graticule on the inside of the tube, as in figure 2.15, called an 'internal graticule'. However, it is considerably more costly, it is not so easy to illuminate and, of course, it cannot be changed.

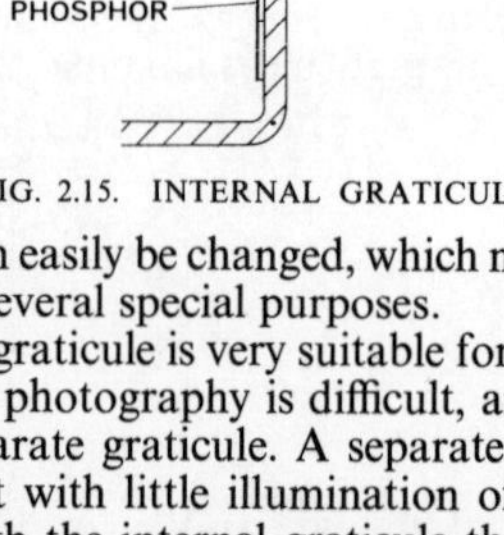

FIG. 2.15. INTERNAL GRATICULE

The external graticule can easily be changed, which may be necessary when the oscilloscope is used for several special purposes.

Although the internal graticule is very suitable for normal viewing, illuminating it satisfactorily for photography is difficult, and a better result may be obtained by using a separate graticule. A separate graticule has its edge lit and the marks stand out with little illumination of either the screen or the rest of the graticule. With the internal graticule the lighting illuminates the graticule marks, but some light falls on the phosphor screen, so reducing the contrast. One manufacturer (Hewlett Packard) uses an internal flood gun for graticule illumination which gives a uniform background on photographs.

The internal graticule requires some type of trace alignment. For example, if the deflection system on a normal circular tube does not correspond exactly to the marking on an external graticule (say the horizontal trace is sloping) this is easily corrected for by simply turning the tube until the tube and graticule are in alignment. This cannot be done with an internal graticule, but the trace can be rotated through a small angle by an axial coil placed as in figure 2.16. The order of 50 AT may be required for a 5° rotation, but depends on the tube construction. Tektronix use two coils. One is near the screen which

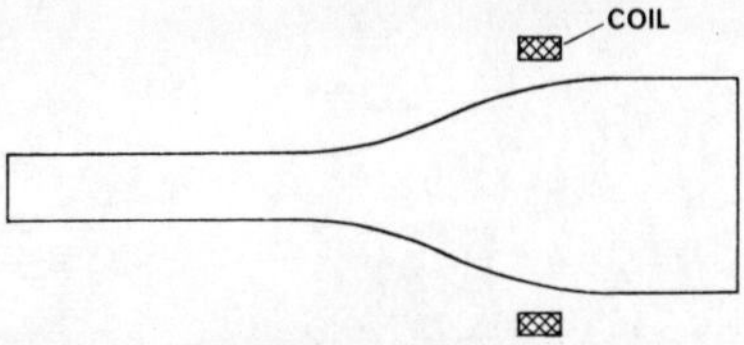

FIG. 2.16. USE OF ALIGNMENT COIL

rotates the whole trace and is used to adjust the alignment along the horizontal lines of the graticule. The other coil is placed in the region of the Y-plates, rotates the vertical trace only, and is used to align the vertical deflection with the graticule.

MAGNETIC SCREENING

Because the beam is easily deflected by stray magnetic fields produced either by the oscilloscope itself (*e.g.* mains transformer) or externally, the tube is magnetically screened by a mumetal screen usually extending the full length of the tube. This is much easier and cheaper to make for a circular tube than for a rectangular one which, of course, is rectangular only at the screen end.

BEAM BLANKING AND MODULATION

In modern oscilloscopes it is usual to cut off or blank the beam when it is not producing the required trace, *e.g.* during flyback. One may consider that the beam is suppressed during periods when not required, *i.e.* the beam is blanked. Alternatively, one may consider that the beam is produced only when required, commonly called 'unblanking'. Blanking or unblanking can be done by the application of a suitable voltage to the grid of the tube, or by blanking electrodes. When blanking electrodes are used it is called 'deflection plate unblanking' (or blanking). A basic arrangement is shown in figure 2.17, where two deflecting plates D_1 and D_2 are placed between two discs forming the first anode A_1. Plate D_2 is connected to A_1. With no potential applied to D_1 the beam passes through the holes in the two anodes and eventually strikes

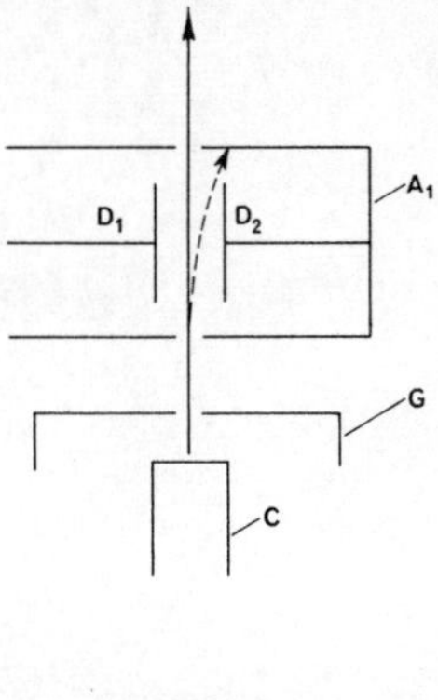

FIG. 2.17. DEFLECTION PLATE BLANKING (OR UNBLANKING)

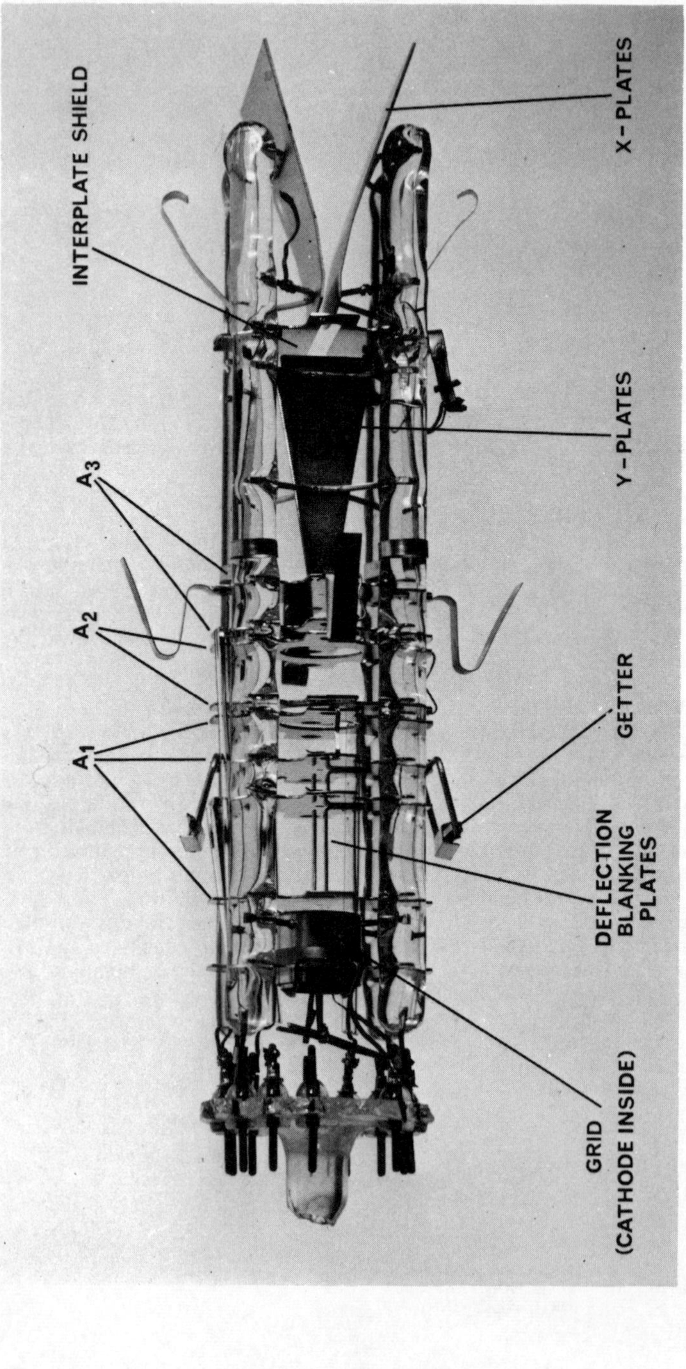

FIG. 2.18. TYPICAL ELECTRON GUN

the screen. If a negative voltage is applied to D_1 the beam is bent, as shown, and is intercepted by the second disc of the anode, so that so far as the screen is concerned the beam is cut off. The arrangement can only be used for cutting off the beam and not for variable intensity modulation as can be done by the grid.

Grid blanking requires more power and can be overriden if the brightness control is advanced too far (with the possibility of screen burning). However, grid blanking does cut off the beam, and therefore lengthens the life of the cathode of the tube. With deflection blanking the cathode current is flowing all the time.

Since the first anode is usually at the same potential as the final anode and at approximately the same voltage as the deflecting plates, it is near earthy potential (see later chapters). Hence it is easy to apply a suitable voltage or pulse to the deflecting plate of the deflection blanking system. The grid is, of course, at approximately the cathode potential, and hence at a large negative voltage (1 kV or more) relative to the earthy connection. This causes problems when unblanking (or blanking) signals have to be fed to the grid.

More details are given in Chapter 7 dealing with cathode-ray tube circuits.

A typical electron gun is shown in figure 2.18. This has a deflection blanking system, but using two pairs of plates. The second anode comprises a series of discs rather than a tube.

TRACE WIDTH

In general, the finer the trace the better, and tube manufacturers often quote the beam width. This is difficult to measure in the normal way as the brightness varies over the width, in the way shown in figure 2.19 for example. It is usually quoted in mm and determined by what is called the 'shrinking raster

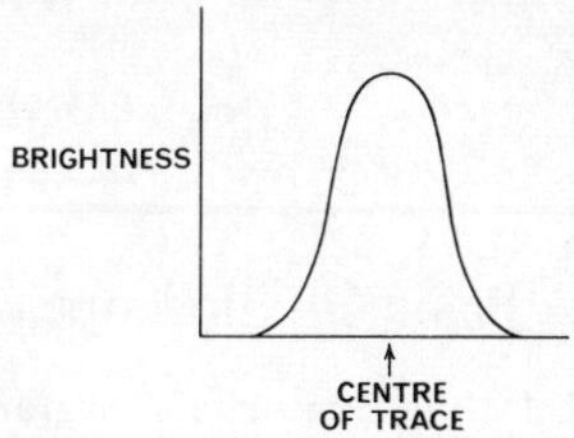

FIG. 2.19. VARIATION OF BRIGHTNESS OF TRACE OVER ITS WIDTH

method'. The idea is to produce a raster consisting of a number of parallel lines, as in figure 2.20, similar to a television raster (which is shown for simpli-

Fig. 2.20. RASTER PRODUCED BY SPOT

city as having only 14 lines). When the lines are spaced apart there will be dark areas in between, as in figure 2.21(a). The lines are now brought nearer together by reducing or shrinking the size of the raster until the dark lines disappear and the raster looks uniformly bright. This occurs when the 50% brightness

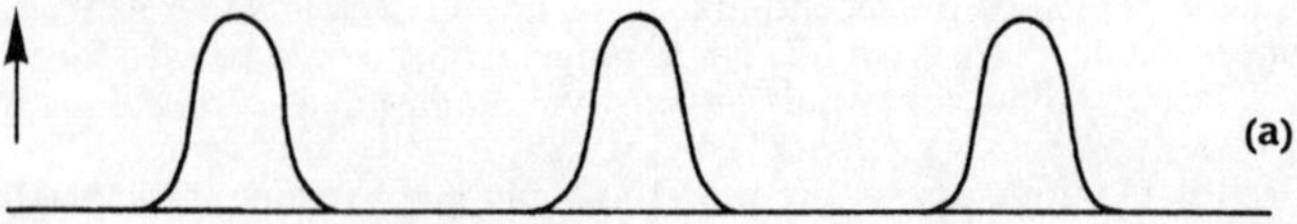

FIG. 2.21(a). LINES SPACED APART

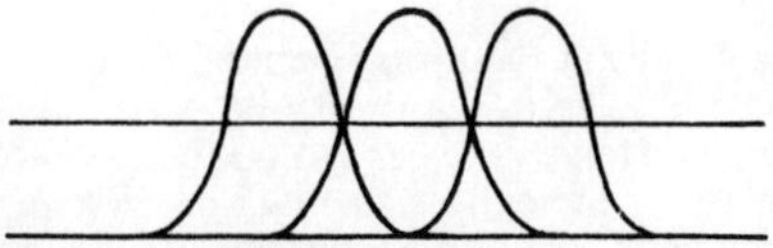

FIG. 2.21(b). LINES AT A DISTANCE WHICH PRODUCES UNIFORM ILLUMINATION

level of one trace just coincides with the 50% brightness level of the other, as shown at (b). To obtain the trace width the height of the raster in mm is now divided by the number of lines.

Dual-beam and split-beam tubes are considered in Chapter 10.

3

The Basic Oscilloscope

In this chapter we shall deal with the block diagram of a simple basic oscilloscope, such as that given in figure 3.1. The first point to make is that the deflecting plates must be at approximately the potential of the final anode A_3 of the gun. The deflecting plates, both X and Y, are normally fed from amplifiers; one side of the amplifier must be at earthy potential. Thus the deflecting plates will be near earth potential (the output voltage of the amplifier), and A_3 is approximately at the same voltage. Hence, the cathode of the tube will be at a relatively high negative voltage with respect to earth (say 1 kV) and this is provided by the negative e.h.t. supply. If a PDA tube is fitted then a positive e.h.t. supply (say 4–12 kV) is required to feed the PDA electrode. A power supply (or supplies) is also required to operate the amplifiers and timebase circuits.

Although it is sometimes possible to connect directly to the X and Y plates of an oscilloscope such a connection is used only for special applications, the plates usually being fed from the X and Y amplifiers as shown. The signal under investigation is normally fed to the Y-plates through the Y-attenuator and Y-amplifier as shown. This arrangement of a fixed-gain amplifier and variable attenuator is almost universal and is used because direct calibration of the vertical scale on the screen is required. The attenuator is a passive device and should maintain its calibration. If, therefore, the calibration is checked or adjusted on one range it should be correct on all ranges. Calibration

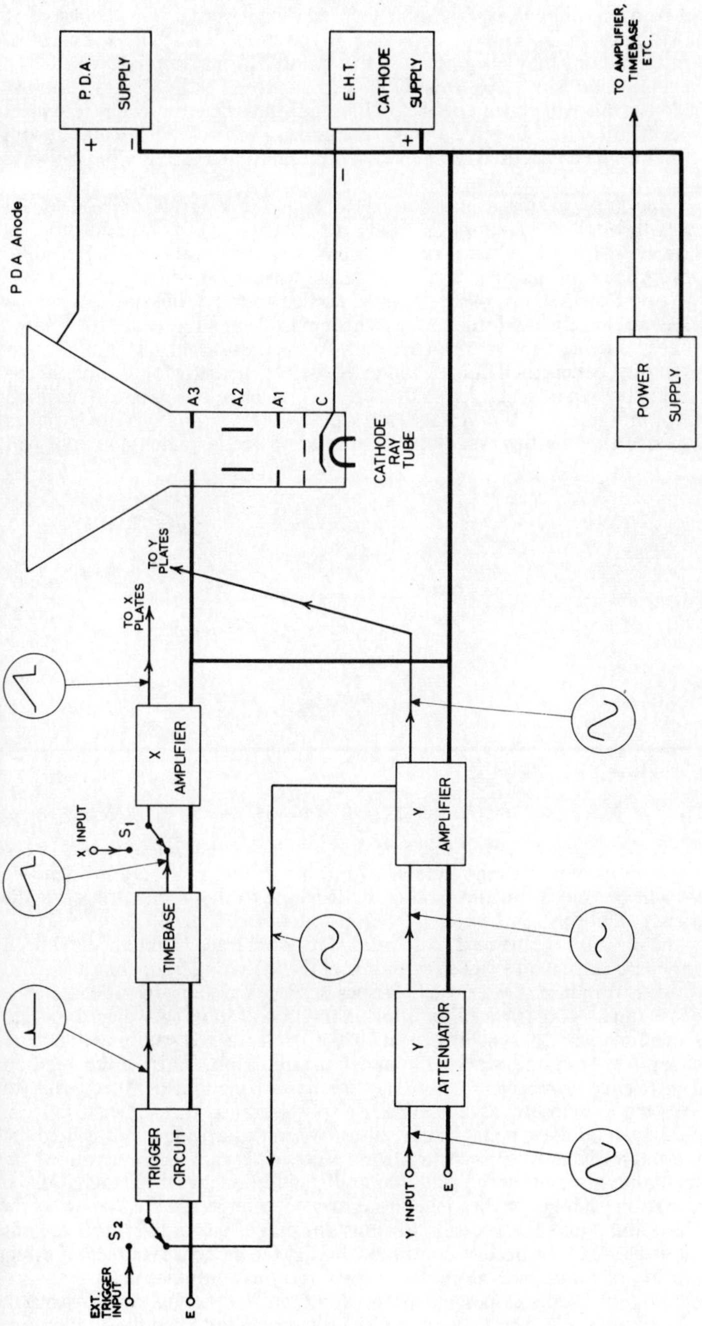

FIG. 3.1. BLOCK DIAGRAM OF SIMPLE OSCILLOSCOPE

would be more difficult if a variable gain amplifier were used. The range of input voltages to be covered is large and the input sensitivity (also called 'deflection factor') may range from 10 mV/cm to 20 V/cm (a range of 2000/1). The attentuation is in steps, usually in 1, 2, 5 ratios. The frequency response of the Y-circuits will depend on the oscilloscope, but must be designed to meet the specification and to have a small rise time. The problems of bandwidth and rise time are considered in Chapter 4. A d.c. amplifier is now almost always used.

In some applications it is necessary to apply an external voltage to the X-plates through the X-amplifier using the switch S_1. In some, but not all, oscilloscopes, there is direct access to the X-amplifier in this way. For most purposes the oscilloscope is used to examine waveforms, the horizontal axis being TIME. For this purpose a device is needed to move the spot across the screen from the left-hand side to the right-hand side at a uniform rate, known as a 'linear timebase' or 'sweep generator'. At the right-hand side of the screen the spot must be returned quickly to the left-hand side ready for another linear movement (sweep or scan) across the screen. The shape of waveform required is shown in figure 3.2. If a voltage (say sinusoidal) is applied to the Y-plates at the same time as this waveform is applied to the X-plates, the spot will

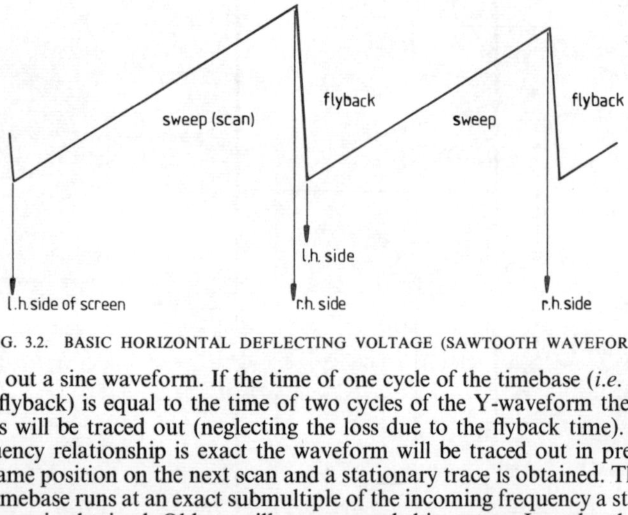

FIG. 3.2. BASIC HORIZONTAL DEFLECTING VOLTAGE (SAWTOOTH WAVEFORM)

trace out a sine waveform. If the time of one cycle of the timebase (*i.e.* sweep plus flyback) is equal to the time of two cycles of the Y-waveform then two cycles will be traced out (neglecting the loss due to the flyback time). If the frequency relationship is exact the waveform will be traced out in precisely the same position on the next scan and a stationary trace is obtained. Thus, if the timebase runs at an exact submultiple of the incoming frequency a stationary trace is obtained. Older oscilloscopes used this system. In order that the timebase ran at EXACTLY a submultiple, a fraction of the Y-waveform was fed into the timebase to synchronize it. Thus it was a free-running timebase synchronized with the waveform under examination. This is the method used in television receivers where the timebases (line and field) self-run (to obviate the possibility of a stationary spot burning the screen) and are synchronized by the synchronizing pulses. When this arrangement is used the FLYBACK is initiated by the synchronizing pulses or Y-signal. The arrangement is unsuitable for a modern oscilloscope. If the frequency of the signal under examination changes, it becomes necessary to change the frequency of the timebase and it may be necessary to adjust the magnitude of the synchronizing signal. Because of the need to continuously vary the timebase frequency, direct calibration of time-scale, along the X-axis, is almost impossible.

All modern oscilloscopes use an arrangement consisting of a triggering circuit and timebase. The signal under examination is fed from the Y-amplifier through S_2 to the trigger circuit; or an external signal may be fed to the trigger circuit as shown in figure 3.1. The trigger circuit generates a pulse to operate the timebase. The timebase may operate in two basic ways, automatic or triggered. (To simplify matters the automatic operation will be described later). The timebase does not self-run and the START of the trace is now initiated by the pulse from the trigger circuit. The speed at which the spot travels across the screen can be varied (usually in steps). It is independent of the triggering pulse and is determined by the SPEED SETTING control. This can be directly calibrated so that the time/division is known. With this method of operation the pulse from the trigger unit starts the trace. When the spot reaches the right-hand side of the screen it returns rapidly to the left-hand side and *remains there until the next triggering pulse*. If the trigger circuit operates at the same point of the waveform in each case, the trace always starts at the same part of the waveform and a stationary trace results. It is important to note that the repetition rate of the timebase is made a submultiple of the Y-waveform by varying the waiting time of the spot on the left-hand side of the screen. This is shown in figure 3.3. It will be seen that a whole number of cycles are not necessarily seen on the screen. If the frequency of the Y-signal is varied then,

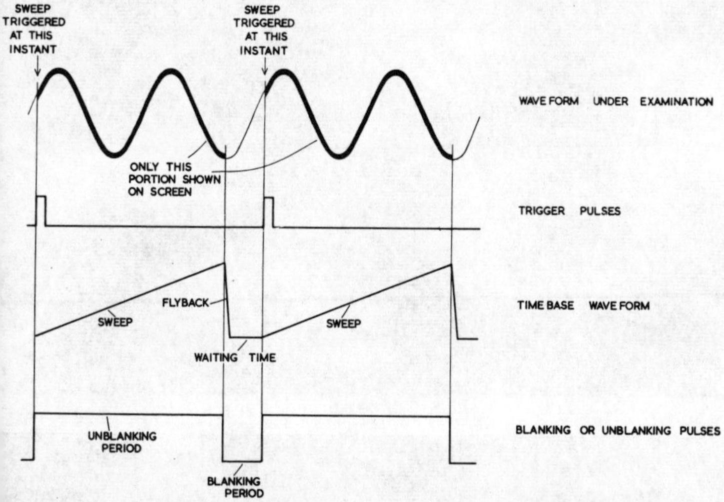

FIG. 3.3. OPERATION OF TRIGGERED TIMEBASE OR SWEEP GENERATOR

or a given timebase speed setting, the number of cycles will change. This is hown overleaf in figure 3.4.

In order that the flyback of the spot and the stationary spot are not visible n the screen, these are blanked out by pulses as shown in figure 3.3. The ulses may be applied to the grid to cut off the beam or to the deflection lanking plates described in the last chapter. Or, put another way, the spot is ormally suppressed and is unblanked by the pulses shown during the trace eriod only.

As no trace is visible until the timebase is triggered by suitable pulses, utomatic triggering' is often used (the two basic systems are described in hapter 6). With this arrangement, in the absence of triggering pulses the mebase free-runs so that a trace is visible on the screen but will lock only when igger pulses are produced.

Details of the various blocks will be considered in succeeding chapters.

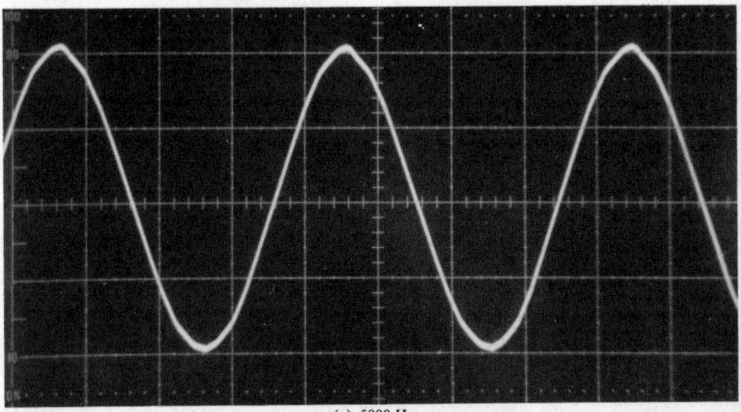

(a) 5000 Hz

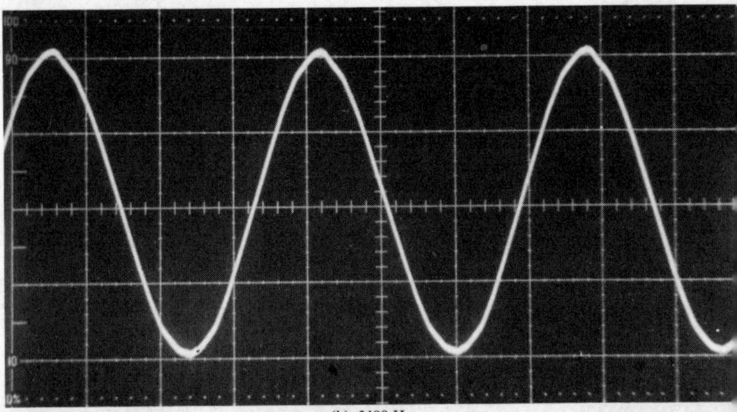

(b) 5400 Hz

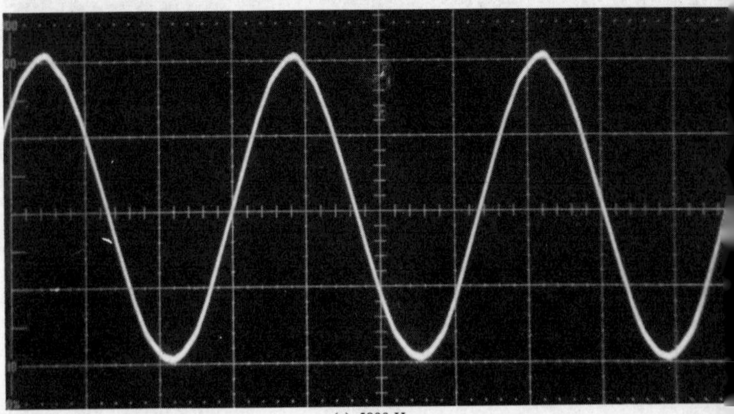

(c) 5800 Hz

FIG. 3.4. EFFECT OF INCREASING THE FREQUENCY FOR A FIXED SWEEP SPEED

4

The Y-attenuator and Amplifier

It is not intended to go into the design of the blocks that follow. Some of the problems will be indicated; the type of circuit will be explained; and, in particular, explanations given of the use of the various controls. The detailed arrangements will vary between type of oscilloscope and manufacturers, so only basic circuits will be given. Also, only transistor circuits will be considered.

To prevent distortion the MEAN potential of the two Y-deflecting plates must be about the same as that of the final gun anode A_3 (the only difference in potential is that for astigmatism correction as described in Chapter 2). For this reason push-pull deflection is required, *i.e.* one plate will change in voltage by $+V$ while the other plate changes by $-V$, so that the mean or average voltage is constant. The input to the amplifier is usually single-sided, *i.e.* earthy and live only, hence there must be some means of converting this input to the push-pull output required.

The input impedance of an oscilloscope should normally be high, say at least 1 MΩ (an exception may occur at very high frequencies when 50 Ω is used), so that connecting the oscilloscope to a piece of equipment does not upset its operation unduly. The input impedance is settled by the attenuator design, but the attenuator can have a high impedance only if the amplifier input impedance is high. When valves were used this was easy, since the input impedance of a valve is very high and largely settled by the grid resistor required; 2 MΩ was quite easily obtained. A bipolar transistor has a relatively low input impedance, hence field effect transistors (FETs) are commonly used. The change from valves to transistors brought its own problem. A large voltage may be fed to the grid of a valve without causing damage. The grid current can be limited by, say, a 100 kΩ resistor in series. Consequently, valve oscilloscopes would usually stand a large input, even with the attenuator set to its most sensitive position. Users were in the habit of applying a signal to the oscilloscope without checking the attenuator setting. If the input was too large the trace went off the screen, no damage was done, and the attenuator was adjusted to produce a suitable trace. This is exactly the opposite to what one should – and in most cases must – do with a non-electronic multirange meter (such as an Avometer). However, mistakes are always possible and oscilloscopes have built-in protection against overloading. To appreciate how much overload there might be suppose that 60 volts peak (42 volts r.m.s.) are applied (intended to give 3 cm deflection in one direction, a total peak-to-peak deflection of 6 cm) on the 20 V/cm range and the oscilloscope is set to the 20 mV/cm ranges. The signal input is now 1000 times or 100,000% above its correct value. Not many devices will withstand such overloads.

A block diagram of a typical Y-channel is given in figure 4.1. In some oscilloscopes the Y-amplifier may be a plug-in unit. If so, part of the block diagram will be in the plug-in and part may be in the main frame, *e.g.* the output amplifier. The various blocks will now be considered.

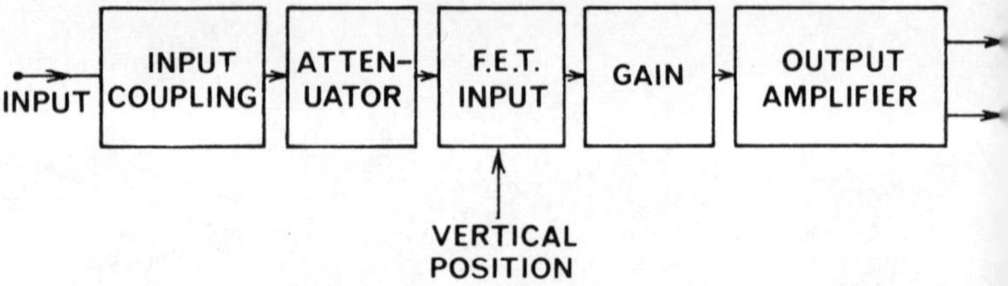

FIG. 4.1. BLOCK DIAGRAM OF Y-CHANNEL

INPUT COUPLING

A common arrangement of the input coupling is given in figure 4.2. Ir position 1 there is a direct connection from the input to the attenuator anc so to the amplifier. In this position both d.c. and a.c. inputs are fed to th

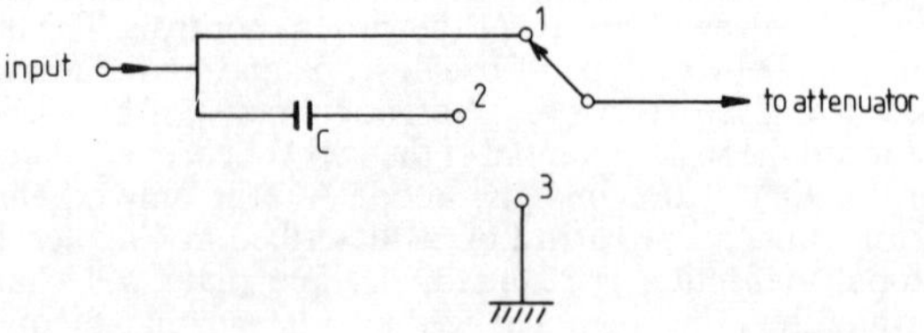

FIG. 4.2. INPUT COUPLING (Y-CHANNEL)

oscilloscope tube. In position 2 a capacitor C is added so that any d.c. com ponent is removed. This enables one to view an alternating voltage super imposed on a larger direct voltage. The frequency response will depend on th value of the capacitance, but usually, say, 3 dB down at 5 Hz. In position the attenuator is connected to the earthy line, the input, of course, not bein short circuited. This facility (not always provided) is useful for setting th zero input position of the trace without disconnecting the signal, *i.e.* th position of the trace for no input, corresponding to the zero line.

ATTENUATOR

This is a stepped attenuator normally going in 1, 2, 5 ratio. The total attenua tion required is large since the input sensitivity may vary from 10 mV/divisio to 20 V/division (much higher sensitivity of 10 μV/division may also b available on some instruments). On the 10 mV/division setting the attenuat is cut out so that on the 20 V/division setting the attenuation is 2000 time This cannot be obtained on a single switch bank, and commonly a number c fixed attenuators are used connected in cascade as required. An example i given in figure 4.3. The attenuators are potential dividers, and a typical circu is shown in figure 4.4. At low frequencies this is a resistance attenuator, th values of resistors R_1 and R_2 being chosen to give the required attenuatio The input impedance of the oscilloscope should be the same on all range therefore the input impedance of the ÷10, ÷100 and ÷1000 attenuato must be the same as the input impedance of the amplifier. These attenuato sometimes feed the amplifier and, in other positions, the ÷2 and ÷5 attenua tors. Again, these must also have the same input impedance as the amplifie The values of R_1 and R_2 must allow for the fact that the attenuators are feedin

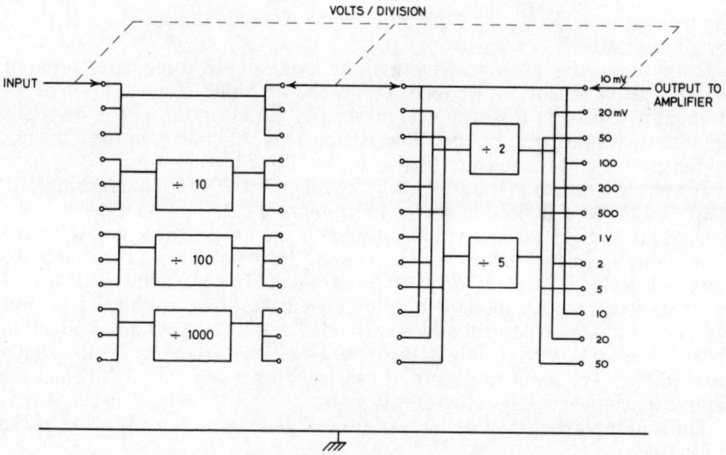

FIG. 4.3. ONE ATTENUATOR ARRANGEMENT

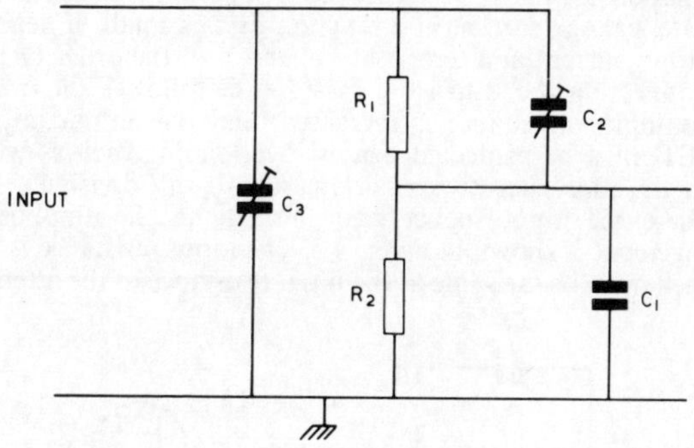

FIG. 4.4. TYPICAL ATTENUATOR CIRCUIT

the amplifier or other attenuators. If, for example, the input impedance of all attenuators is 1MΩ, then the input impedance of the amplifier must also be 1 MΩ. If an attenuator is 10/1 then R_1 = 900 kΩ and the effective lower arm must be 100 kΩ. This is composed of R_2 and the 1 mΩ input resistance of the device being fed from it, or

$$\frac{1}{R_2}+\frac{1}{1000}=\frac{1}{100} \text{ (}R_2\text{ being in kilohms).}$$

Thus, $R_2=\dfrac{1000\times 100}{1000-100}=111$ kΩ.

To offset the effect of stray capacitance at high frequencies, the potential divider becomes a capacitance potential divider consisting of C_1 and C_2 and stray capacitances. C_2 is adjusted to give the required attenuation at high frequencies (the same as that given by R_1 and R_2 at low frequencies). Again, C_2 must allow for the fact that C_1 includes the input capacitance of the attenuator or amplifier being fed from the attenuator. The input capacitance of the oscilloscope should be the same at all settings (so that a potential probe can be used: see Chapter 8). C_3 is adjusted to give the same input capacitance for all attenuators.

Other attenuator arrangements may be used, and in some cases some of the change in sensitivity is done within the amplifier itself. Although an attenuator is used in this way it is marked in VOLTS/DIVISION. The divisions are usually centimetres, but not always, and refer, of course, to the divisions on the graticule. If the sensitivity is set to, say, 10 volts/division, a d.c. voltage of 10 volts will move the trace vertically by 1 division. If 10 volts sinusoidal a.c. is applied the total deflection will be much more. First, because it displays the peak value and the normal voltage quoted is the r.m.s. (for a sine wave the peak value is $1{\cdot}414 \times$ r.m.s. value); secondly, the polarity changes and the trace is deflected in both directions. The trace would be of magnitude $1{\cdot}414 \times 2 = 2{\cdot}828$ divisions. All modern oscilloscopes have direct calibration in this way. A variable gain control may be provided, but it is important to note that the calibration is correct only *in one position* of the variable control, usually marked CAL. The input resistance of oscilloscopes is generally 1 MΩ and the input capacitance is, say, 11 to 50 pF.

Thick film planar resistors may be used in the attenuators in place of the ordinary types.

INPUT CIRCUITS

The most common arrangement is to use a junction FET as a source-follower. The gate leakage current of a junction FET is small; it depends on drain voltage, drain current and temperature, and is of the order of 10^{-5} to 10^{-4} μA. By connecting the transistor as a source-follower the current is reduced (*i.e.* the input impedance is increased) and the output impedance reduced. An FET must be protected against overloading such as when the attenuator is in the most sensitive position, say 10 mV/division (*i.e.* zero attenuation) where the input socket feeds directly to the amplifier. One protective arrangement is shown in figure 4.5. The input resistance is R_1 and R_2 in series, which must be the same as the input resistance of the attenuators.

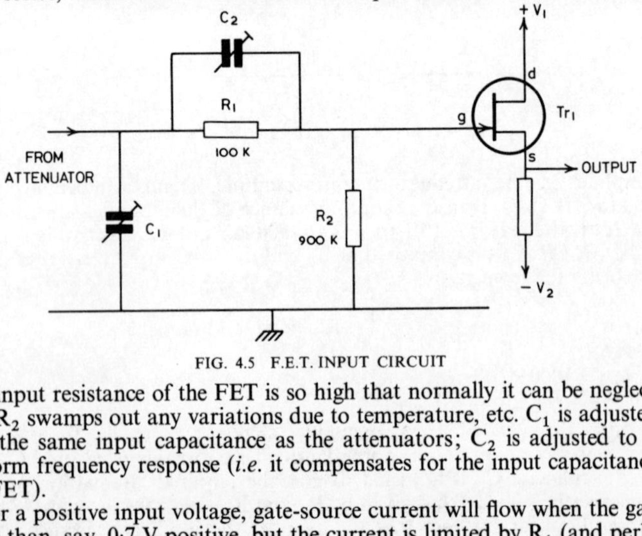

FIG. 4.5 F.E.T. INPUT CIRCUIT

The input resistance of the FET is so high that normally it can be neglected, and R_2 swamps out any variations due to temperature, etc. C_1 is adjusted to give the same input capacitance as the attenuators; C_2 is adjusted to give uniform frequency response (*i.e.* it compensates for the input capacitance of the FET).

For a positive input voltage, gate-source current will flow when the gate is more than, say, 0·7 V positive, but the current is limited by R_1 (and perhaps to some extent by the resistance in the FET source circuit). With an input of 100 V the current is limited by R_1 to 1 mA. The FET will ordinarily carry this current (or larger) without damage because the dissipation is low (the gate-source voltage being low). When a negative voltage is applied the gate-source voltage rises until avalanche breakdown occurs, the current again being limited by R_1. In this case the dissipation is greater and the FET must be capable of withstanding these conditions without damage. Owing to the dissipation, damage may be done if the overload is maintained. Better protection can be obtained by increasing R_1 and decreasing R_2. However, this results in some loss of signal, and necessitates more gain in the amplifier.

Some further protection may be incorporated by the use of diodes, as in figure 4.6, the input circuit being as figure 4.5. These diodes protect the FET, and also the transistors fed from the FET. For a positive gate input the FET

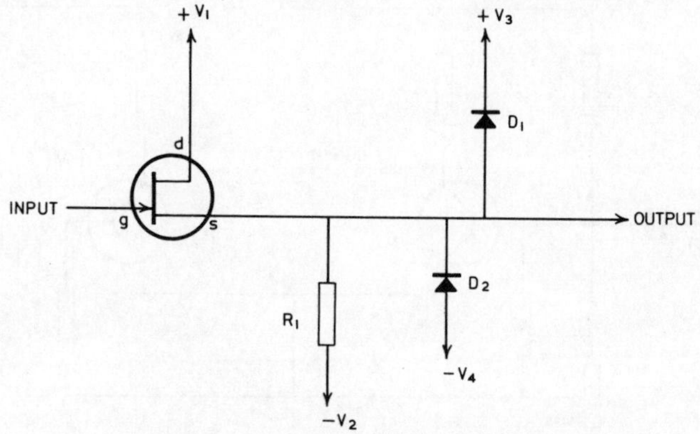

FIG. 4.6 F.E.T. INPUT CIRCUIT WITH DIODE PROTECTION ON F.E.T. SOURCE CIRCUIT

source voltage will rise, but limited by D_1 to a little over $+V_3$. With a negative gate input the FET source voltage will decrease, but this will be limited by D_2 to a little over V_4 and so protect later transistors. The gate-drain junction will break down as previously. An alternative is to place the diodes on the input as in figure 4.7. R_2 gives the required value of input resistance and R_1 limits the current flow. C_1 is for frequency compensation. For a positive input voltage D_1 will conduct and clamp the gate voltage to a value slightly above $+V_3$ (*i.e.* V_3 plus the drop across D_1). In the negative direction D_2 will clamp the gate voltage at a value slightly higher than V_4. In this way the gate voltage can be

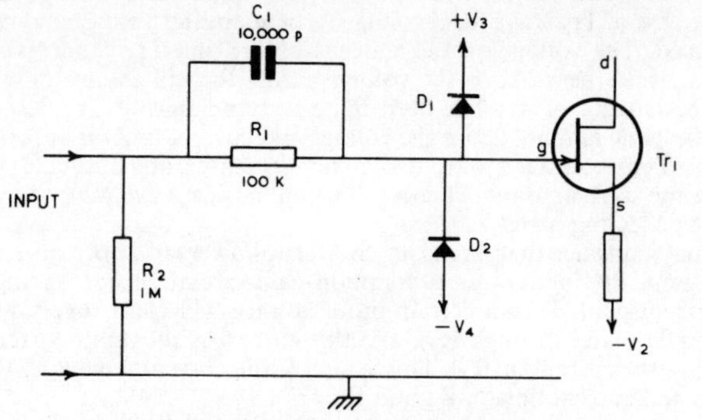

FIG. 4.7. F.E.T. INPUT CIRCUIT WITH DIODE PROTECTION ON THE GATE

limited to safe values. The disadvantage is the leakage currents of D_1 and D_2 and their capacitancies, which are now across the input of the FET.

The maximum voltage input to oscilloscopes is commonly given as 400 volts, but the conditions are not often stated. Generally, oscilloscopes will stand 400 volts d.c. or peak a.c. (or d.c. + peak a.c. if both present) on the most sensitive range, *i.e.* direct to the amplifier. However, this overload condition should not be maintained or damage may result.

As already stated, the input is single ended, but push-pull signals are required for feeding to the deflecting plates. This is almost always done by the use of a long-tailed pair. The circuit shown in figure 4.8 uses bipolar transistors, and will now be described.

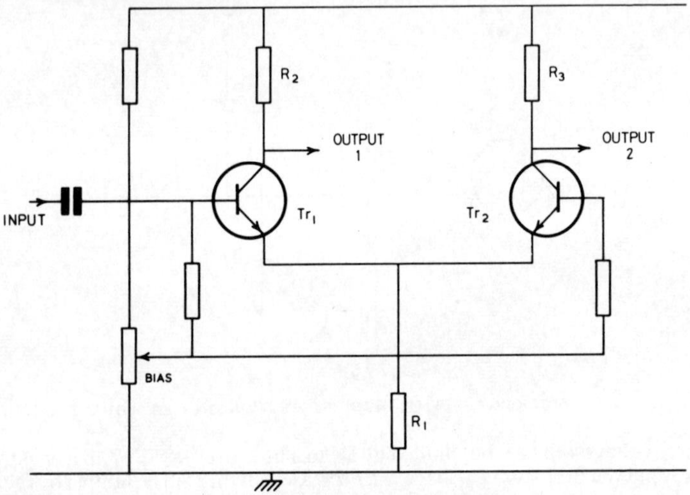

FIG. 4.8. BASIC LONG-TAILED PAIR CIRCUIT

Tr_1 and Tr_2 have a common emitter resistor R_1 (which may be replaced by a constant current circuit). The bases of Tr_1 and Tr_2 are fed with a suitable bias voltage and the signal is applied to the base of Tr_1. If the transistors have the same bias the circuit is symmetrical and equal currents will flow in Tr_1 and Tr_2, producing equal steady voltages at outputs (1) and (2). Suppose that the input signal goes in a positive direction. This will cause the current of Tr_1 to increase and this current will flow in R_1 causing a rise in voltage across R_1. Since the base of Tr_2 is at a fixed voltage its base-emitter voltage and its current are reduced. The voltage on the emitters of Tr_1 and Tr_2 cannot vary much without cutting them off, so the voltage across R_1 will change only slightly. Hence the current of Tr_1 has been increased and that of Tr_2 decreased by almost the same amount (since the voltage and current in R_1 is approximately constant). The potential of output (1) hence decreases, and output (2) increases by about the same amount. Thus a push-pull output is available to eventually drive the deflecting plates.

One may consider that Tr_1 is an emitter-follower, of approximately unity voltage gain, driving Tr_2 as a common-base circuit, hence having almost unity current gain. Thus a certain input voltage will cause a certain current to flow in Tr_1 (and through R_2), and this current is the input current to Tr_2 (neglecting the current in R_1). This is almost the current flowing in R_3. Thus, almost equal current flow in R_2 and R_3.

A circuit is shown in figure 4.9 where C_1, R_1 and R_2 form the protection circuit already described. Tr_1 and Tr_2 are junction FETs, commonly a

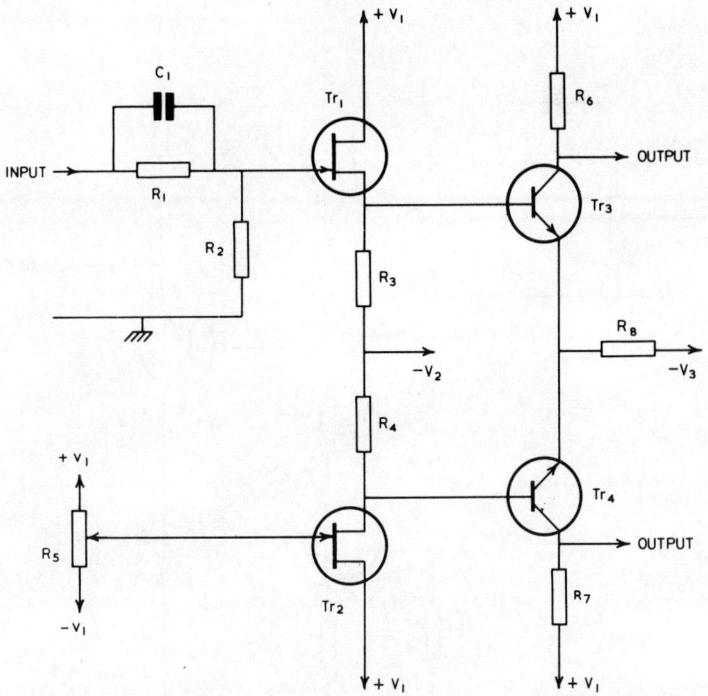

FIG. 4.9. TYPICAL INPUT CIRCUIT USING FETs

matched pair to reduce drift due to temperature changes. Tr_2 is connected similarly to Tr_1 but fed with a variable bias from R_5, which may form the SHIFT or VERTICAL control. R_3 and R_4 are the source-resistors connected to a suitable negative supply. The signal voltage is developed only across R_3, but if there is no signal and the circuit is balanced (*i.e.* no shift voltage is applied), the voltages across R_3 and R_4 will be equal. Thus, equal steady or bias voltages are fed to bipolar transistors Tr_3 and Tr_4, which form a long-tailed pair with the common emitter resistor R_8 (which may be replaced by a constant current circuit). R_6 and R_7 are the collector-load resistors and from these resistors push-pull signal outputs are available, as explained earlier. Varying R_5 is equivalent to an equal and opposite signal fed to Tr_1, and causes a shift of the beam so that the trace can be moved vertically to any part of the screen. There are many variations of the basic circuit; emitter-followers may be placed between the FETs and Tr_3 and Tr_4.

An alternative is to maintain a single-sided circuit until later in the amplifier. A basic circuit is given in figure 4.10. The two FETs, Tr_1 and Tr_2 carry the same current since they are in series. The bias voltage on Tr_2 is the voltage across R_3. Since $R_3 = R_1$ the voltage across R_1 must equal that across R_3. If the transistors are assumed to be the same then the drop across R_1 must be the bias voltage of Tr_1. If the output is taken from the source of Tr_1 it will be more positive than the gate voltage by this bias voltage, but by taking the output from the lower end of R_1 the voltage at this point must equal the input or gate voltage. R_2 together with Tr_2 form the source load. Tr_3 acts as an emitter-follower with emitter-load resistor R_4. In the oscilloscope (Advance Electronics) this output is taken to a common-emitter amplifier with switched

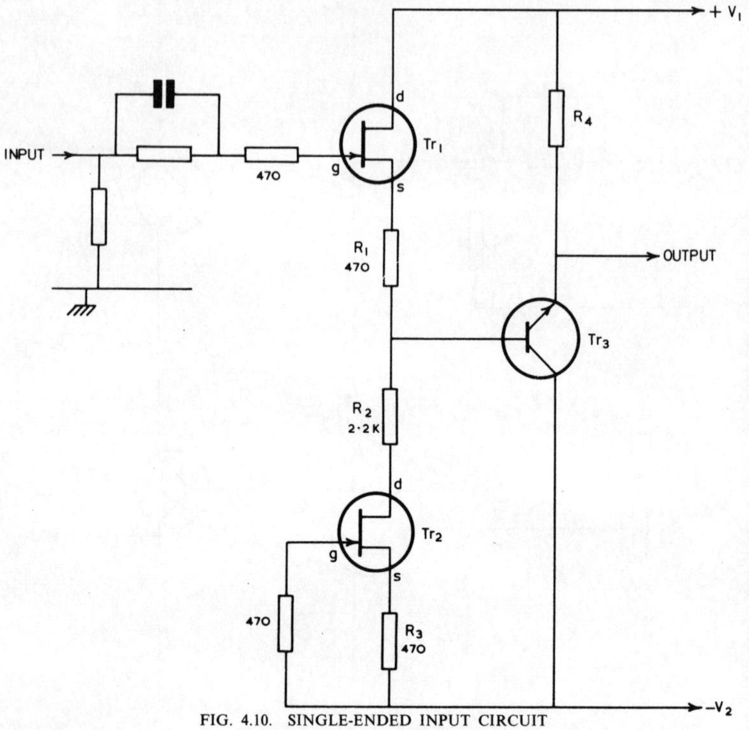

FIG. 4.10. SINGLE-ENDED INPUT CIRCUIT
(Simplified Advance Electronics)

feedback so the gain can be varied (this takes the place of some of the attenuators, only a ÷100 attenuator being used). This stage then feeds a long-tailed pair from which a push-pull output is obtained to drive the output stage consisting of a push-pull cascode stage.

In all these circuits, since they are d.c. amplifiers, drift occurs hence the circuits are designed to reduce it as much as possible, particularly with temperature. They are basic circuits; the methods used to reduce drift are outside the scope of this book.

In place of junction FETs for the input stage MOSFETs (metal oxide semiconductor field effect transistors) may be used. As the input resistance of a MOSFET is so high (leakage current, say, 1 nA) it can be connected as a common-source amplifier instead of a source-follower. The feedback capacitance between drain and gate can be reduced to an extremely small value by the use of dual gate MOSFETs (say 20 fF where fF is a femto Farad = 10^{-3} pF). They act like a cascode circuit, the second gate being at earthy potential as regards a.c. A simplified circuit used by Scopex is given in figure 4.11.

Protection is applied by the use of diodes D_1 and D_2, the current being limited by R_2 (the circuit is similar to one already described). Excess voltage must not be applied to a MOSFET as the insulation will break down and the transistor will be destroyed. The second gates of Tr_1 and Tr_2 are fed from R_3 and R_4, and a suitable bias voltage is fed to Tr_2 from R_5 which forms the Y SHIFT control. These two transistors act as a long-tailed pair with the common-emitter resistor R_6. The drain-load resistors are R_7 and R_8 which feed another long-tailed pair, Tr_3 and Tr_4. The common emitter resistor is now R_{10}.

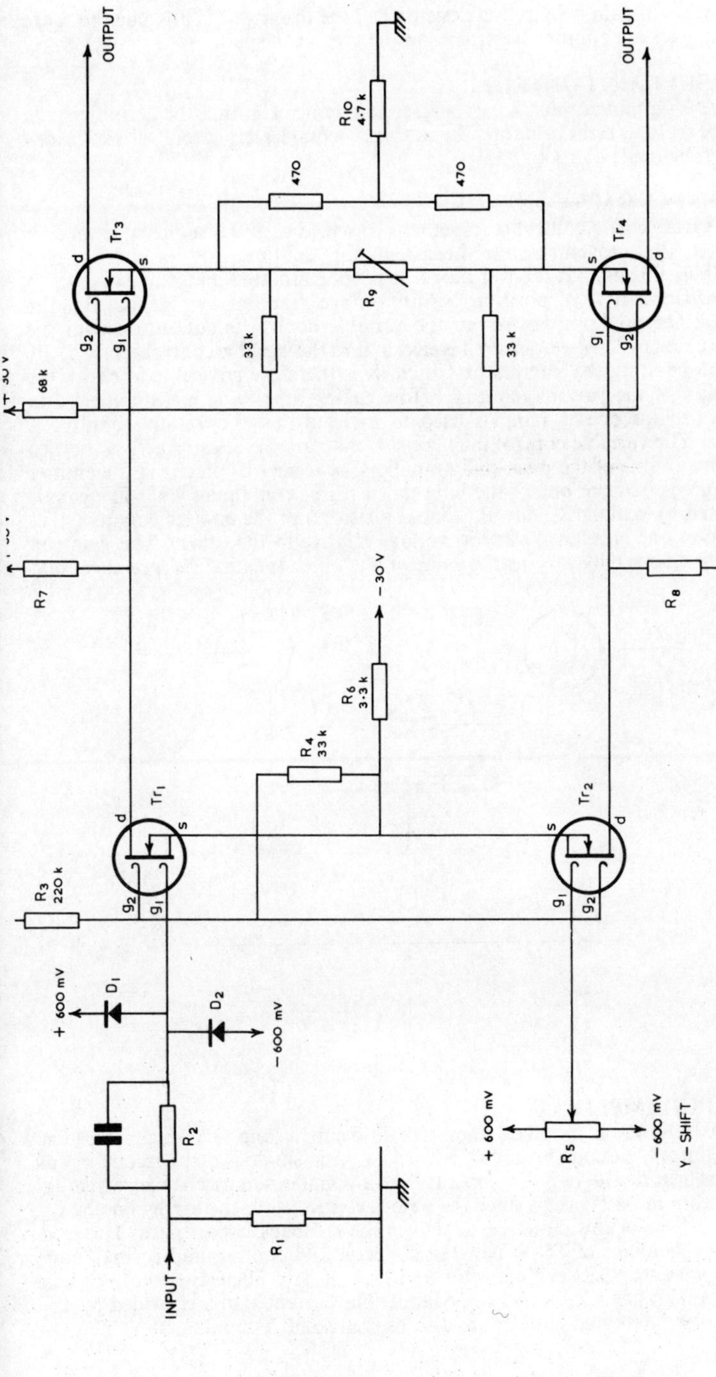

Fig. 4.11. INPUT CIRCUIT USING MOSFETs
(Simplified Scopex Instruments)

Variation of gain is by preset resistor R_9; the lower this is in value the more the sources are coupled together and the greater the gain.

PRESET GAIN CONTROL

The Y-amplifier must have a preset gain control so that the amplifier gain can be set to its right value and the calibration of the attenuator (volts/division) switch be made correct.

VARIABLE GAIN CONTROL

A variable gain control is sometimes provided to give a sufficient change of gain so that, together with the attenuator, continuously variable gain is available throughout the full range. It is not calibrated and it must be set to the *calibrate* or CAL position before voltage readings are taken from the screen. On oscilloscopes where the variable control is concentric with the attenuator switch (VOLTS/DIV) it is easy to turn the variable control accidentally when operating the attenuator switch. It is therefore advisable to check the position of the variable control before taking amplitude measurements. In some oscilloscopes a lamp is used to indicate that the variable control is in use. The variable control may consist of a variable resistor across the two outputs of one of the push-pull amplifiers, or it may be placed in the emitter or source circuit of one of the long-tailed pairs, as in figure 4.12. The gain is reduced by adding R_4 and R_5 (equal resistors) in the emitter circuits (since it allows one emitter to vary in voltage relative to the other). The gain can then be changed by varying the value of R_6, which eventually shorts them out.

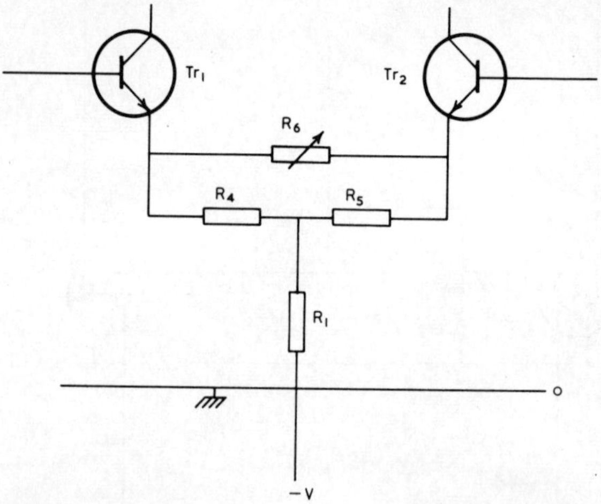

FIG. 4.12. VARIABLE GAIN CONTROL CIRCUIT

OUTPUT AMPLIFIER

This follows on the same lines, usually being a long-tailed pair amplifier. Sometimes a cascode amplifier is used; a basic single-ended cascode circuit is shown in figure 4.13. Tr_1 operates as a common-emitter stage, while Tr_2 is a common-base circuit since the base is connected to the earthy line by C_1. The purpose of this circuit is to reduce the feedback capacitance. The feedback capacitance of Tr_2 is between collector and emitter, and is small compared with that between collector and base of Tr_1. There is no current gain in Tr_2 and it has a low input impedance. The current gain is provided by Tr_1.

Bipolar integrated circuits are also used in some Y-amplifiers.

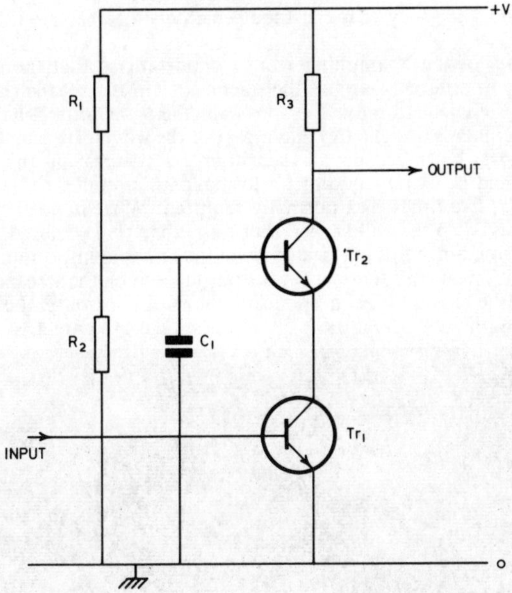

FIG. 4.13. BASIC CASCODE AMPLIFIER CIRCUIT

DELAY LINE

In many cases the timebase will be triggered by the signal itself; details are given in Chapter 6. The triggering of the timebase takes a finite time. Therefore, if the same signal is fed to the timebase as to the Y-plates then the first part of a signal will not be displayed when it occurs at high speed. To overcome this the signal can be delayed in the Y-amplifier by a time approximately equal to the triggering time. This is normally produced by a length of coaxial cable connected to a suitable point (*after* the trigger take-off point) in the amplifier. The cable must of course be suitably terminated at both ends to prevent distortion of the waveform being examined. The delay time required will depend on the oscilloscope, but may be 0·1–0·25 μs. A thin film circuit is sometimes used.

Obviously, frequency compensation is necessary in the amplifier, particularly when the Y-amplifier goes up to high frequencies. A bandwidth of 10 MHz now appears to be a minimum figure; 50 MHz is fairly common and there are amplifiers available going up to 500 MHz. The detailed analysis of these circuits and means of compensation are beyond the scope of this book.

Some amplifiers have facilities to increase the gain by, say, a factor of 10, but at reduced bandwidth. This is often useful but is becoming less important with the greater sensitivity of the amplifiers. When two Y-amplifiers are used (in double-trace oscilloscopes) it is sometimes possible to cascade two so as to get limited increased gain, say $\times 10$.

Since the Y-amplifier sensitivity is now calibrated it is important that it is stable and that there should be little drift on d.c. Variations in mains voltage can be prevented from altering the gain by the use of stabilizers for the various voltages. These may be conventional electronic stabilizers or a constant-voltage transformer (used in one model by Advance Electronics).

Variation in mains voltage will vary the e.h.t. The use of stabilized supplies (which are costly) can be eliminated by making the gain of the amplifier increase in the same way that the tube sensitivity decreases with increased e.h.t. voltage.

The linearity of the Y-amplifier is also important so that the calibration does not vary at different parts of the screen. In most cases the shift control is more than sufficient to move the spot over the total vertical height of the tube. The amplifier is designed so that a part of the waveform may be examined, say a $\frac{1}{2}$ to $\frac{1}{5}$, by increasing the sensitivity *i.e.* decreasing the VOLTS/DIV setting. It would be extremely difficult to make an amplifier to have a linear output of, say, five times that normally required. What happens is that the amplifier limits when the voltage is rather more than that required to give full screen deflection. Since this part is not seen, limiting is unimportant, provided the amplifier is linear over that part of the output seen on the screen. The effect of increasing the sensitivity, and operating the shift control on the waveform (as seen on another c.r.o.) on the Y-plates, is shown in figure 4.14.

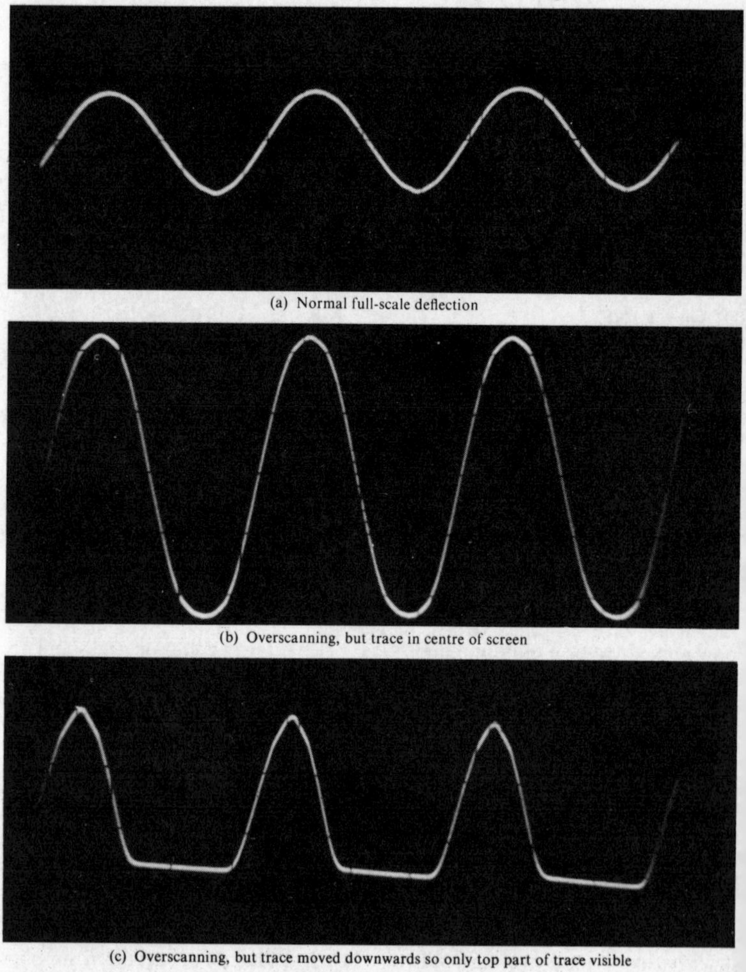

(a) Normal full-scale deflection

(b) Overscanning, but trace in centre of screen

(c) Overscanning, but trace moved downwards so only top part of trace visible

FIG. 4.14. VOLTAGE ON DEFLECTING PLATE OF OSCILLOSCOPE

BEAM LOCATE BUTTON

When an oscilloscope is used with large X and Y shifts it is sometimes difficult to find the beam, and many oscilloscopes have a BEAM LOCATE or BEAM FINDER button. This reduces the gain of both X and Y amplifiers (including the amount of the shift) so that a trace will appear on the screen under all conditions.

The trace may be distorted and will not be in the centre of the screen. The X and Y shift controls should be operated to bring the trace to the centre of the screen, both horizontally and vertically. If the button is now released the trace will appear approximately in its correct position. It may be too large in amplitude, in which case the VOLT/DIV controls must be moved to a less sensitive (more volts/div) position.

Some oscilloscopes have neon indicators, one at the top and bottom and one at each side. If the trace is off screen then the neon lamp(s) lights, corresponding to the position the trace is off screen.

In the case of a triggered timebase no trace will appear when the BEAM LOCATE button is pressed unless the timebase is triggered as the beam will be cut off. This difficulty can be overcome by arranging that the BEAM LOCATE button makes the timebase self-run; or a lamp can be used to indicate whether the timebase is triggered. In the latter case one must get the timebase running either by adjustment of the trigger level or setting it to AUTO before using the locate button.

BANDWIDTH AND RISE TIME

At one time a frequency response of 1 MHz was considered good, but today the 'normal' oscilloscope goes up to 10 MHz, and there are instruments readily available going up to at least 500 MHz. In this section we will discuss what is meant by bandwidth and its importance.

First, the definition of bandwidth. Assuming a d.c. amplifier, the response curve, *i.e.* the amplitude of deflection at different frequencies for a constant input amplitude, is as shown in figure 4.15. At high frequencies the amplitude of the trace will decrease and the frequency at which the amplitude has

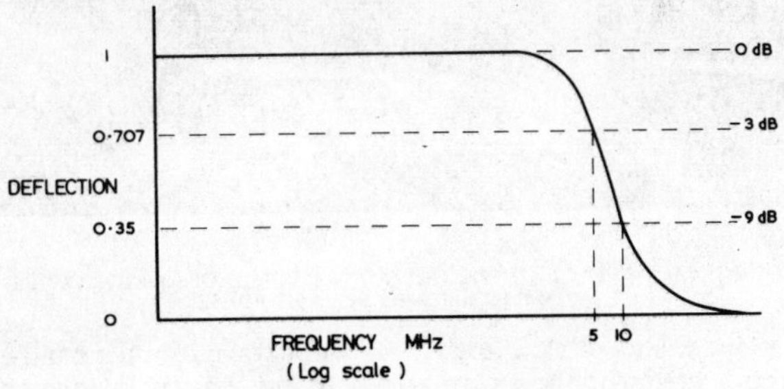

FIG. 4.15. RESPONSE OF OSCILLOSCOPE

decreased to 0·707 (−3 dB) of its low-frequency value (often taken as 50 kHz) is called the bandwidth. The way in which the response drops can vary, *i.e.* it may be a slow roll off or a rapid decrease; the ideal is as shown, *i.e.* it falls to 0·35 (−9 dB) at a frequency of twice that at the 0·707 (−3 dB) point. In other words, the rate of fall over this region is 6 dB per octave (6 dB per doubling of the frequency). The frequency response required obviously depends on the range of frequencies to be examined, but *most important* on the shape of the waveforms of the voltage to be examined. Sine waveform voltages up to a maximum frequency of 1 MHz would need an oscilloscope with a bandwidth of 1 MHz (or a little more). If one wishes to examine 1 MHz square waveforms then the situation is quite different. A square waveform of 1 MHz contains a fundamental of 1 MHz and a large number of harmonics – in theory an infinite number. If it is considered necessary to reproduce the 3rd, 5th, 7th and 9th harmonics without reduction in magnitude, an oscilloscope with a bandwidth of 10 MHz is essential. A square waveform on an oscilloscope having a bandwidth of 1 MHz would be very distorted and look almost like a sinewave. Thus it is necessary to make sure that the bandwidth of the oscilloscope is adequate or there will be misleading results. This is illustrated in figure 4.16.

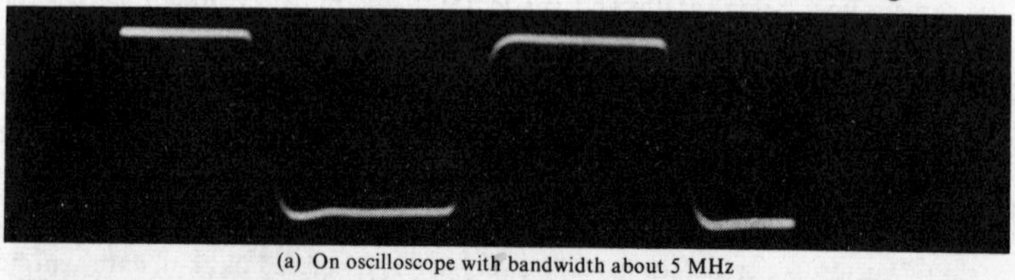

(a) On oscilloscope with bandwidth about 5 MHz

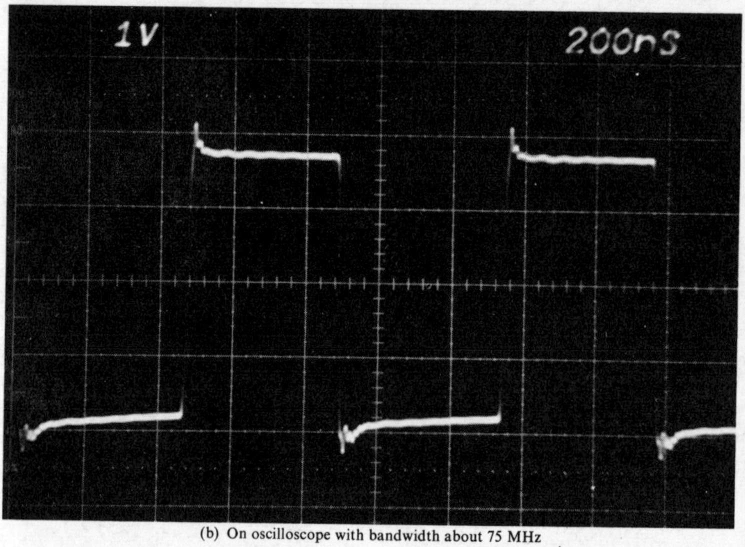

(b) On oscilloscope with bandwidth about 75 MHz

FIG. 4.16. EFFECT OF FREQUENCY RESPONSE OF OSCILLOSCOPE. APPROX. 1 MHz SQUARE PULSE

Thus if one wishes to view any type of pulse waveform the bandwidth must be large compared with the repetition rate or frequency of the pulse waveform. Instead of specifying the bandwidth the rise time of the oscilloscope is commonly quoted as this is more important when dealing with pulses having fast rise times. The rise time is defined as shown in figure 4.17. It is the time t taken for the voltage to rise from 10% to 90% of its maximum amplitude. There is a relationship between the rise time and bandwidth as follows:

$$\text{Rise time } t \times \text{bandwidth } f = 0{\cdot}35 \qquad (4.1)$$

where t is in seconds and f in Hertz.

This is an approximate relationship and depends on how rapidly the response falls off at high frequencies. If it falls more rapidly than that shown in figure 4.15, some overshoot in the waveform will result, as in figure 4.18(a); or a

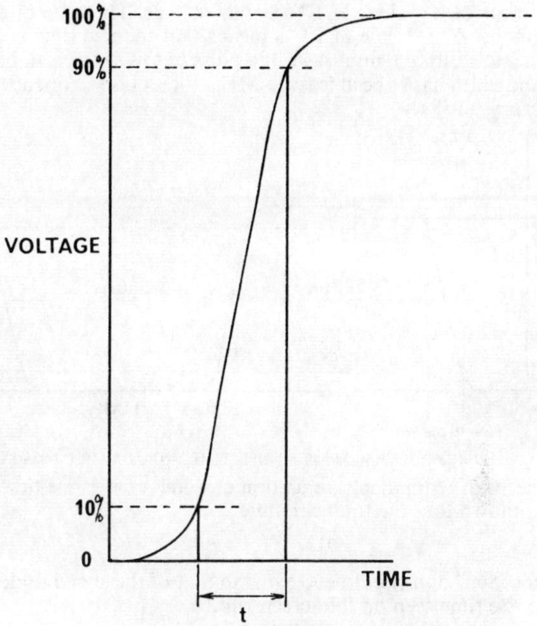

FIG. 4.17. MEASUREMENT OF RISE TIME

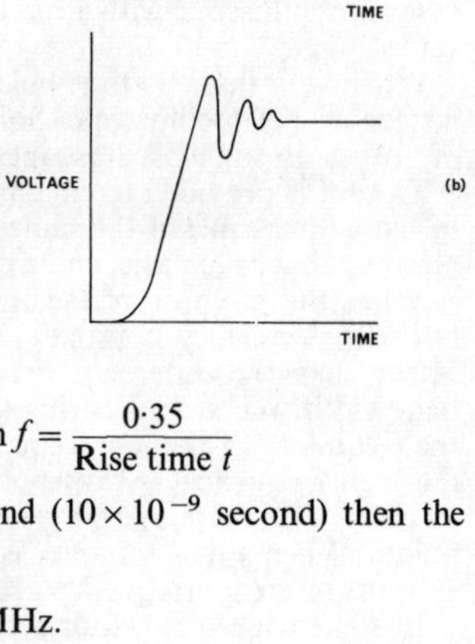

FIG. 4.18. EFFECT OF RAPID DECREASE OF GAIN WITH FREQUENCY
(a) Overshoot; (b) Damped oscillation

damped oscillation may occur, as at (b).

Turning equation (4.1) around $\text{Bandwidth } f = \dfrac{0{\cdot}35}{\text{Rise time } t}$

Thus, if the rise time is to be 10 nanosecond (10×10^{-9} second) then the bandwidth must be at least

$$\frac{0{\cdot}35}{10 \times 10^{-9}} = 35\text{ MHz.}$$

To see what this means consider pulses at 1 MHz having equal mark-space ratios. The time of a cycle is $1/10^6$ seconds or 1 μs. The time of the pulse is 0·5 μs or 500 ns. A rise time of 10 ns means that the rise time is $\frac{1}{50}$th of the period of the pulse, as in figure 14.19. If a pulse of this shape is to be examined then the bandwidth has to be at least 35 MHz, for a pulse frequency of 1 MHz, for satisfactory results.

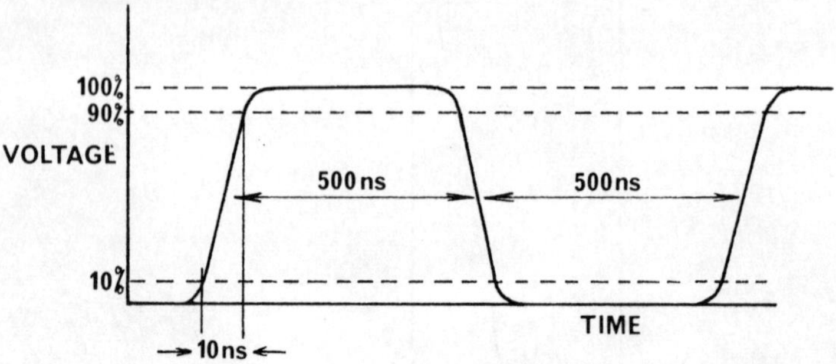

FIG. 4.19. 1 MHz PULSE WITH 10 ns RISE TIME (NOT DRAWN TO SCALE)

If two pieces of equipment are used in cascade having rise times of t_1 and t_2 it can be shown that the total rise time

$$T = \sqrt{t_1^{\,2} + t_2^{\,2}}$$

If an oscilloscope has a rise time t_1 equal to that of the signal under examination t_2, the rise time seen on the screen will be

$$T = \sqrt{t_1^{\,2} + t_1^{\,2}} = 1{\cdot}41\, t_1$$

i.e. it will be 40% longer than it should appear. To prevent this error the rise time of the oscilloscope should be small compared with that of the signal being examined. If the rise time of the oscilloscope is five times less than that of the signal the error is only 2%. Thus, if we wish to measure pulse rise times of 10 ns the rise time of the oscilloscope should be less than 2 ns, which corresponds to a bandwidth of some 175 MHz. Rise-time measurements are important in many pulse applications. When very fast rise-time measurements are required comparisons are better than direct measurements, and in this case an oscilloscope with a rise time approximately the same as the signal will suffice.

When very fast rise-time pulses are to be examined great care is necessary in feeding the oscilloscope. Some details are given in Chapter 8 on the use of probes. In some oscilloscopes designed for measuring high rise-time pulses a Y-input is provided having an input impedance of 50 ohms. This can then be fed with a cable of the same characteristic impedance from a source of 50 ohms so that reflections and distortion are reduced to a minimum.

When the response of the amplifier drops off at low frequencies the 0·707 (−3 dB) frequency is usually given, together with the high frequency limit rather than the difference between them, which is the usual definition of bandwidth. When considering the reference frequency this should be 20 times the frequency of the low frequency −3 dB point, and $\frac{1}{20}$ of the frequency at the high frequency −3 dB point. In giving the bandwidth in this way it is assumed that there are no great variations of gain between the two −3 dB points. When a d.c. amplifier is used the response at d.c. should be the same as at the reference frequency and not −3 dB from it.

In some wide-bandwidth oscilloscopes the bandwidth can be reduced (say from 200 MHz to 20 MHz) when the wide bandwidth is not required. This reduces the noise and gives a clearer trace.

5

X-deflection Amplifier

THE main purpose of the X-amplifier is to apply the signal from the timebase or sweep generator to the X-deflecting plates. The design will depend on whether this is its only purpose or whether it is to be used for external signals and of what type. In many oscilloscopes the X-amplifier simply amplifies the timebase output, and little amplification may be required. It is necessary to change the single-ended (unsymmetrical) output of the sweep generator to a balanced or push-pull output to feed the deflecting plates. This change is normally done by the use of a long-tailed pair, as described in Chapter 4. The deflecting voltage required is usually larger than that on the Y-plates since the X-plates sensitivity is less than that of the Y-plates. The ratio of X to Y sensitivity varies, but commonly the Y-plates are about twice as sensitive as the X-plates. The required frequency response of the amplifier will depend on the timebase frequency.

The maximum speed of the sweep generator is often about 100 ns/division and, assuming 10 divisions across the screen, this means the time for the complete sweep is 1000 ns or 1 μs. There will be some time for the flyback and hold off, but the repetition rate of the timebase is approaching 1 MHz. This is a sawtooth waveform, and one might expect that the bandwidth would have to be high to reproduce the sawtooth correctly. It is unnecessary to reproduce the sawtooth waveform correctly provided the sweep is maintained linear. What happens during flyback is of little importance. Thus, the bandwidth of the X-amplifier is usually less than that of the Y-amplifier and may, in fact, be only a tenth. The general design follows that of the Y-amplifier already described.

It is usual to provide some magnified sweep so that the trace can be expanded. This may be by means of a VARIABLE MAGNIFICATION CONTROL (variable gain control) giving a maximum magnification of say, 5 or 10 times. When this is in use it MUST be borne in mind that the calibration of the X-scale has been lost and that the *calibrated horizontal time-scale can be used only when the* VARIABLE MAGNIFICATION *control is in the* CALIBRATE *position*. Alternatively, a switch may be provided to give either 5 or 10 times magnification (in some cases 100 times). With this the calibration is maintained and the SWEEP SPEED setting is multiplied by the magnification used.

As in the Y-amplifier an X-SHIFT or POSITION control is required in order that the waveform may be moved horizontally. This is essential when SWEEP MAGNIFICATION is used so that any part of the waveform can be examined. The POSITION control must be able to move the trace from one end to the other, *e.g.*, 10 times the screen width. The magnitude of the X-deflecting voltage for normal scan may be 60 V p-p and it might be inferred that, on 10 TIME MAGNIFIED SWEEP, the amplifier must provide 600 V p-p. This is not the case as the change of voltage is not much more than 60 V p-p, say 100 V p-p, due to limiting by the circuit.

This is shown in figure 5.1, where it is seen that a slice is taken from the normal sweep waveform in the MAGNIFIED SWEEP setting, the position of the slice being determined by the setting of the X-POSITION control. The amplifier

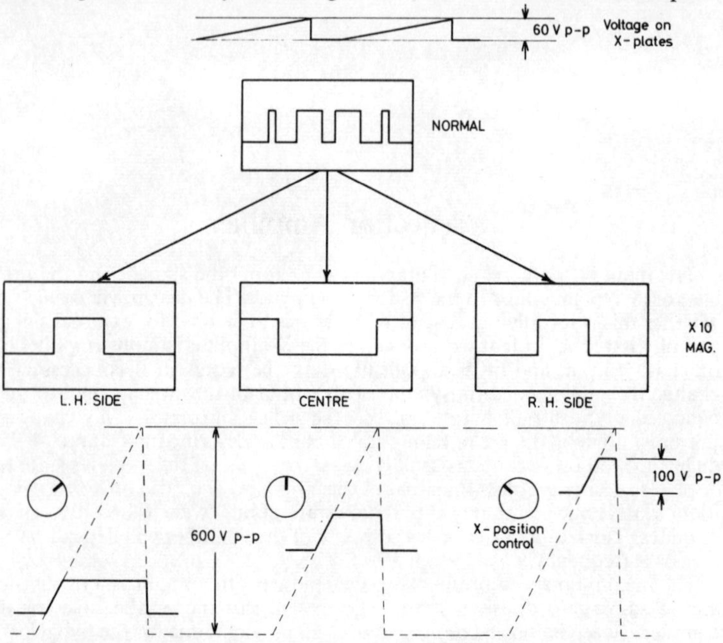

FIG. 5.1. USE OF X-EXPANSION CONTROL AND POSITION CONTROL

must be designed so that this limiting does not upset the accuracy of the sweep when operating under all conditions.

The effect of using the × 10 MAGNIFICATION switch is shown in figure 5.2. The × 10 magnification does, of course, increase the sweep speed by this factor and hence, if the normal maximum sweep speed is 100 ns/division, the maximum with sweep magnification is 10 ns/division. This is a very fast sweep speed, particularly if one realizes that at this speed a Y-signal at 100 MHz (assuming the oscilloscope will operate at this frequency) will produce a trace with 1 cycle corresponding to 1 division.

When a TRACE LOCATE or SPOT FINDER button is fitted the X-gain is reduced, and also the amount of shift, so that the trace always appears on the screen.

Where plug-in timebases are fitted part of the X-amplifier is situated in the plug-in unit and part in the main frame.

Turning now to external X-deflection facilities, these vary greatly depending on the oscilloscope, many only making a very modest provision. A switch may be provided as in figure 3.1, so that an external signal can be fed to the X-amplifier provided for the timebase. In this case no provision is made for differing magnitudes of input signal, although some variation of gain is available by the use of the SWEEP MAGNIFICATION control. The sensitivity of the amplifier will generally be limited and, as already stated, the bandwidth may be quite small compared with that of the Y-amplifier.

For best results the X and Y amplifiers should be similar so that there are equal phase shifts on the two axes. If this is not the case then when the oscillo-

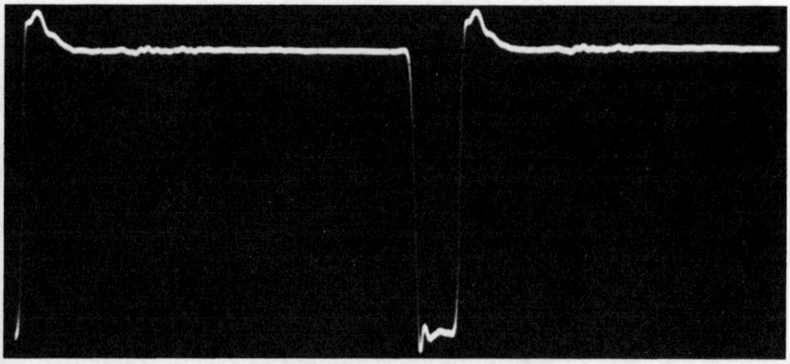

(a) Sweep speed set to 0·1 μs/div and no X expansion

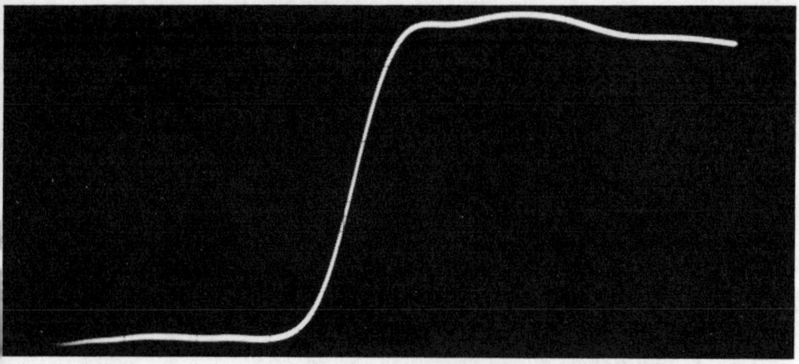

(b) Sweep speed set to 0·1 μs/div and × 10 horizontal expansion corresponding to sweep speed of 0·01 μs/div or 10 ns/div.

FIG. 5.2. PULSE WAVEFORMS

scope is used to display characteristics of devices, loops tend to form due to the differing phase shifts. Some oscilloscopes are made with identical X and Y amplifiers, which increases the cost.

When dual-trace oscilloscopes are used (see Chapter 10) two Y-amplifiers are provided, one for each trace. On some of these equipments it is possible to switch one of these amplifiers into the X-circuit so making identical amplifiers available in both X and Y directions, but, of course, only a single trace can then be used. In one oscilloscope (Tektronix) it is possible to plug a Y-amplifier into the position normally occupied by the plug-in timebase unit. This gives the advantage of having two high-quality amplifiers, one for each direction of deflection. Both arrangements afford full facilities of the normal Y-amplifiers for both X and Y direction, *e.g.* attenuator, a.c./d.c. switch, full bandwidth, etc.

6

Trigger Circuits and Timebase Circuits

THE trigger and timebase circuits can be extremely complex and are very variable as regards circuits. Accordingly, only the general principles and block diagrams can be considered with some basic circuits. Almost all oscilloscopes now use a triggered timebase or sweep generator. This timebase does not run continuously (except in the automatic mode to be described later) and, in the absence of trigger pulses, the spot remains on the left-hand side of the screen. The purpose of the trigger circuit is to produce suitable pulses normally from the signal under examination (but may be from another external signal). When a pulse occurs the timebase starts its scan or sweep and travels across the screen at a speed settled by the timebase SWEEP SPEED control. When the beam reaches the right-hand side of the screen it rapidly returns to the left-hand side (the flyback) and waits for another triggering pulse.

A block diagram of the basic arrangement is shown in figure 6.1. The TRIGGER SELECTOR block determines from where the triggering waveform is to be

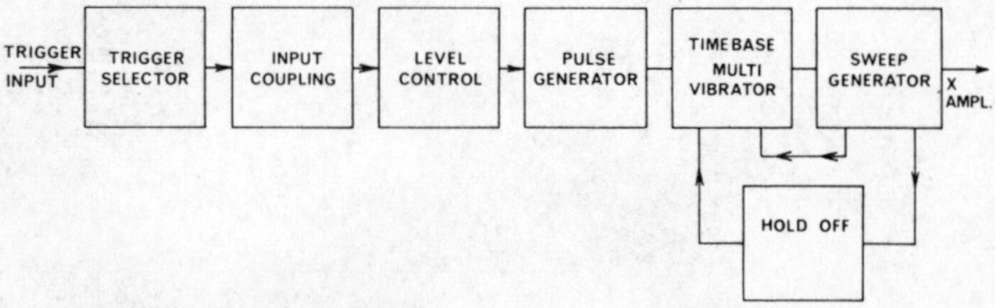

FIG. 6.1. BLOCK DIAGRAM OF TRIGGER AND TIMEBASE CIRCUITS

obtained. The INPUT COUPLING determines in what way the selected trigger waveform will be modified. The LEVEL CONTROL block determines at what level of input signal the timebase shall be triggered, and also whether it is to be triggered by a positive- or negative-going signal. The PULSE GENERATOR gives a pulse of fixed magnitude to operate the TIMEBASE MULTIVIBRATOR which, in turn, operates the RAMP GENERATOR or SWEEP GENERATOR to produce the scan *via* the X-AMPLIFIER. Other blocks may be added, as will be seen later.

The trigger circuit will be considered first.

TRIGGER SELECTOR BLOCK

This block establishes from where the triggering waveform shall come; there are three general sources:

(a) Internally from some part of the Y-amplifier.
(b) Externally from the external trigger socket.
(c) Mains or line, the waveform being obtained from a suitable transformer connected to the mains.

(a) The output from the Y-amplifier must be taken from such a point that a reasonable amplitude is obtained even with a small trace height, say for a signal of 10%, or less, of maximum. The circuit must be designed so that the connection of the triggering circuit does not upset the operation of the Y-amplifier, and so a buffer stage may be necessary. If a Y-delay line is used then the triggering waveform *must* be taken before the delay line.

(b) The external trigger socket should have a high resistance, say 1 MΩ, and a small capacitance, say 25 pF or less, so that its connection to a circuit has a minimum effect on the circuit. Obviously, there will be a minimum signal for satisfactory triggering; this should be fairly small, say 100 mV to

500 mV. The figure will vary with frequency, the higher the frequency the greater the voltage for triggering. There will also be a maximum input. In some oscilloscopes a simple attenuator can be switched into circuit to decrease the triggering sensitivity, usually marked ÷10.

(c) Although not always provided, this facility can be extremely useful. When working with mains frequency equipment the internal trigger may be used, but if the voltage varies, it may be necessary to alter the trigger level control from time to time. This can be overcome by feeding a mains frequency voltage into the external trigger socket, but it is more convenient if it can be done internally by the operation of a switch.

Multitrace oscilloscopes require additional switching so that the trigger waveform can come from any of the inputs. This is fully discussed in Chapter 10.

INPUT COUPLING BLOCK

This block determines how the selected triggering waveform shall be modified before being applied to the LEVEL CONTROL block. The facilities available will depend on the type and intended use of the oscilloscope. The facilities that may be provided are as follows:

(1) D.C. In this position there is a direct d.c. connection from the trigger waveform source to the LEVEL CONTROL block. The timebase will now trigger when the waveform reaches a certain d.c. level as set by the level control. Suppose that we have a waveform as shown by the full line of figure 6.2, which has both a d.c. component and an a.c. component, *i.e.* it is a varying

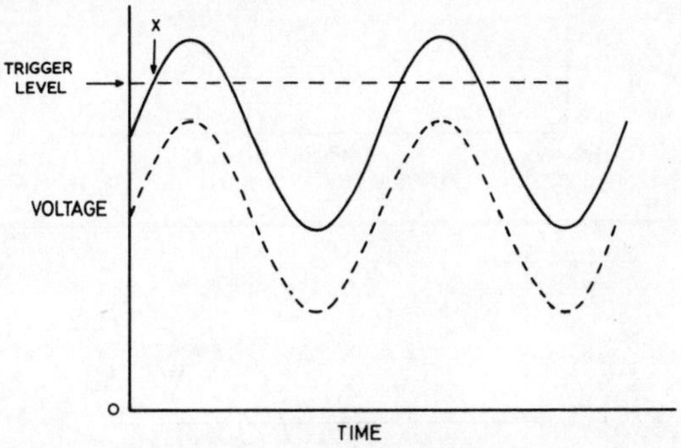

FIG. 6.2. EFFECT OF D.C. COMPONENT ON TRIGGERING

d.c. voltage. For the trigger level shown the timebase will trigger at point X. If the d.c. component is decreased (but the a.c. component remains at the same value), as shown by the dotted curve, the timebase will not be triggered. Figure 6.3 shows an application where this is useful. Suppose now that we have a television waveform as shown, which has its full d.c. component. At (a) is the case where the picture is mainly white, while at (b) it is mainly black. However, if the d.c. component of the waveform is present, then the tips of the synchronizing pulses will stay at the same voltage. Thus, triggering will always occur at point X (for the particular trigger level setting shown) independent of the picture content. Thus stable triggering will result.

(2) A.C. In this position a blocking capacitor is placed in series to the LEVEL CONTROL block so that any d.c. component is eliminated. The response is commonly uniform down to a frequency of, say, 5 Hz. Thus, considering the waveform of figure 6.2, *as regards the trigger circuit* (and not that seen on the screen), the waveform will be as in figure 6.4 and, with the trigger level shown, the timebase would trigger at point X. If the d.c. component were changed,

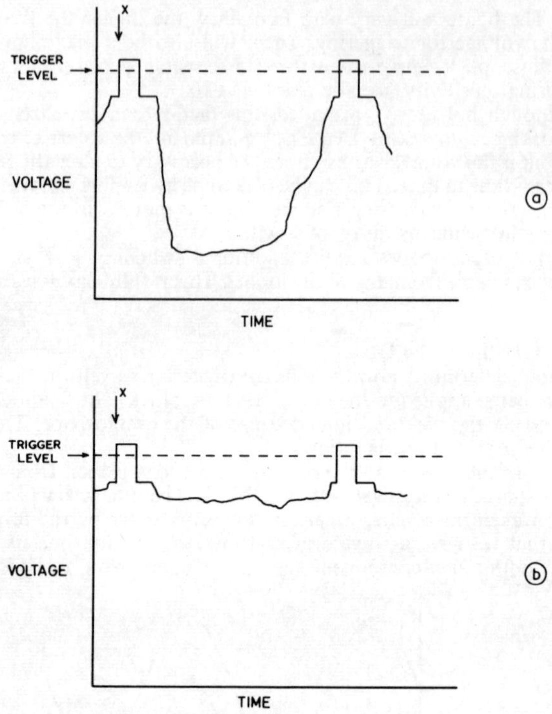

FIG. 6.3. USE OF D.C. TRIGGER SETTING WITH TELEVISION WAVEFORM

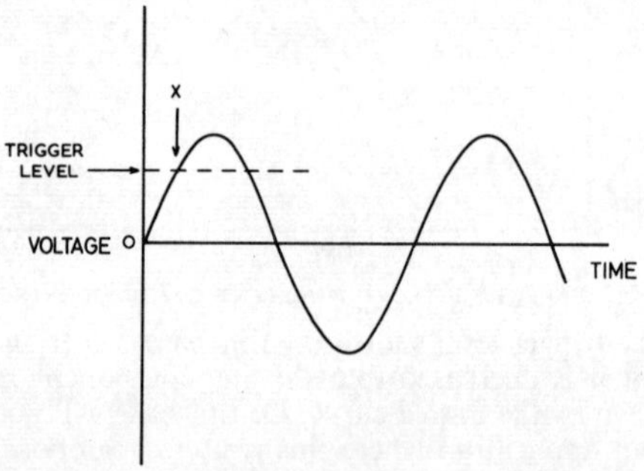

FIG. 6.4. TRIGGER WAVEFORM WHEN INPUT COUPLING SET TO A.C.

as shown dotted in figure 6.2, then, *as regards the trigger circuit* (but not as regards the display on the screen), the waveform will be as figure 6.4 and will not change (assuming the a.c. component to be the same in both cases). Thus the triggering would remain constant independent of the d.c. component. It must be emphasized that it is the triggering circuit that is being discussed and not the selection of the waveform to the Y-plates. If the Y-selector were on the d.c. position the waveform would appear as figure 6.2 in both cases.

Consider now the television waveform as in figure 6.3 but with the trigger selector in the a.c. position. As regards the LEVEL CONTROL block the wave-

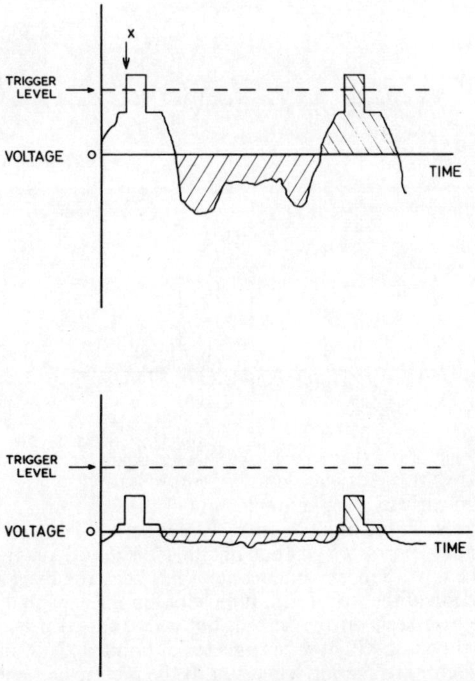

FIG. 6.5. EFFECT OF CHANGE OF TELEVISION WAVEFORM WHEN TRIGGER INPUT COUPLING SET TO A.C.

forms would now appear as in figure 6.5, with no d.c. component, *i.e.* the area above the zero line must be equal to the area below, as indicated. With a picture of mostly white content, as at (a), the synchronizing pulse will cross the trigger level shown and the timebase will be triggered at point X. However, when the picture is mainly black, as at (b), the synchronizing pulse does not reach the trigger level and the timebase will not be triggered. Therefore, to maintain the timebase triggering on a synchronizing pulse it is necessary to keep changing the trigger level as the picture content changes. Hence, for this waveform the d.c. position is the correct one to use.

If a low frequency signal is being examined it is better to use the d.c. position due to the attenuation in the a.c. position at very low frequencies. This is particularly true in the case of a sine waveform where the rate of change of voltage will be small in relation, say, to a pulse waveform.

(3) L.F. reject (low frequency reject), also called a.c. fast or h.f. (high frequency). In this case a small capacitor is used between the TRIGGER SELECTOR block and the LEVEL CONTROL block. This, of course, removes the d.c. component as in (2), but greatly attenuates the low frequencies, or, one may say that it passed only high frequencies. The frequency at which the input is attenuated will depend on the oscilloscope; it may be below 10 KHz or much lower, say below 200 Hz.

An application is shown in figure 6.6. At (a) is a television waveform with superimposed low frequency hum with the input coupling to a.c. (To simplify the drawing the hum frequency and line frequency are not in the correct ratio). It is assumed that we wish to trigger the timebase on every alternate pulse. At (a) the timebase would be triggered at P, but not again until R, while at S

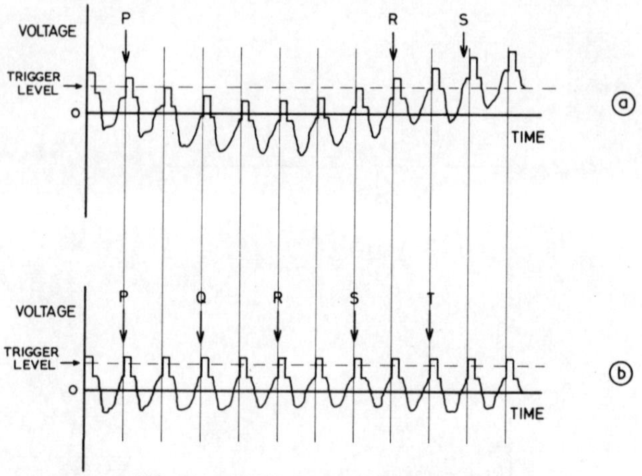

FIG. 6.6. EFFECT OF HUM ON WAVEFORM
(a) Input waveform; (b) Trigger waveform when input coupling set to l.f. reject

it would be triggered early by the picture portion of the waveform rather than the synchronizing pulse. If the input coupling is now set to l.f. reject the waveform *seen by the level control block* (but not that displayed on the screen) will be as at (b). The low frequency hum content has been removed and there may be some distortion of the waveform. It must be emphasized that this does not alter the waveform seen on the screen, but only that seen by the level control block. The timebase will now be triggered at point P, Q, R and S, as is required, and satisfactory triggering will result. If the picture content changes then, of course, the same problem will arise as described in (2) because the d.c. component has been removed. An interesting effect of this control is described in Chapter 13.

(4) H.F. reject (high frequency reject) or L.F. (low frequency) a.f. (audio frequencies). In this position the high frequency components fed to the LEVEL CONTROL block are reduced. The frequency at which attenuation occurs will depend on the oscilloscope; at, say, 15–40 kHz or much less, say 1 kHz. The coupling may be d.c. or a.c. but where the d.c. is removed it is more correctly called 'a.c. h.f. reject'. This position can be used to remove noise or any high frequencies from a low frequency waveform to get more stable synchronizing. This is shown in figure 6.7. At (a) is the input waveform, and that fed to the LEVEL CONTROL block if the INPUT COUPLING is a.c. It will be seen that the timebase is triggered at instants X and Y, but these are different in relation to the low frequency waveform due to the high frequency component. If the input coupling is on h.f. reject then the waveform *seen by the level control block* will be as at (b), and correct triggering will take place, both points X and Y being the corresponding points on the low frequency waveform.

(5) T.V. line. It is frequently required to trigger from a composite (*i.e.* video plus synchronizing pulses) television waveform. Often this can be done using the d.c. or a.c. position of the input coupling and by suitable setting of the trigger level, as already explained. However, it is better if the video signal is first removed. In this position of the input coupling a television synchronizing separator is brought into circuit which produces only the composite synchronizing pulses and these trigger the timebase. The exact arrangement depends on the oscilloscope; in some instruments the synchronizing separator will only

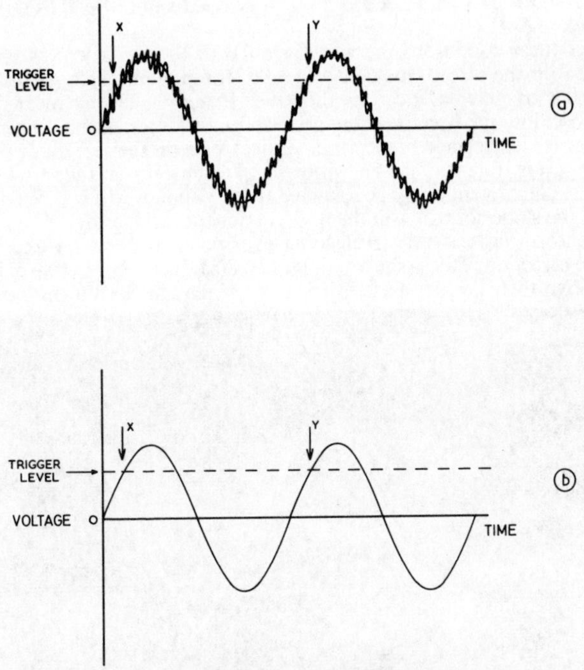

FIG. 6.7. EFFECT OF NOISE ON WAVEFORM

(a) Input waveform; (b) Trigger waveform when input coupling set to h.f. reject

operate with one polarity of video input.

(6) T.V. field. There are numerous occasions when it is necessary to trigger the timebase by the field pulses, which is not usually satisfactory with the other settings of the input coupling. In this position a synchronizing separator is used, as in (5), but a partly integrating circuit is added to produce a pulse corresponding to the field synchronizing pulses (*i.e.* a field synchronizing separator). These are fed to the LEVEL CONTROL block to trigger the timebase. Again, the exact arrangements vary and the synchronizing separator may only operate with one polarity of video signal. A simpler method sometimes used is a low-pass filter (similar to h.f. reject but at a lower frequency) which will produce an output corresponding to the field pulses but may need more careful adjustment of the triggering level.

(7) H.F. Although this is not part of the INPUT COUPLING block its inclusion here is for completeness. Its purpose is to assist triggering at high frequencies and its action will be discussed later. It is NOT the same as L.F. Reject; unfortunately the term h.f. is used for both purposes but not on the same oscilloscope.

LEVEL CONTROL BLOCK

The purpose of this block is to select the voltage at which the triggering takes place. It has a slope or polarity switch which decides whether the timebase is triggered on a positive-going or a negative-going voltage. The exact action of the level control will depend on the setting of the TRIGGER SELECTOR block and INPUT COUPLING block. Various settings will be considered.

(a) TRIGGER SELECTOR block set to Internal and INPUT COUPLING block to D.C.

Under these conditions the waveform fed to the LEVEL CONTROL block is that seen on the screen, including the d.c. component. This means that the timebase is triggered at a certain d.c. level. Exactly what this means depends on the oscilloscope because there are two possibilities. It may mean (i) that the timebase is triggered at a certain vertical level on the screen; or (ii) it may mean that it is triggered at a certain vertical distance from the zero line of the trace. For the present it will be assumed that (i) applies; (ii) will be considered later. If the slope control is in the positive position triggering will occur when the waveform crosses this level going in a positive direction and not in a negative direction. This is shown in figure 6.8(a). If the slope control is turned to negative then the timebase is triggered when the waveform crosses this

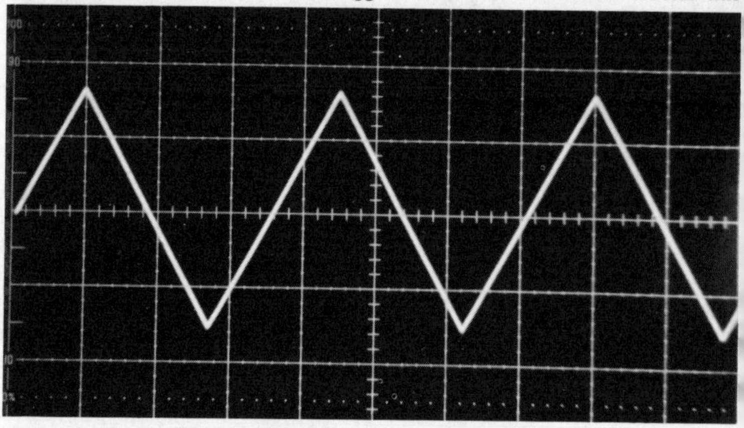

FIG. 6.8(a). POSITIVE TRIGGER SLOPE SETTING

level going in a negative direction, but not when going in a positive direction. This is shown in figure 6.8(b) for the same trigger level setting and same amplitude of trace.

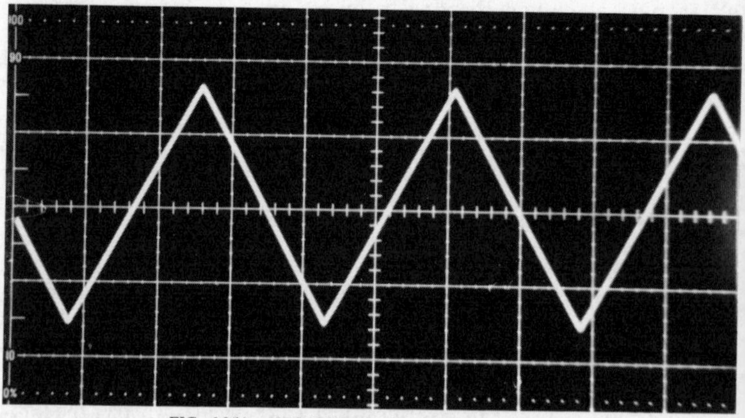

FIG. 6.8(b). NEGATIVE TRIGGER SLOPE SETTING

If the trigger level control is varied then the instant of triggering will be varied. With the slope control at positive, increasing the trigger level moves

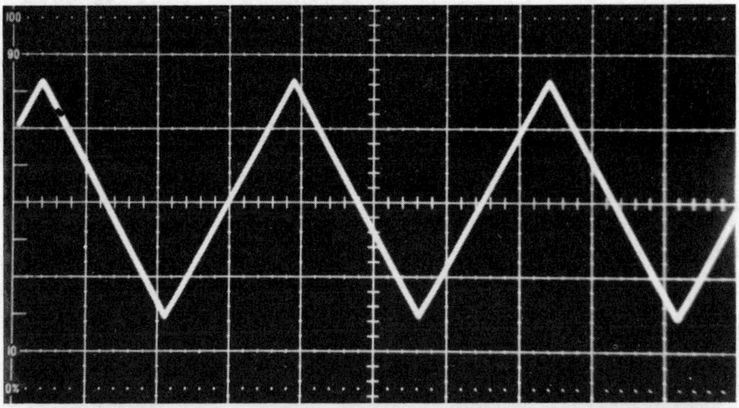

(a) Effect of increasing trigger level

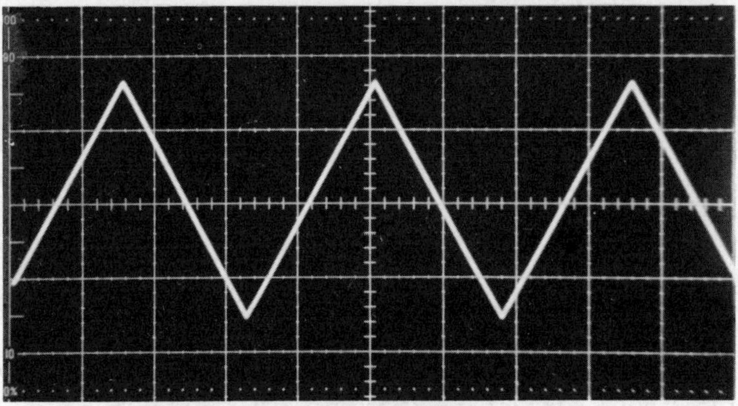

(b) Effect of decreasing trigger level

FIG. 6.9. POSITIVE TRIGGER SLOPE SETTING

the instant of triggering up the trace, as at figure 6.9(a) compared with 6.8(a), and triggering occurs later. If the trigger level is decreased then the timebase will be triggered earlier, as at 6.9(b). When the slope control is negative, increasing the trigger level causes the waveform to be triggered earlier, as shown in figure 6.10(a) compared with figure 6.8(b). Decreasing the trigger level moves the instant of triggering in the opposite direction, as at 6.10(b).

If the vertical shift or position control is operated so as to move the whole trace up and down it must be remembered that the trigger level on the screen remains at the same point, and so changes in relation to the waveform. This is shown in figure 6.11. At (a) is with the waveform in the centre, with the trigger level on the zero (centre line), and with the SLOPE CONTROL on positive. At (b) the waveform is shifted upwards, and at (c) it is moved downwards. Using the vertical shift or position control varies the instant of triggering. The effect of varying the amplitude of signal, with the trigger level and slope control at the same setting is shown in figure 6.12. At (a) the amplitude is small and the triggering level has been set to a low value and the slope control to positive. At (b) and (c) the amplitude is increased; it will be seen that the level at which the timebase triggers remains the same but, because of the

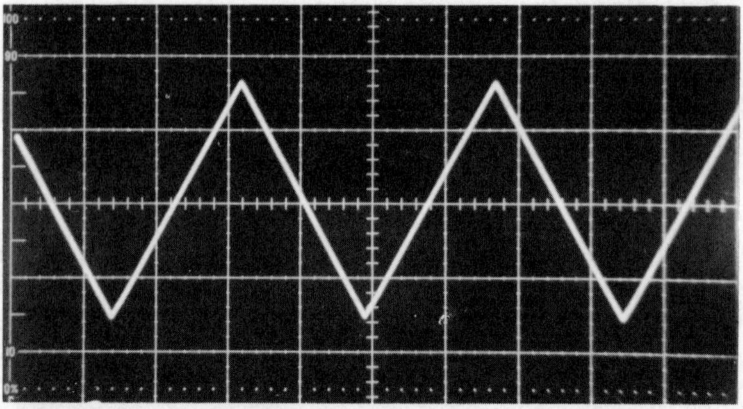

(a) Effect of increasing trigger level

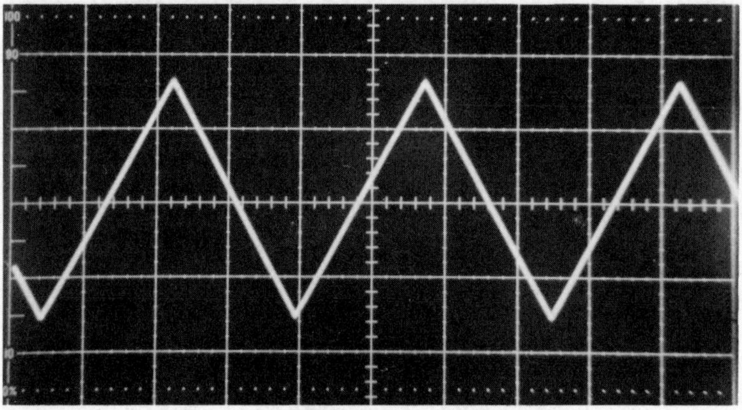

(b) Effect of decreasing trigger level

FIG. 6.10. NEGATIVE TRIGGER SLOPE SETTING

change in amplitude of the trace, the instant of triggering on the waveform changes.

If, of course, the triggering level is greater (either in a positive or negative direction) than the maximum amplitude of the trace then the timebase will not be triggered and *there will be no trace on the screen* (unless the timebase is set to the AUTO position, to be described later).

If the Y-amplifier input is set to d.c. the waveform on the screen will contain any d.c. component. This will therefore have an effect on the triggering, since the trace will move up and down the screen (and relative to the trigger level). If, however, the input is set to a.c., then the d.c. level will have no effect on the position of the trace. This does not alter the fact that the trigger level control is still d.c. coupled, hence the trigger level is constant, *i.e.* the trigger level control still sees what is on the screen, including a d.c. level such as produced by the vertical shift or position control. It is important to distinguish between the effect of changing the Y-amplifier input circuit and the input coupling setting of the trigger circuit.

Now consider the other possibility where the trigger level is at a certain vertical distance from the zero line. The only difference is that the trigger

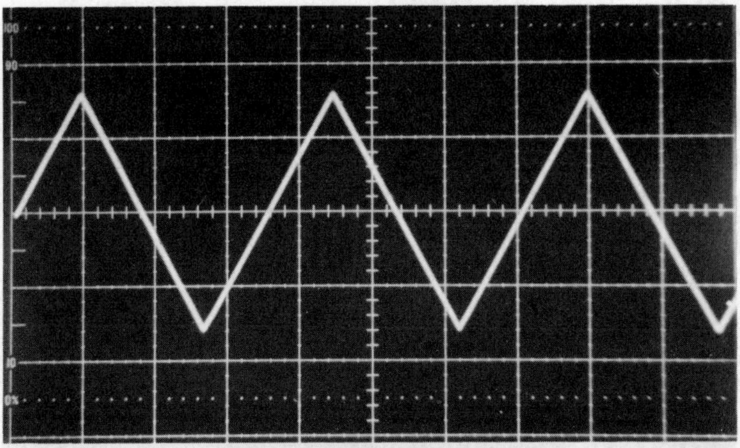

(a) Waveform at centre

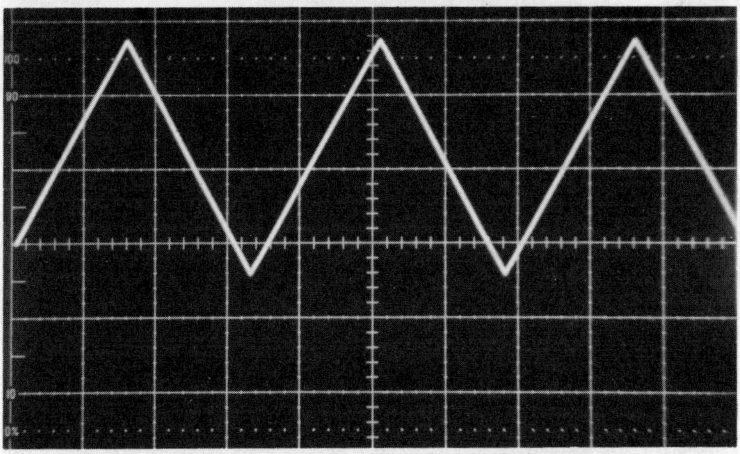

(b) Waveform shifted upwards

FIG. 6.11(a) & (b) EFFECT OF VARYING Y-SHIFT CONTROL WITH D.C. TRIGGER INPUT COUPLING (on some oscilloscopes)

level now moves up and down with the vertical position control (so that it always remains at a certain distance from the zero line). With this type of oscilloscope figure 6.11 does not apply, and the instant of triggering is not varied by the vertical position control. Apart from this the oscilloscope behaves as described for the first possibility, *i.e.* figures 6.8, 6.9, 6.10 and 6.12 still apply. The difference is caused by the fact that in one case the position control operates before the trigger output point from the Y-amplifier and in the other case after it.

In the examples shown the setting of the trigger level is perhaps not important, but if one considers a television waveform, as shown in figure 6.13, (pp 55 & 56) then it is of importance. If the triggering is not to vary with the picture content then the timebase should be triggered by the synchronizing pulses, as shown at (a). If the trigger level is set as shown and the slope set to

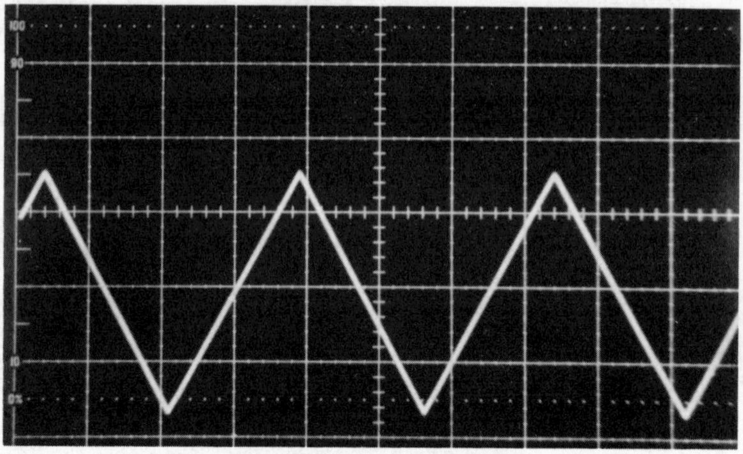

(c) Waveform shifted downwards

FIG. 6.11(c). EFFECT OF VARYING Y-SHIFT CONTROL WITH D.C. TRIGGER INPUT COUPLING (on some oscilloscopes)

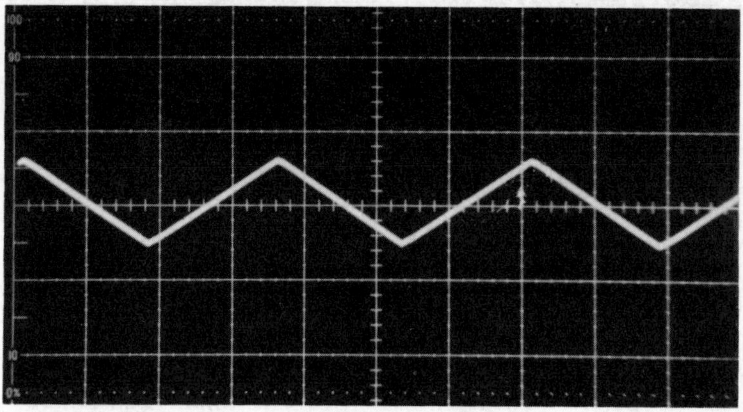

(a) Small amplitude

FIG. 6.12(a). EFFECT OF VARYING AMPLITUDE OF SIGNAL, TRIGGER LEVEL BEING FIXED

positive the timebase will be triggered consistently on the pulses and the picture content will have no effect. In order that the pulse remains at the same level the Y-input should be set to D.C. and the full d.c. component of the signal must be present, otherwise the waveforms move up and down with picture content, as explained. At (b) is shown a television signal synchronized in this way. This is the waveform of an actual picture, hence the vision portion is different on each line, and results in the complex video portion. At (c) is shown the result of lowering the trigger level so that it triggers on the video content. The result is unsatisfactory triggering and a multiple trace of line pulses.

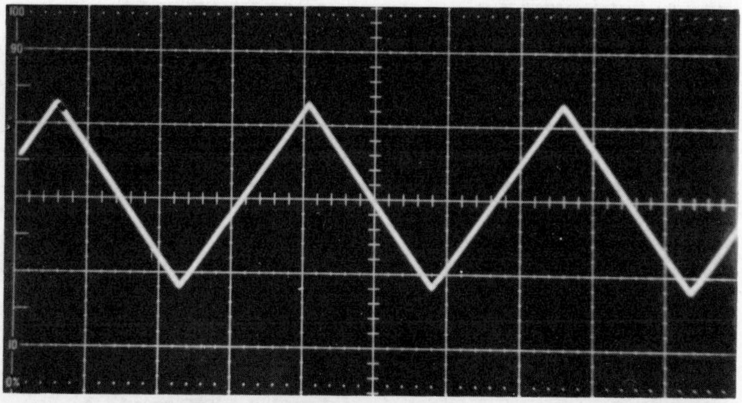

(b) Medium amplitude

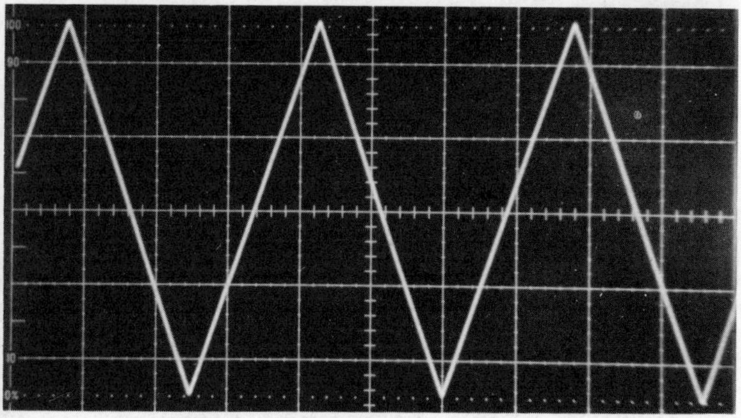

(c) Large amplitude

FIG. 6.12(b) & (c). EFFECT OF VARYING AMPLITUDE OF SIGNAL, TRIGGER LEVEL BEING FIXED

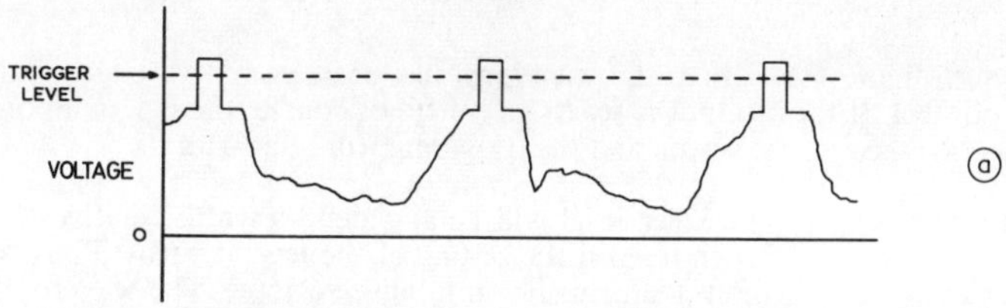

FIG. 6.13(a). CORRECT SETTING OF TRIGGER LEVEL

(b) TRIGGER SELECTOR block set to Internal and INPUT COUPLING block to A.C.

The waveform now fed to the trigger level control is that seen on the screen, but without any d.c. component. Thus, if the Y-input is set to D.C. and there is a d.c. component to the signal this will be displayed on the screen *but will have no effect on the triggering*. Similarly, if the shift control is operated there will be no effect, whatever the type of oscilloscope (see earlier). This is

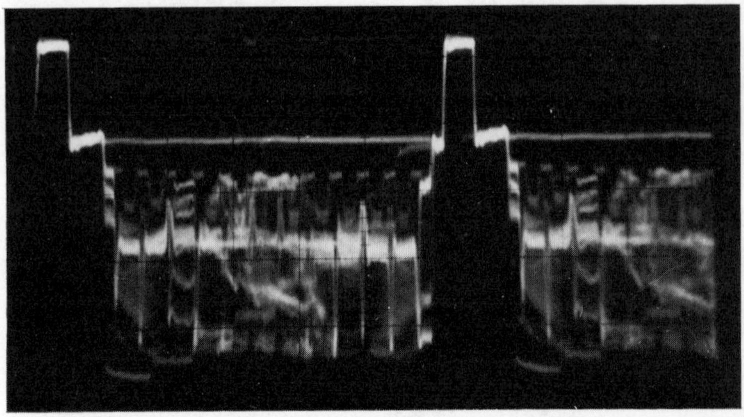

FIG. 6.13(b). DISPLAY WITH CORRECT SETTING OF TRIGGER LEVEL. TRIGGERING ON SYNCHRONIZING PULSES AND POSITIVE SLOPE

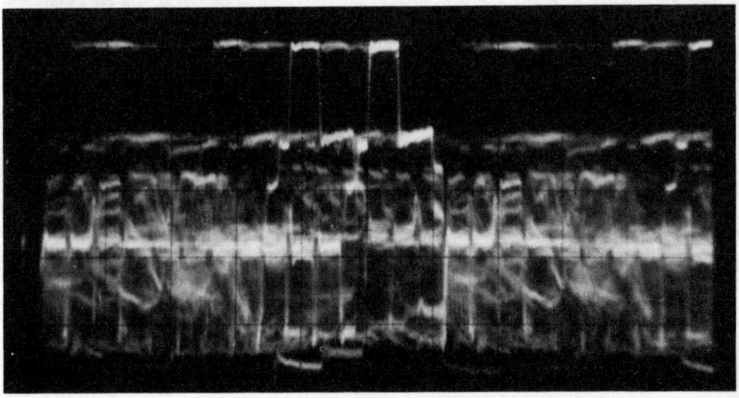

FIG. 6.13(c). DISPLAY WITH INCORRECT SETTING OF TRIGGER LEVEL. TRIGGERING NOW ON PICTURE PORTION

shown in figure 6.14 where the waveform has been moved vertically by the shift control. If the Y-input is set to A.C. then, of course, the d.c. component will not be seen on the screen and the triggering will be as with the Y-input set to D.C.

In this position the trigger level will be at a definite value relative to the zero line, the value depending on the setting of the level control. Thus, suppose the input is of the waveform shown in figure 6.15(a). The waveform to the trigger circuit will settle down so that the area above the zero line is equal to the area below, as shown. If the trigger level is set as shown, and the polarity or slope set to negative, then the timebase will be triggered at instant X. If the mark-space ratio of the waveform changes to that at (b) then the waveform to the trigger circuit will move upwards, as shown, and with the same trigger level setting the timebase will not be triggered.

Varying the trigger level and slope controls have a similar effect as with the input coupling set to D.C. shown in figures 6.8, 6.9, 6.10 and 6.12, but not as figure 6.11.

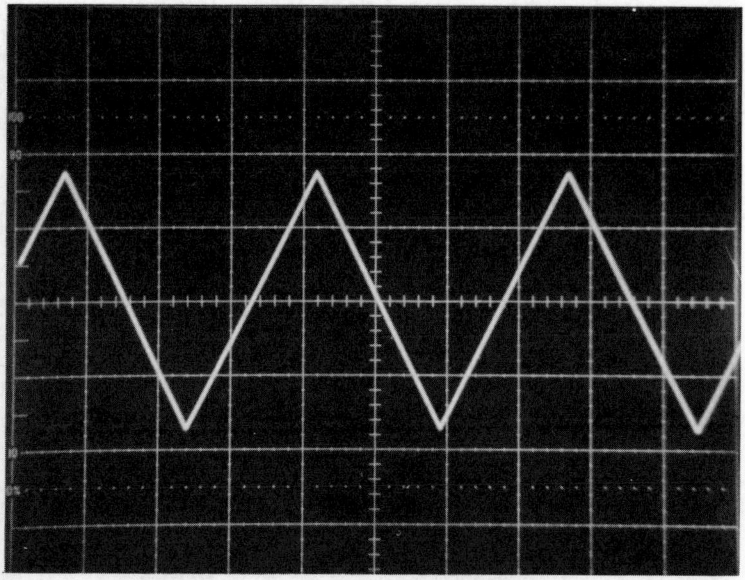

(a) Waveform in centre

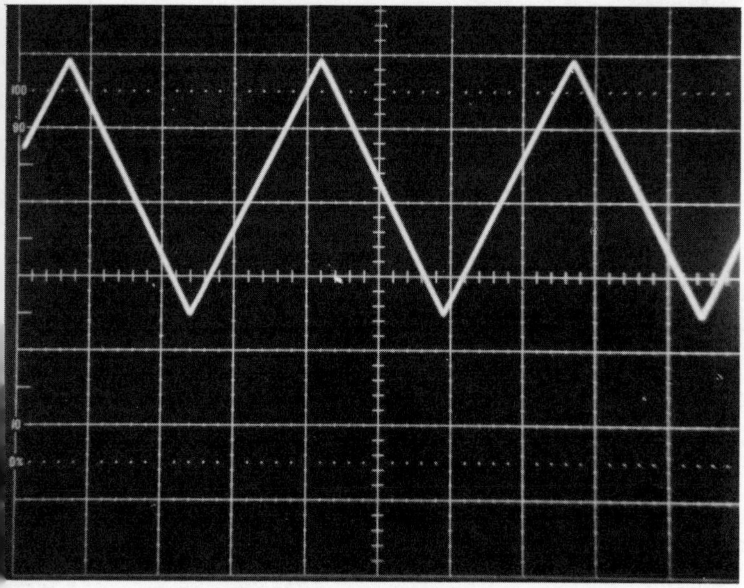

(b) Moved upwards

FIG. 6.14(a) & (b). EFFECT OF Y-SHIFT CONTROL WITH INPUT COUPLING SET TO A.C.

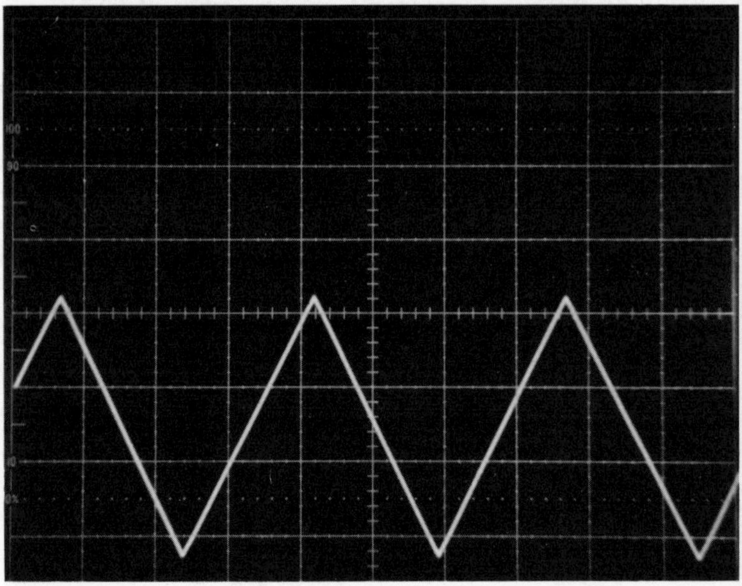

(c) Moved downwards

FIG. 6.14(c). EFFECT OF Y-SHIFT CONTROL WITH INPUT COUPLING SET TO A.C.

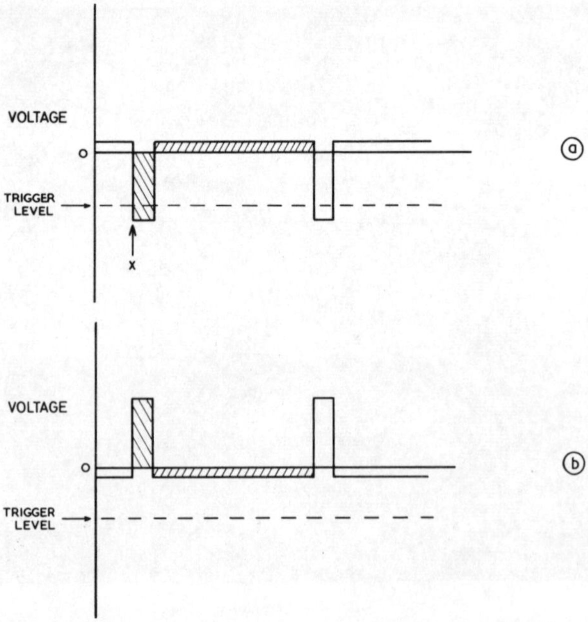

FIG. 6.15. EFFECT OF CHANGE OF MARK-SPACE RATIO

(a) Correct triggering; (b) No triggering

(c) TRIGGER SELECTOR block set to Internal and INPUT COUPLING block to either L.F. reject or A.C. H.F. reject

In both these cases the d.c. component is removed, hence the general effect is as (b). As already mentioned the L.F. reject may be used to reject hum or low frequency interference, and the A.C. H.F. reject for removing the effect of noise or high frequency interference on triggering. If the oscilloscope is fitted with a D.C. H.F. reject position, the general effect is as (a) since the d.c. component is fed to the LEVEL CONTROL block.

Where TV triggering facilities are in the form of a synchronizing separator the effect will depend on the arrangement used and is too varied to be considered in this book.

(d) TRIGGER SELECTOR block set to External and INPUT COUPLING block to D.C.

The signal now fed to the trigger circuit is not that fed to the Y-amplifier (it could be but there is little point in doing so). Whether or not the d.c. component is present on the screen depends ONLY on the setting of the Y-amplifier selector. The signal now fed to the LEVEL CONTROL block is that fed to the trigger input socket and will include the d.c. component. The general operation is the same as (a) bearing in mind that it is the *waveform fed to the trigger input socket which is controlling the triggering*.

(e) TRIGGER SELECTOR block set to External and INPUT COUPLING block to A.C.

This is similar to (d) except that the d.c. component of the signal fed to the Trigger Input Socket is not now fed to the LEVEL CONTROL block. The general operation is as (b), keeping in mind that the trigger signal is not usually the same as that seen on the screen.

(f) TRIGGER SELECTOR block set to Line and INPUT COUPLING block to D.C. or A.C.

In this case a sine wave is fed to the TRIGGER LEVEL CONTROL at mains frequency and as it is a.c. only it is unimportant as to whether the INPUT COUPLING block is set to D.C. or A.C. The level control will select the part of the waveform at which triggering occurs. This is of little importance in general, although it will affect the phase of the triggering relative to a mains voltage fed to the Y-input.

BASIC LEVEL CONTROL CIRCUIT

A basic level control circuit using a long-tailed pair is given in figure 6.16. This is not the only possible circuit, but one that is commonly used. Suppose that the base of Tr_2 is set to zero voltage and that the base of Tr_1 is negative. Under these conditions Tr_2 will be conducting, and the current flowing will be such that the drop across R_3 results in the emitter voltage of Tr_2 (and Tr_1) being almost zero. It will, in fact, be negative by the base-emitter drop of Tr_2 – a fraction of a volt. The voltage on the collector – the output voltage – will be low. Tr_1 is cut off. Suppose now that a ramp voltage is applied to Tr_1 (as the triangular waveform used in earlier examples). At some input voltage, approximately zero in this particular case, Tr_1 will start to conduct. As the current in Tr_1 increases, the current of Tr_2 will decrease, since the current in R_3 must be approximately constant if Tr_2 is conducting; the base-emitter voltage can be only a fraction of a volt and the base voltage is constant. As the input voltage rises, the current of Tr_1 also rises, so tending to increase the voltage across R_3 and so reduce the current in Tr_2. Eventually, the current in Tr_2 is reduced to zero and the emitters then move in a positive direction following the voltage applied to the base of Tr_1. When Tr_2 is cut off the col-

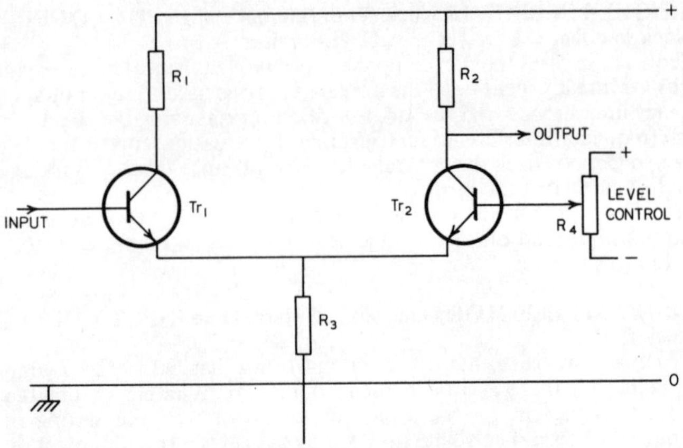

FIG. 6.16. LEVEL CONTROL CIRCUIT

lector voltage rises and therefore the output voltage rises to the voltage of the positive line. If the input voltage now decreases, a similar action occurs when the input is approximately equal to that from the level control R_4 (zero volts in this example). As the input voltage goes through this value the current in Tr_2 rises and the collector voltage and output voltage drop to a low value.

The level at which the switch-over occurs and a change of output voltage occurs can, of course, be varied by varying the voltage fed to the base of Tr_2 from the level control R_4.

The level control circuit feeds a pulse generator and is designed so that it will produce a triggering pulse only when the input moves in one direction, say when the input voltage rises in a positive direction. Thus, in the circuit described, a synchronizing pulse will only be produced for a voltage to Tr_1 which is rising, and no triggering pulse when it is going in a negative direction. To be able to switch the circuit so that triggering will occur with the input going in the opposite direction (*i.e.* opposite slope) the base connections of Tr_1 and Tr_2 may be reversed as shown in figure 6.17.

With the slope control switch in position 1 the operation is as before. In position 2 the input is fed to Tr_2 and the trigger level voltage to Tr_1. Assume that the trigger level control is at zero, that the input voltage is negative then Tr_1 is conducting, and Tr_2 is cut off. When the input reaches approximately 0 volts Tr_2 becomes conducting and Tr_1 is cut off. The output voltage is reduced, but this will not result in a triggering pulse. When the input voltage decreases again (*i.e.* going in a negative direction or has negative slope), Tr_2 becomes cut off (when the voltage goes through 0 volts). This results in a positive output and a triggering pulse. Thus the timebase is triggered by a negative slope but not by a positive slope. Looking at waveforms, suppose that the input is a triangular waveform, as shown in figure 6.18(a). The output will vary as at (b) if the slope switch is in the negative position. Thus when the input voltage reaches the trigger level (set by R_4) at A the output goes in a negative direction to a low voltage. This does not produce a trigger pulse to operate the timebase. As the input voltage decreases and becomes equal to the trigger level voltage at B the output rises to positive line voltage and the pulse generator produces a pulse to trigger the timebase. Instead of switching the inputs in this way the same effect can be obtained by taking the output from one or other of the collectors.

We have now selected a suitable source for triggering, processed it and obtained a voltage change, the instant of which depends on the setting

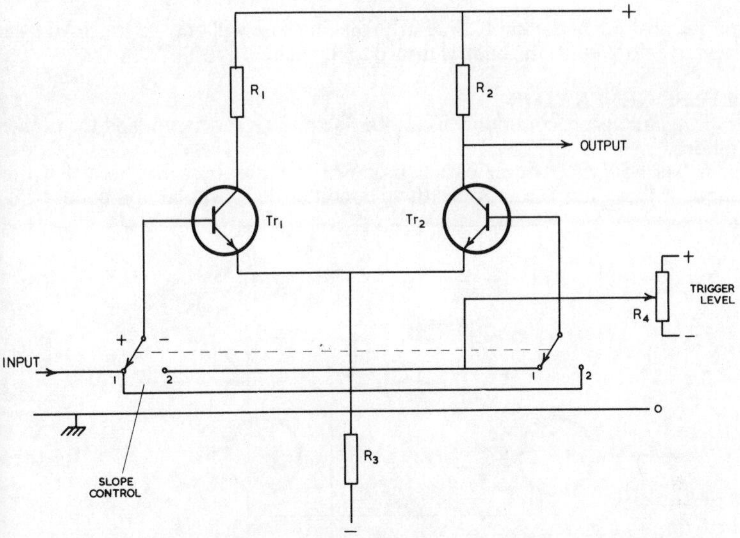

FIG. 6.17. LEVEL CONTROL CIRCUIT WITH MEANS FOR TRIGGERING ON POSITIVE OR NEGATIVE SLOPES

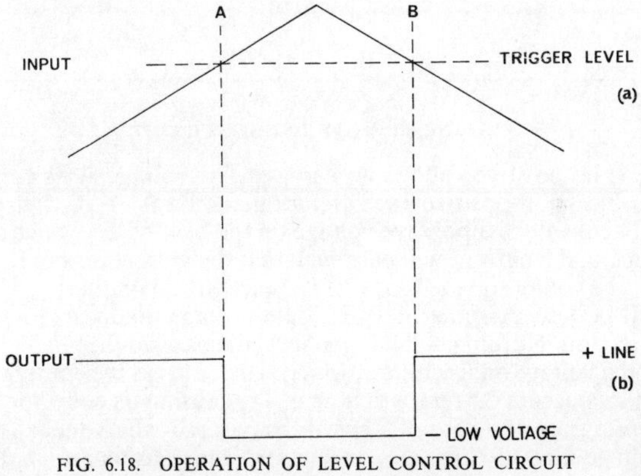

FIG. 6.18. OPERATION OF LEVEL CONTROL CIRCUIT

(a) Input; (b) Output

of the level control and the slope control. The trigger level and slope control may be combined, as shown in figure 6.19. If the knob is moved to the left the timebase will trigger only on a positive slope and the trigger level will depend

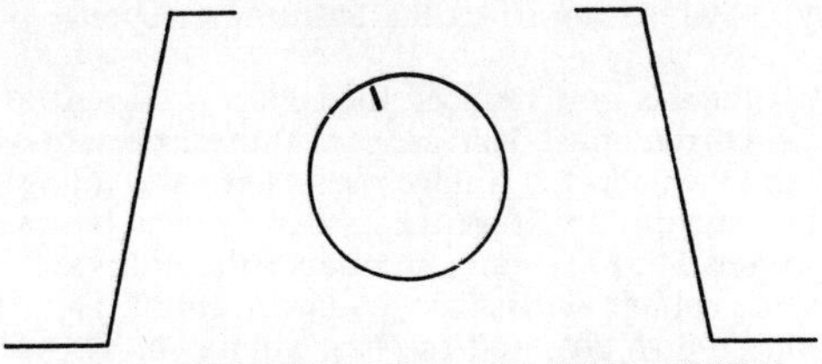

FIG. 6.19. COMBINED LEVEL AND SLOPE CONTROL

on the position of the knob. Similarly, the timebase will only be triggered by a negative slope when the knob is turned to the right.

PULSE GENERATOR

Two circuits are commonly used: the Schmitt trigger circuit, and the tunnel diode.

A basic Schmitt trigger circuit is shown in figure 6.20. Suppose that the input voltage V is zero. Under these conditions Tr_1 will be non-conducting

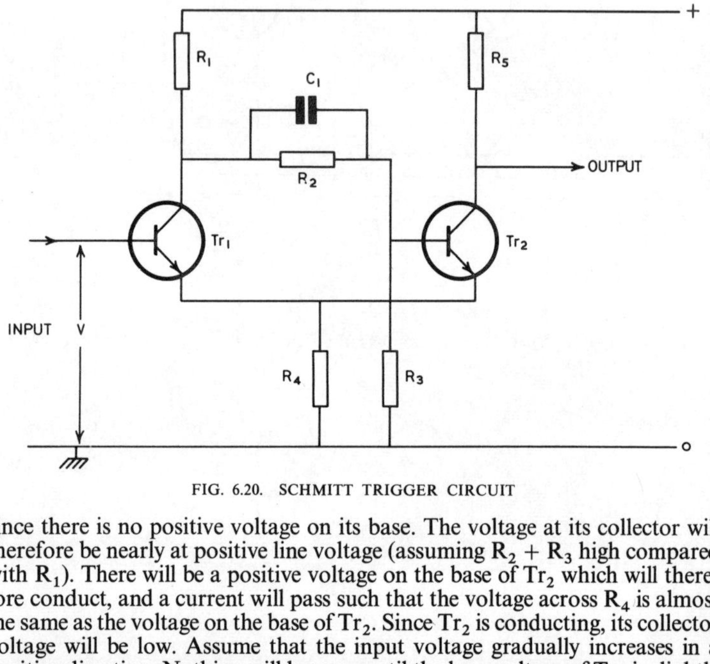

FIG. 6.20. SCHMITT TRIGGER CIRCUIT

since there is no positive voltage on its base. The voltage at its collector will therefore be nearly at positive line voltage (assuming $R_2 + R_3$ high compared with R_1). There will be a positive voltage on the base of Tr_2 which will therefore conduct, and a current will pass such that the voltage across R_4 is almost the same as the voltage on the base of Tr_2. Since Tr_2 is conducting, its collector voltage will be low. Assume that the input voltage gradually increases in a positive direction. Nothing will happen until the base voltage of Tr_1 is slightly more than the voltage on its emitter, which is the same as the emitter voltage of Tr_2. When this occurs current will flow in Tr_1 causing its collector voltage to fall. The voltage on the base of Tr_2 will be reduced which tends to drop the emitter voltage and so cause Tr_1 to conduct more to maintain the voltage across R_4. Thus there is cumulative action (*i.e.* positive feedback), Tr_2 is rapidly cut off and Tr_1 made conducting. When Tr_2 is cut off its collector voltage rises to positive line voltage. If the input voltage is made more positive Tr_1 conducts more so that the drop across R_4 almost equals the input voltage. The changeover from one transistor to the other is very rapid, being speeded up by C_1 which improves the frequency response of the potential divider R_2R_3.

If the input voltage is now reduced the collector current of Tr_1 decreases (since the emitter current must decrease so that the drop across R_4 follows the input voltage) and the collector voltage rises. Hence the voltage on the base of Tr_2 rises and at some input voltage the base of Tr_2 will become positive with respect to its emitter. Thus Tr_2 starts to conduct, the additional current flowing in R_4 increases the voltage across it so tending to cut off Tr_1. This still further increases its collector voltage and the base voltage of Tr_2. This cumulative action rapidly results in Tr_1 being cut off and Tr_2 conducting – the original state. The input voltage at which the circuit switches back is less than the voltage required to switch Tr_1 on. The difference is known as the 'hysteresis voltage'. Within limits this can be varied by the choice of component values and can be made variable by splitting the common-emitter resistor as in figure 6.21. This results in the non-conducting transistor having a slightly lower emitter voltage than that of the conducting transistor, so causing changeover

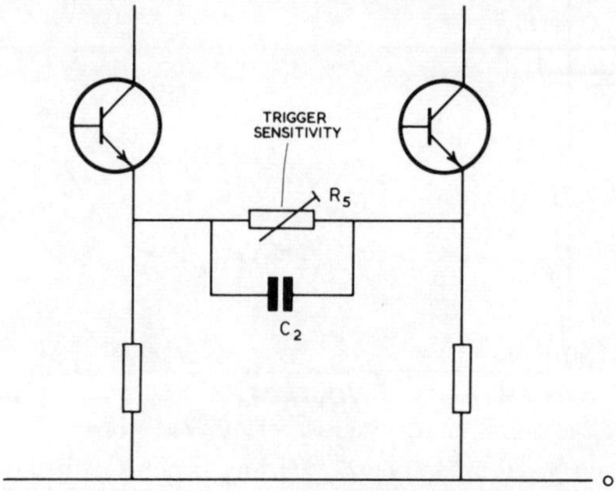

FIG. 6.21. METHOD OF VARYING HYSTERESIS VOLTAGE

to occur sooner. The circuit becomes unstable if R_5 is made too large. C_2 is used to maintain the full loop gain at high frequencies so as not to slow down the speed of changeover.

The action can be seen more clearly from figure 6.22. At (a) is shown the voltage from the trigger level circuit where the rate of voltage change is

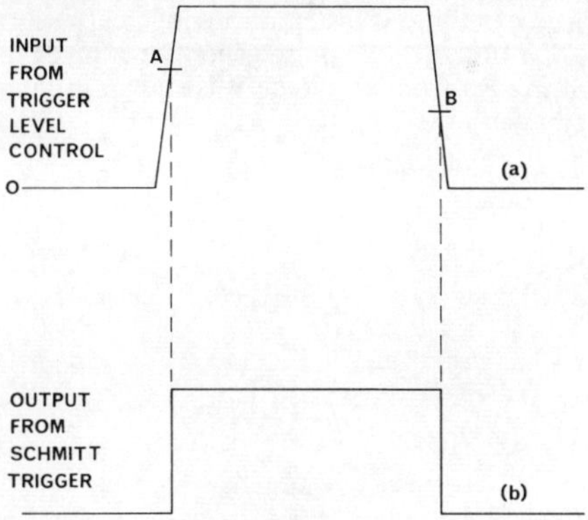

FIG. 6.22. WAVEFORMS OF SCHMITT TRIGGER CIRCUIT
(a) Input; (b) Output

relatively slow. (There is no positive feedback or trigger action in the level control). At point A the Schmitt trigger operates and the output voltage rises to positive line voltage. Due to the positive feedback or cumulative action this rate of rise of voltage is high. When the voltage from the trigger level control decreases, the Schmitt trigger will operate at point B, which is at a lower voltage than A. Again, a sudden drop in output voltage now occurs.

For high-speed timebases a tunnel diode is used as the pulse generator because a Schmitt trigger circuit is not fast enough. The characteristic of a

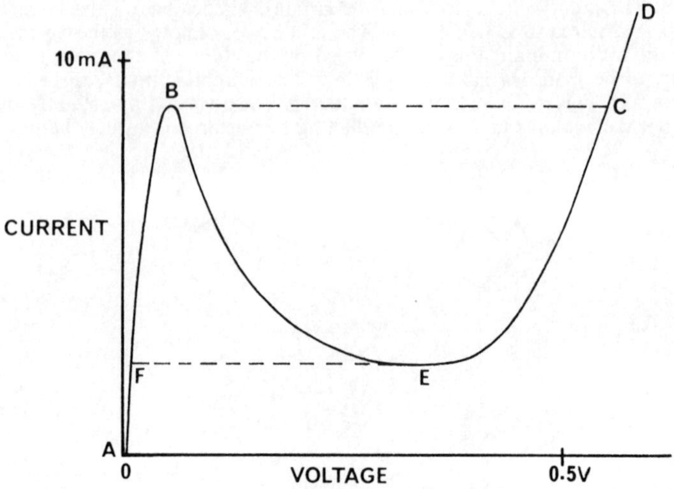

FIG. 6.23. CHARACTERISTIC OF TUNNEL DIODE

tunnel diode is given in figure 6.23. Starting at point A, if the current is increased the diode operates along the characteristic AB and the voltage drop is low. If, however, the current reaches point B then the operating point suddenly changes to point C with a greater voltage drop. If the current is increased further then the operating point follows the line CD. If the current is now decreased the operating point moves along the characteristic to E and then there is a sudden change to point F at a much reduced voltage. Thus the device switches rapidly between two voltage levels in a similar way to a Schmitt trigger but the speed of change is faster, occurring in about 1 ns.

A basic circuit of a tunnel diode pulse generator is given in figure 6.24. The load line representing R_2 from a voltage V is shown in figure 6.25, and cuts

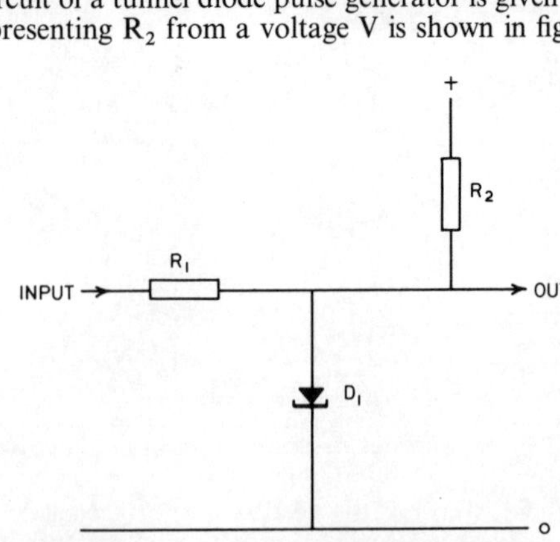

FIG. 6.24. TUNNEL DIODE CIRCUIT

the characteristic at three points. Suppose that the operating point is A with a low output voltage. If a positive input voltage is now applied a current will flow in R_1 so increasing the diode current. When point B is reached the diode will suddenly switch to point C, the position of C depending on the values of R and R_2. The voltage output has now suddenly increased. If the input voltage i

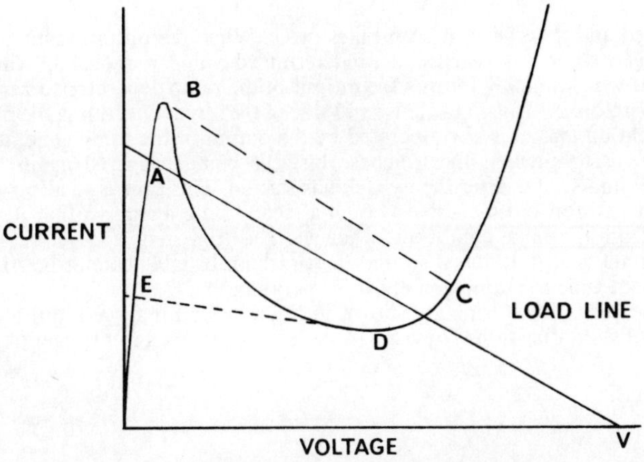

FIG. 6.25. OPERATION OF THE CIRCUIT OF FIGURE 6.24

reversed the current will be reduced until D is reached, when there will be a sudden change to point E and a rapid drop in voltage occurs. If the input voltage is now returned to zero the operating point becomes A. The action of the circuit is similar to that of figure 6.22 except that the d.c. level of the input waveform must be displaced so as to give a positive and negative input. The rate of change of voltage is now very rapid but small because of the small voltage drop across the tunnel diode.

By the use of one of these circuits we now have a pulse which has a rapid rate of rise and fall which can be used to trigger the timebase accurately. This is essential: if the instant of triggering of the timebase varies relative to the signal then jitter of the waveform will occur.

SWEEP GENERATOR

We must now consider that section which actually produces the sweep and flyback. This section, as shown in figure 6.1, is reproduced in figure 6.26. The detailed arrangements vary, but these are the basic blocks. The ramp or sweep

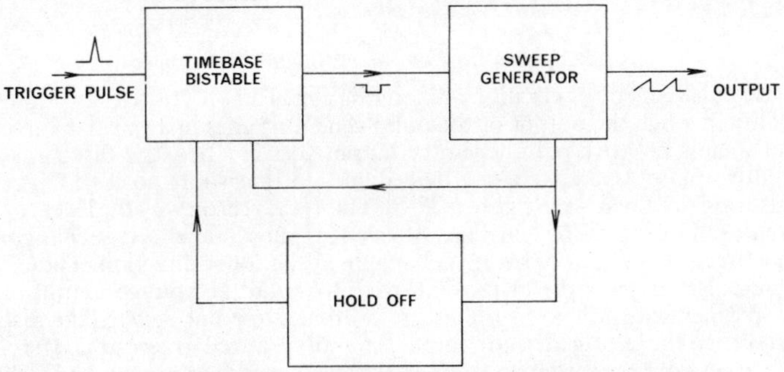

FIG. 6.26. SWEEP GENERATOR BLOCK DIAGRAM

generator is a device that will produce a voltage changing in a linear way with time (*i.e.* a voltage proportional to time). Under rest conditions, *i.e.* with the spot on the left-hand side, the timebase bistable is in such a state that the ramp generator is clamped so that its ouput is fixed. When a trigger pulse is applied to the timebase bistable from the Schmitt trigger or tunnel diode the bistable is switched over, so allowing the ramp generator to operate. Thus, an output is obtained and the sweep or scan takes place. When the output voltage of the ramp generator has reached a predetermined value it resets the timebase bistable which quickly returns the output of the ramp generator to zero. The spot therefore returns to the left-hand side of the screen, *i.e.* this is the flyback. The hold-off circuit is also operated by the output of the ramp generator. Its purpose is to prevent the timebase bistable being triggered again until a predetermined time after the flyback has started. This time is to allow for the ramp generator and bistable to reach a steady state again, so that it always starts from the same conditions whenever it is triggered. This reduces jitter. The circuit is also designed so that the timebase bistable cannot be triggered during the time the ramp generator is operating.

Some basic waveforms are shown in figure 6.27. First, we require trigger pulses of short duration. These are obtained from the Schmitt trigger or tunnel

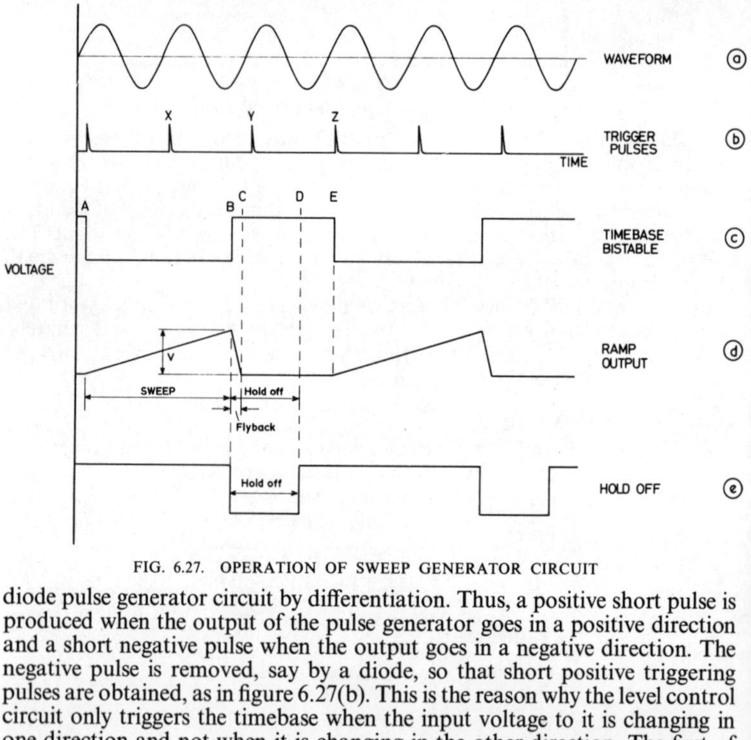

FIG. 6.27. OPERATION OF SWEEP GENERATOR CIRCUIT

diode pulse generator circuit by differentiation. Thus, a positive short pulse is produced when the output of the pulse generator goes in a positive direction and a short negative pulse when the output goes in a negative direction. The negative pulse is removed, say by a diode, so that short positive triggering pulses are obtained, as in figure 6.27(b). This is the reason why the level control circuit only triggers the timebase when the input voltage to it is changing in one direction and not when it is changing in the other direction. The first of these pulses triggers the bistable at instant A, and the voltage output of the ramp generator starts to rise. This rise continues to point B, when the bistable is reset and the ramp generator output is rapidly reduced to zero at C. The hold-off circuit now prevents any action occurring during the period C to D [shown at (e)]. At D the circuit is free ready to be triggered again. It will now be triggered by the first pulse after instant D, *i.e.* at E. This switches the timebase bistable and the action restarts. As already mentioned, after instant A and until after D the circuit is made to be insensitive to pulses from the pulse generator. Thus, pulses X and Y have no effect. The rate of rise of the ramp voltage (and hence the speed of the sweep in milliseconds/division) will be settled only by the time-constant of the charging circuit of the ramp generator. Since the timebase bistable is reset by a predetermined ramp voltage amplitude, the faster the rise of ramp voltage the shorter the time AB, hence fewer cycles will be displayed on the screen.

It is perhaps important to realize that the period AB does not have to correspond to a complete number of cycles. The only effect of gradually increasing the sweep speed (*i.e.* the rate of rise of ramp voltage) is to cut off part of the last cycle of the waveform, but without altering the width of sweep or scan, since the width is determined by the ramp output V (which is predetermined and independent of the sweep speed). This is shown in figure 6.28 where the sweep speed increases from (a) to (e). The next time that the timebase is triggered must, of course, be a whole number of cycles from the first time, *i.e.* at instant E (figure 6.27). Thus, if the time AB is reduced the time BE must

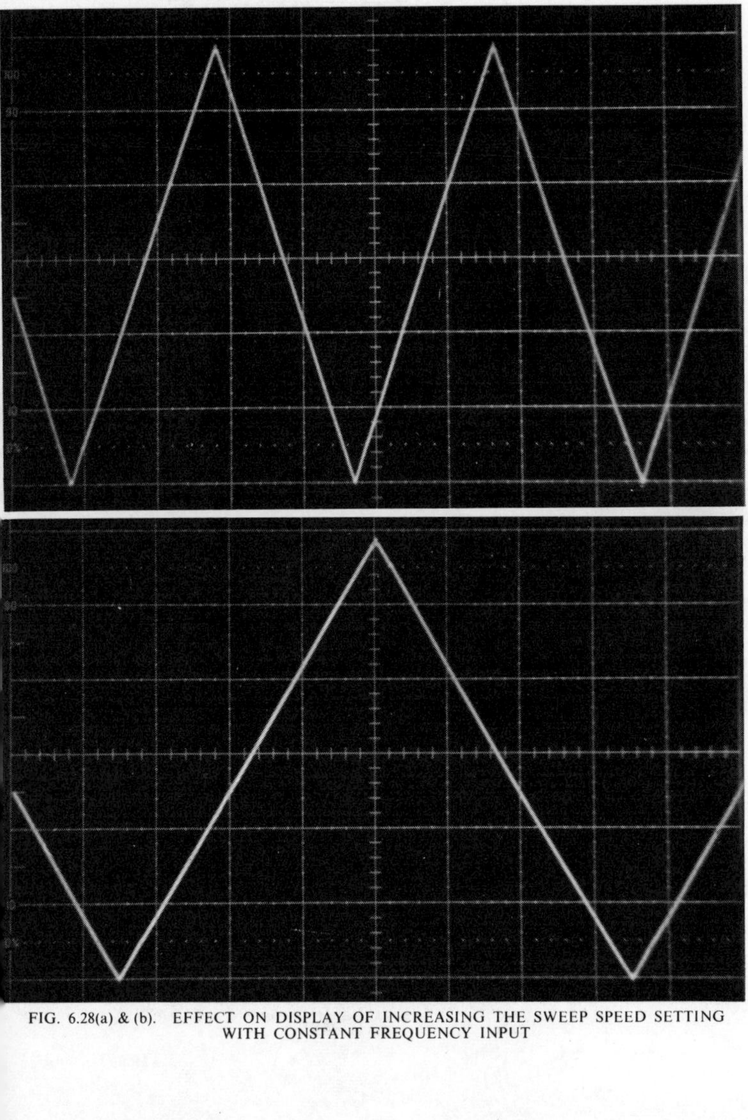

FIG. 6.28(a) & (b). EFFECT ON DISPLAY OF INCREASING THE SWEEP SPEED SETTING WITH CONSTANT FREQUENCY INPUT

be increased by the same amount until such time as the timebase is triggered by pulse Y instead of pulse Z.

In figure 6.28 (f) to (i) is shown the output of the sweep generator under various conditions. At (f) the sweep generator is triggered soon after the flyback, as would be the result in the case of the display shown at (a). At (g) the frequency of the triggering signal was varied so that there was an appreciable time before the sweep generator was triggered again. In the case of (h) this is taken further where the waiting period is as long as the sweep period. This is the sort of thing that would happen in the case of the display shown at (d), since there must be a considerable time before the voltage is going in the correct direction and of suitable amplitude to trigger the timebase again. The effect of increasing the sweep speed still further is shown at (i), where again there is a

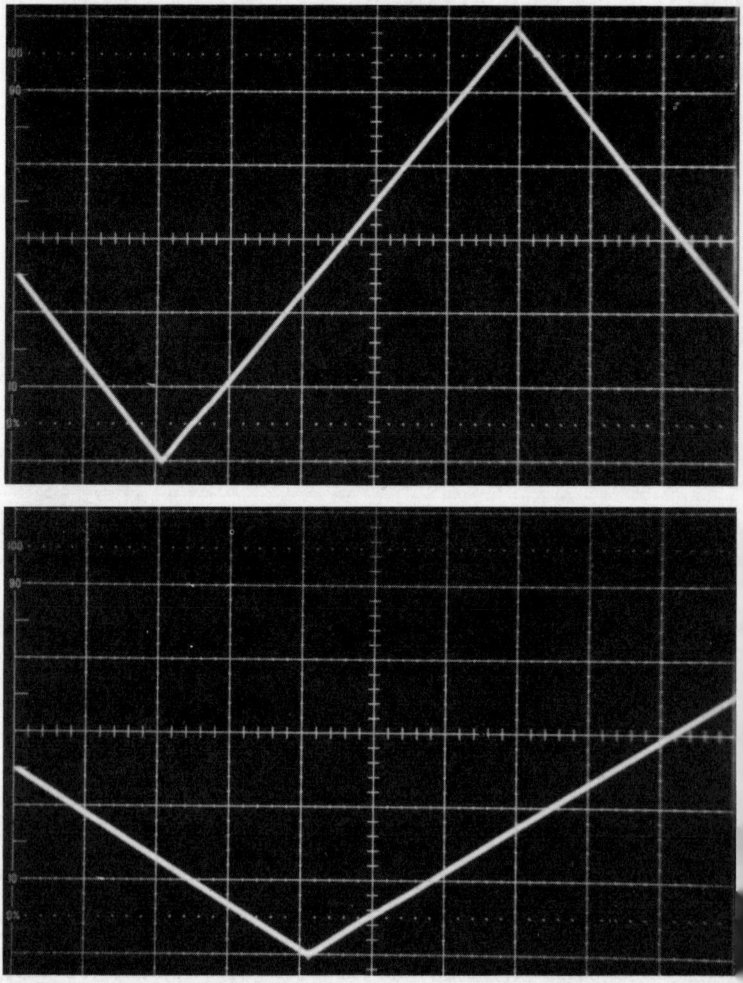

FIG. 6.28(a) & (d). EFFECT ON DISPLAY OF INCREASING THE SWEEP SPEED SETTING WITH CONSTANT FREQUENCY INPUT

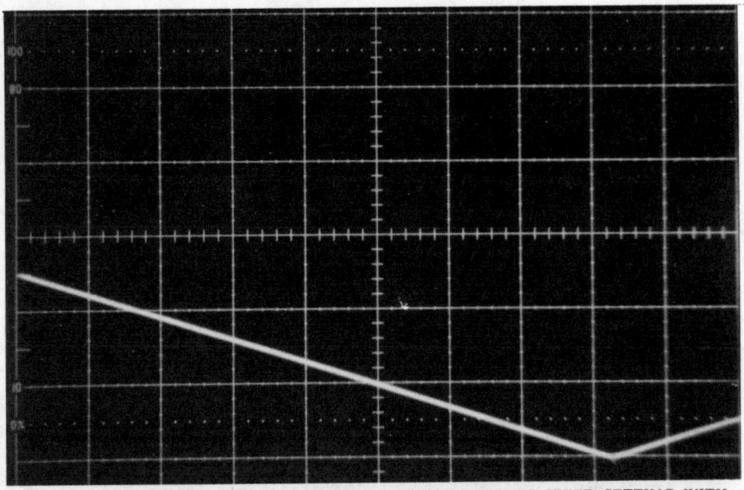

FIG. 6.28(e). EFFECT ON DISPLAY ON INCREASING THE SWEEP SPEED SETTING WITH CONSTANT FREQUENCY INPUT

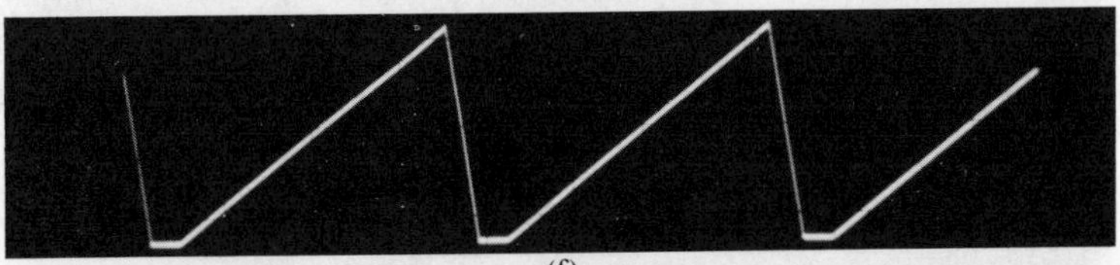

(f)

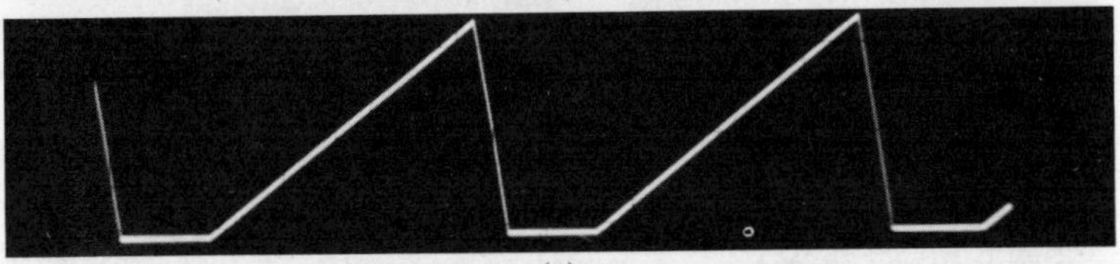

(g)

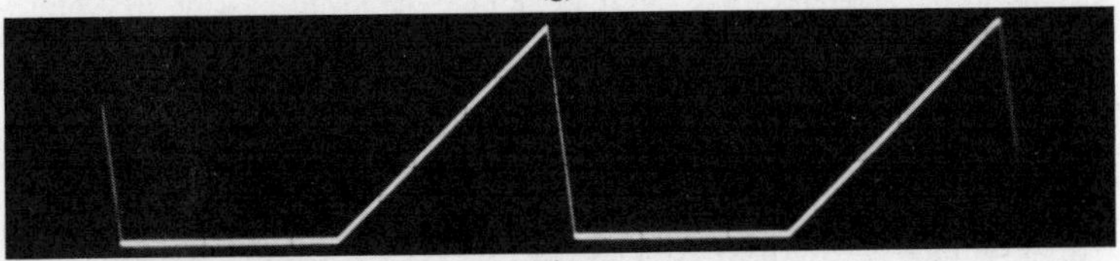

(h)

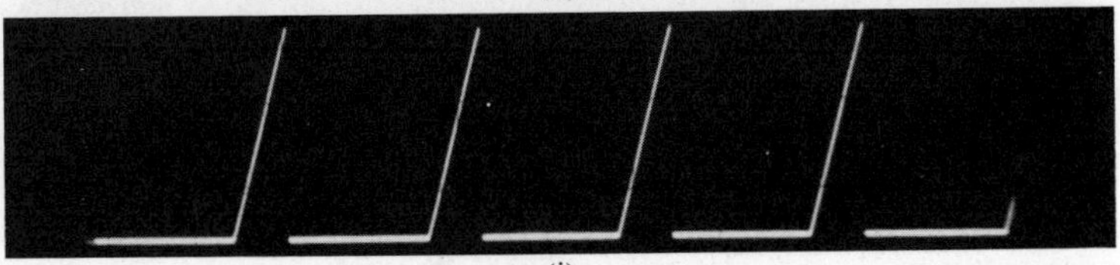

(i)

FIG. 6.28(f)–(i). EFFECT ON TIMEBASE WAVEFORM ON INCREASING THE SWEEP SPEED SETTING WITH CONSTANT FREQUENCY INPUT long waiting time between sweeps. It should be noted that the amplitude of the waveforms is the same in all cases, and so the amplitude of the horizontal scan is the same.

RAMP GENERATOR

We will now consider the basic principles of the ramp generator. Like almost all timebases the ramp voltage is produced by charging a capacitor through a resistor. If a capacitor C is charged through a resistor R from a constant voltage V, as shown in figure 6.29(a), the voltage across the capacitor

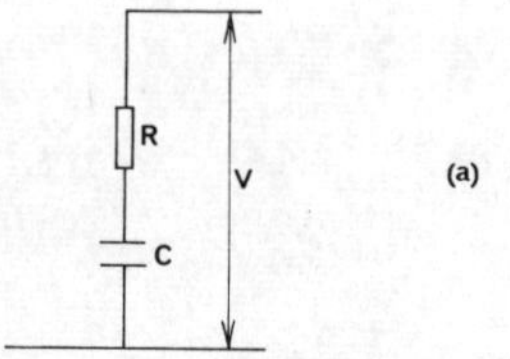

FIG. 6.29. (a) BASIC C-R TIMEBASE CIRCUIT

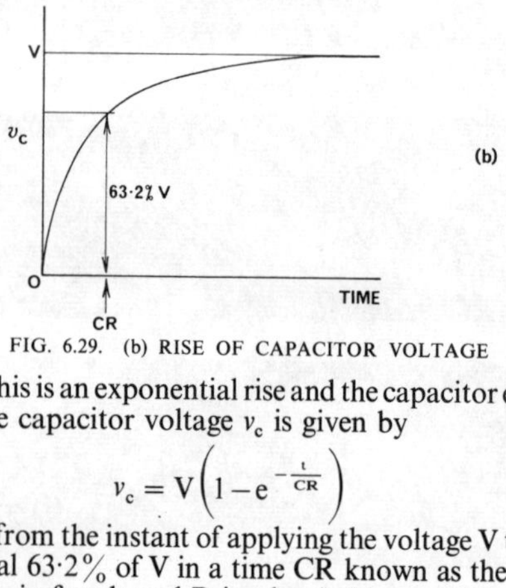

FIG. 6.29. (b) RISE OF CAPACITOR VOLTAGE

will rise as at (b). This is an exponential rise and the capacitor eventually reaches the voltage V. The capacitor voltage v_c is given by

$$v_c = V\left(1 - e^{-\frac{t}{CR}}\right)$$

where t is the time from the instant of applying the voltage V to the circuit. The voltage v_c will equal 63·2% of V in a time CR known as the time-constant of the circuit (C being in farads and R in ohms). The problem with this voltage across C is that it does not rise in a linear way, being rapid at first and the rate of rise decreasing as C becomes charged. The first part of the characteristic (say 10%) is nearly linear. However, if only 10% of the total voltage is used it means that either the output voltage is small or a high supply voltage V is required. To overcome these difficulties a means of making the curve more linear is commonly used. The reason why the rate of rise of voltage decreases is that the voltage across the resistor decreases as the capacitor becomes charged. This therefore reduces the current and hence the rate of charge, since

$$\frac{dv}{dt} = \frac{i}{C}$$

If the voltage across the resistor can be maintained constant then the rate of change of voltage will be constant. Two methods (although others are possible) are commonly used in oscilloscopes: the bootstrap circuit and the Miller integrator.

The speed of the sweep must, of course, be made variable over a large range to cover all requirements. Usually, the sweep speed is varied in steps, commonly in a 1, 2, 5 ratio. In order to cover the large range of speeds required the SWEEP SPEED control changes both R and C of the timing circuit. The range of sweeps covered will depend on the oscilloscope, but may be as large as 5 second/division to 50 ns/division – a range of 10^8 to 1. If the X-magnification facility is used these are then increased by the magnification factor, commonly 5 or 10. In many cases a continuously variable sweep speed control is fitted which covers at least the range from one fixed setting to the other. However, when this is moved from the calibrate position the calibration is upset. It is most important to remember this when taking time measurements.

BOOTSTRAP CIRCUIT

The basic idea is shown in figure 6.30. The voltage across the capacitor C is fed to a unity gain amplifier (with infinite input resistance) so that at all times

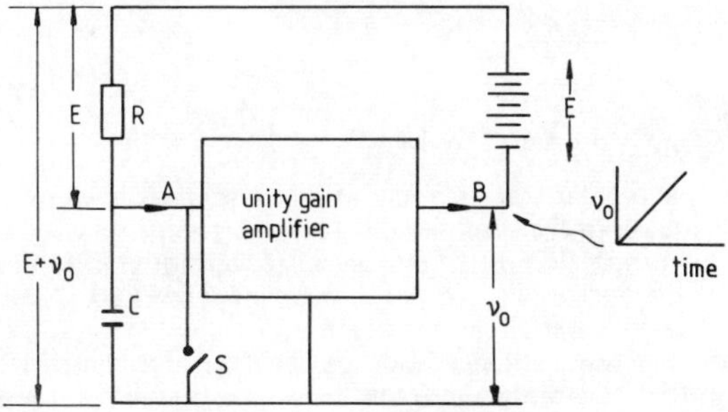

FIG. 6.30. BASIC BOOTSTRAP CIRCUIT

the voltage at B is equal to that at A. The voltage fed to the R–C circuit is now the battery voltage E plus the voltage v_o from the amplifier. On switching on (and assuming that C is initially discharged) voltage v_o will be zero and the voltage fed to the circuit is E. Since there is no voltage across C the voltage across R is equal to E. When C has become partly charged there will be a voltage output v_o from the amplifier, equal to that across the capacitor. The voltage now fed to the RC circuit is $E+v_o$. Since the voltage across C is v_o the voltage across R must remain at E. The voltage across R is constant, equal to the battery voltage E at all times. Thus a constant current flows in R and the rate of rise of voltage across C is constant, resulting in a linear rise of voltage (until limited by the amplifier). This is a positive feedback circuit with 100% feedback, but a gain of unity (hence it does not oscillate). To reset the circuit the switch S is closed so rapidly discharging C.

Obviously a battery or floating supply is inconvenient since the whole supply is varying at the output voltage and frequency of the ramp generator. In practice it may be replaced by a capacitor, and a modified circuit is given in figure 6.31, where the capacitor C_1 replaces the battery of figure 6.30. The value of C_1 must be considerably greater than the value of C – say 20 times. With switch S closed, v_o is zero and C_1 charges through D_1 from the positive line to a voltage V_1. On opening S the voltage v_o from the amplifier equals the voltage v across C. This is applied in series with C_1 and therefore C charges through R and, as before, the voltage across $R = V_1$. As soon as the capacitor C charges, v_o increases and the diode D_1 becomes reverse biased. Assuming that the voltage across C_1 remains constant, the circuit operates exactly as that of figure 6.30. Provided C_1 is large compared with C the voltage across C_1 will remain almost constant during the sweep period. It will be recharged again through D_1 when S is closed. For good linearity the gain of the amplifier should be exactly unity. When using valves a cathode-follower can be used,

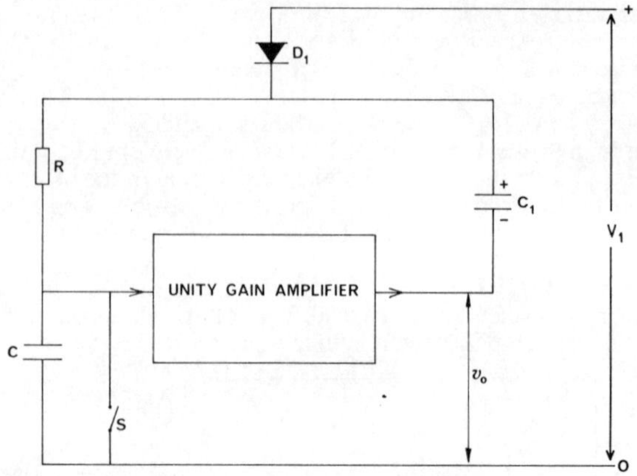

FIG. 6.31. PRACTICAL BOOTSTRAP CIRCUIT

but not a simple bipolar transistor circuit because of its low input resistance. The input resistance of the amplifier must be high compared with R, otherwise an appreciable proportion of the current in R will not go into the capacitor. An F.E.T. connected as a source follower may be used. Alternatively, a multiple emitter-follower circuit may be used.

Instead of capacitor C_1 a zener diode may be used, and a basic circuit used by Advance Electronics is shown in figure 6.32. The timing resistor and capacitor are R and C, which are varied to give the required range of sweep speeds. The voltage across C is applied to the triple emitter-follower circuit comprising Tr_1, Tr_2 and Tr_3. The use of three emitter-followers results in a high input resistance for Tr_1. A current flows from the $+170$ V supply through R_3 to zener diode Z_1, so maintaining a constant voltage across Z_1 and R_2. The voltage v_o (the output of Tr_3) will be about equal to the voltage across C, hence the voltage fed to the C–R circuit is v_o plus a fraction of the voltage across Z_1 selected by R_2, which forms a fine sweep speed control. The circuit therefore operates to give a linear rising voltage in the same way as described. The output to the X-amplifier is taken from Tr_3 and C is discharged by a transistor connected across it (not shown in the figure).

MILLER INTEGRATOR

The basic Miller integrator circuit is shown in figure 6.33. The timing circuit consists of R and C, and C is connected between the output and input of a high gain amplifier having 180° phase shift. In this case the feedback is negative. If the gain of the amplifier is high then point A is almost at a constant voltage, and there is a constant voltage across the resistor R equal to V_1. Assuming an infinite input resistance to the amplifier then the current through R is the charging current of C and hence the rate of change of voltage across C must be constant. As point A tends to go positive point B goes negative and B will move in voltage by an amount such as to maintain A at an approximately constant voltage, the more the gain the more constant the voltage. Thus, point B moves in a negative direction at a linear rate. The effective value of C is $(A+1)C$ where A is the gain of the amplifier without feedback. This is, in fact, an integrating circuit and is used as such in analogue computers.

When valves are used the amplifier can easily be a pentode valve with the capacitor connected between anode and grid, the resistor R then going to a negative supply. Simple bipolar transistor amplifiers cannot be used because of the low input impedance, but FETs may be used. A simple circuit is shown

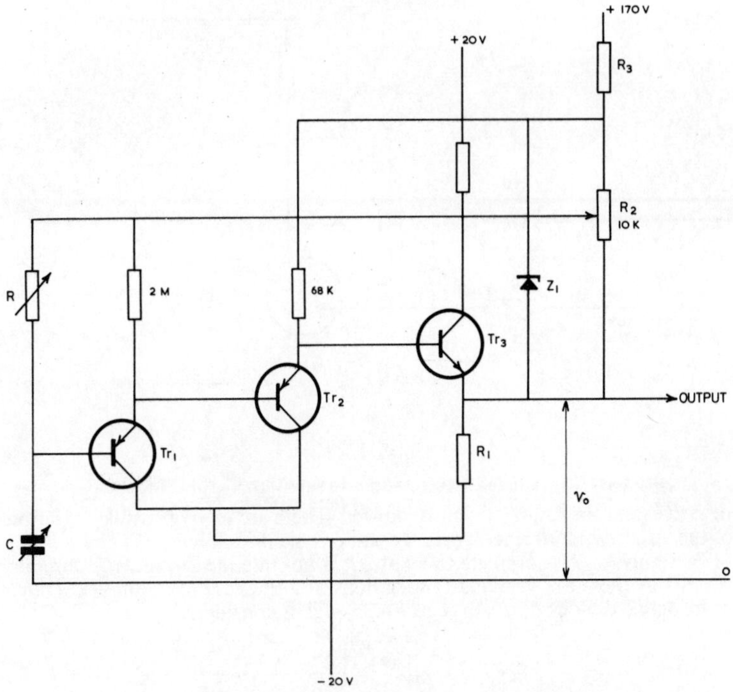

FIG. 6.32. BOOTSTRAP SWEEP GENERATOR CIRCUIT

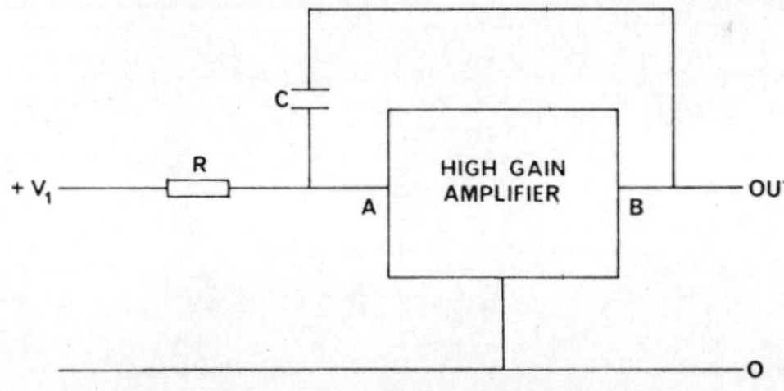

FIG. 6.33. MILLER INTEGRATOR CIRCUIT

in figure 6.34. Suppose that S is closed so that a suitable negative voltage is applied to the gate of the FET so that the drain current is small and hence the voltage at X is high. The capacitor will now be charged in the direction shown. On opening S a current will flow into C from R so discharging it (*i.e.* causing the voltage across C to decrease) the operation being similar to that described earlier. The gate will move slightly in a positive direction as C discharges causing the voltage at point X to decrease in a linear manner. At the end of the sweep S is closed and C is rapidly recharged through R_1. The circuit may operate the opposite way by R being fed with a negative voltage. In this case the initial condition is little bias on the gate so that a large current flows and point X is at a low voltage. On opening the switch current now flows out through R, C charges, the voltage of point X rising as the gate goes more

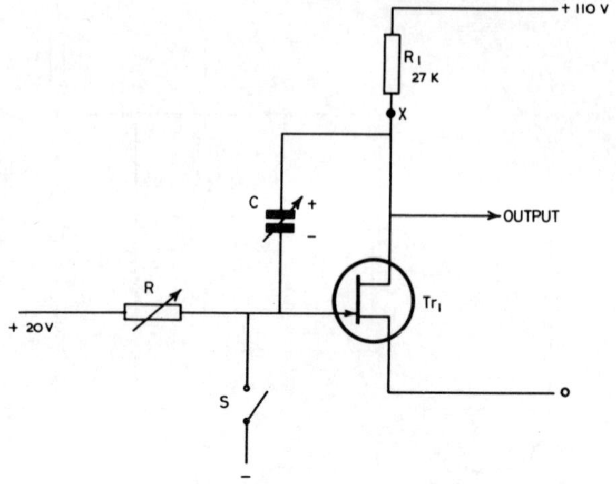

FIG. 6.34. MILLER INTEGRATOR SWEEP GENERATOR CIRCUIT

negative, and the drain current is reduced. There are many possible variations of the circuit, but all operate on the same basic principle.

To return to the complete timebase, it is not intended to include complete circuits as there are very many variations and they are too complex. Only a more detailed block diagram (figure 6.35) will be considered.

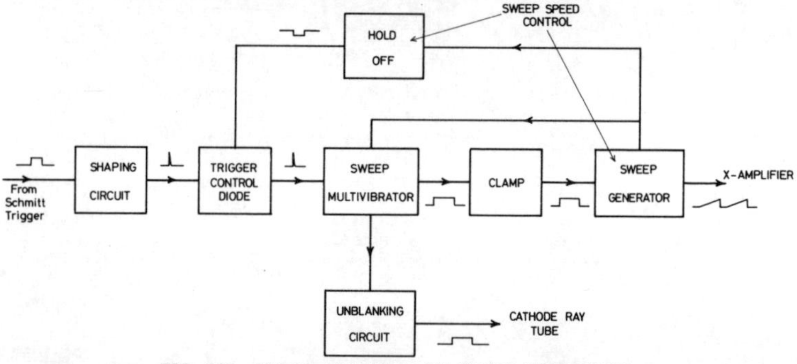

FIG. 6.35. BLOCK DIAGRAM OF SWEEP GENERATOR

There may be some variations from this diagram in some oscilloscopes, but the general principles are the same. The output from the Schmitt trigger or tunnel diode is fed to the SHAPING CIRCUIT which produces a pulse when the Schmitt or tunnel diode suddenly changes output. This is shown as a positive pulse, but the polarity will depend on the circuit arrangements, also the polarity of waveforms shown may be different. If the circuit is in the quiescent state (*i.e.* the beam at rest on the left-hand side of the screen) the pulse from the shaper circuit is fed through the TRIGGER CONTROL DIODE to the SWEEP MULTIVIBRATOR. This switches over the multivibrator (usually bistable but can be monostable), activating the clamping circuit to allow the sweep generator to operate and so produce the sweep. (The sweep generator is usually a Miller integrator or bootstrap circuit). When the SWEEP MULTIVIBRATOR is switched to this position it biases the TRIGGER CONTROL DIODE so that no further pulses can be fed to the sweep multivibrator and upset the sweep. At the same time this multivibrator operates the UNBLANKING CIRCUIT, so producing a trace on the screen. The sweep generator continues to run, but at a certain output voltage it causes the sweep multivibrator to return to its original state. This causes the CLAMP circuit to operate and terminate the sweep and produce the flyback. The switching back of the multivibrator causes the beam to be blanked so that the flyback is not visible on the screen. At the same time the HOLD-OFF circuit is operated, this block feeding a voltage to the TRIGGER CONTROL DIODE so preventing any trigger pulses being applied to the SWEEP MULTIVIBRATOR until the flyback has been completed and the circuit settled down to equilibrium. After a time, settled by the hold-off circuit, the trigger diode is again made conducting, so that the sweep multivibrator can be operated by the next trigger pulse. The hold-off circuit may be a multivibrator or a capacitor-resistor circuit. Its hold-off time will depend on the sweep velocity, its component values being varied by the SWEEP SPEED control. The output from the unblanking circuit may be fed to the grid of the cathode-ray tube or to the unblanking deflection plate (see Chapter 2).

VARIABLE HOLD-OFF CONTROL

Some oscilloscopes are fitted with a variable hold-off control. Considering figure 6.36, the hold-off time is the time after flyback when the timebase cannot be triggered by a triggering pulse. The purpose of this control is to

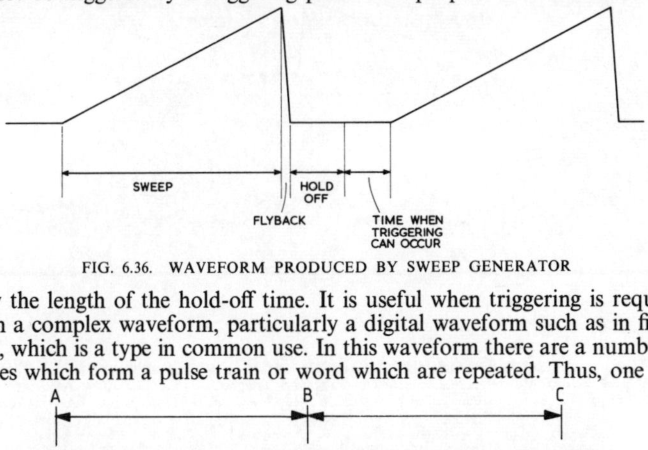

FIG. 6.36. WAVEFORM PRODUCED BY SWEEP GENERATOR

vary the length of the hold-off time. It is useful when triggering is required from a complex waveform, particularly a digital waveform such as in figure 6.37, which is a type in common use. In this waveform there are a number of pulses which form a pulse train or word which are repeated. Thus, one may

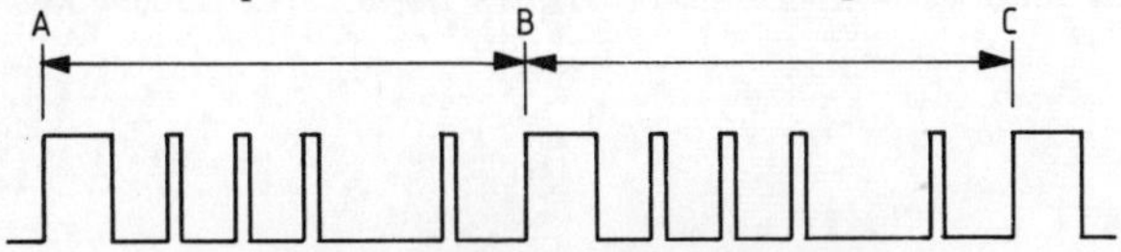

FIG. 6.37. DIGITAL PULSE WAVEFORM

consider that the cycle or word is from A to B or B to C. So that a steady display of the pulse train can be obtained the timebase should be triggered by the pulse at A and B and C, etc., and not by other pulses.

If the sweep speed setting is as shown in figure 6.38(b) the timebase is triggered first at A, a sweep will then take place followed by a hold-off, but the timebase will be triggered next time by pulse P instead of B and a multiple display will be obtained. If the sweep speed adjustment is, say, in 1, 2, 5 steps the next position will make the sweep too long, as shown at (c). If there is a continuously variable sweep speed control it can be adjusted until the hold-off finishes just before pulse B, and then satisfactory triggering will occur as shown at (d).

Not all oscilloscopes have a variable control and if fitted then the time/division (*i.e.* sweep speed) calibration is upset when moved from the calibrate

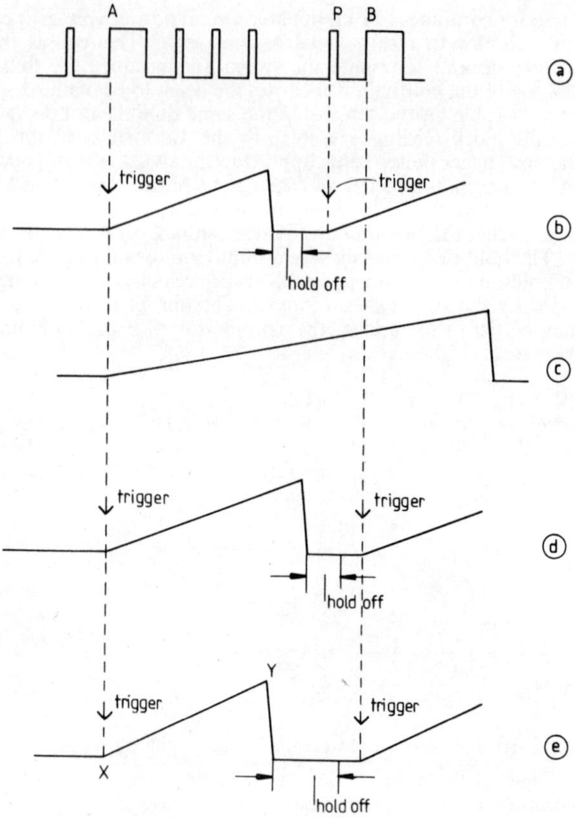

FIG. 6.38. TRIGGERING FROM DIGITAL PULSE WAVEFORM
(a) Pulse waveform; (b) Triggering on incorrect pulses; (c) Triggering on incorrect pulses (slower sweep speed setting); (d) Correct triggering (use of variable sweep speed control); (e) Correct triggering (use of variable hold-off control)

position. To overcome this the sweep speed is set to the value shown at (b) and the hold-off time increased. If the hold-off time is increased as at (e) so that it finishes just before pulse B, triggering will be satisfactory because the waveform will be triggered by the same pulse each time. The complete train will not now be shown, only those pulses during the sweep period X Y which may not be important. By altering the sweep speed and hold-off so that triggering occurs at A and C, etc. (figure 6.37) a whole train and part of a train of pulses will be displayed. Figure 6.39 shows some actual results. At (a) is the pulse train waveform as obtained by external triggering. With a normal sweep setting as shown at (b) a multiple display is obtained because the timebase triggers on different pulses. Operating the fine sweep speed control results in (c), which is satisfactory except that the time calibration has been lost. Resetting the variable sweep speed control to CALIBRATE and operating the variable hold-off control results in (d), which is also satisfactory and the time-scale is still correct.

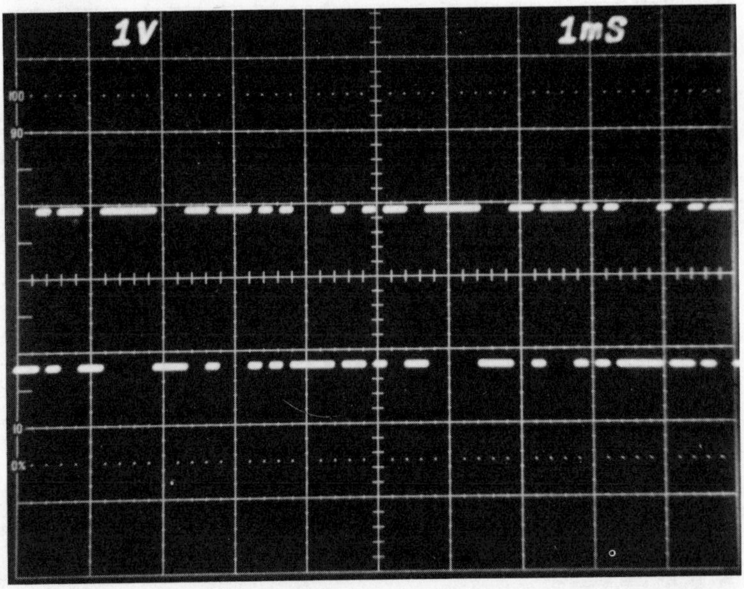

(a) By external synchronizing pulses

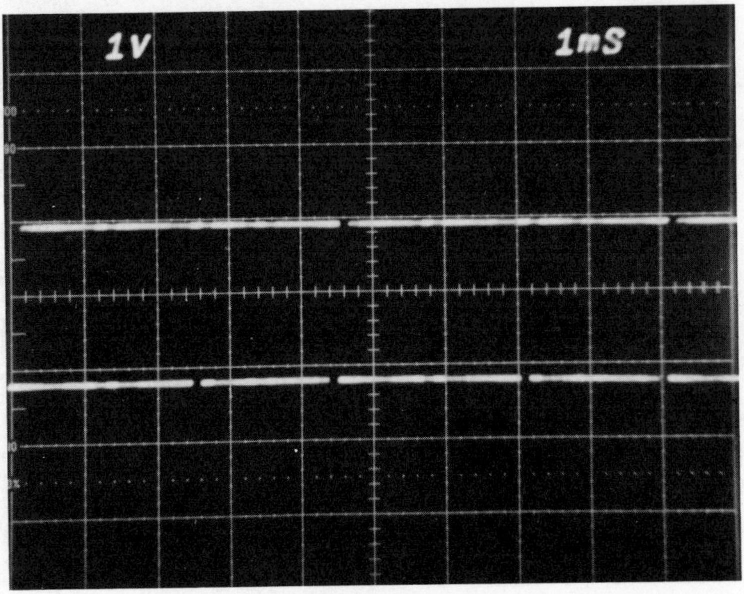

(b) Incorrect triggering

FIG. 6.39(a) & (b). TRIGGERING FROM DIGITAL PULSE WAVEFORM

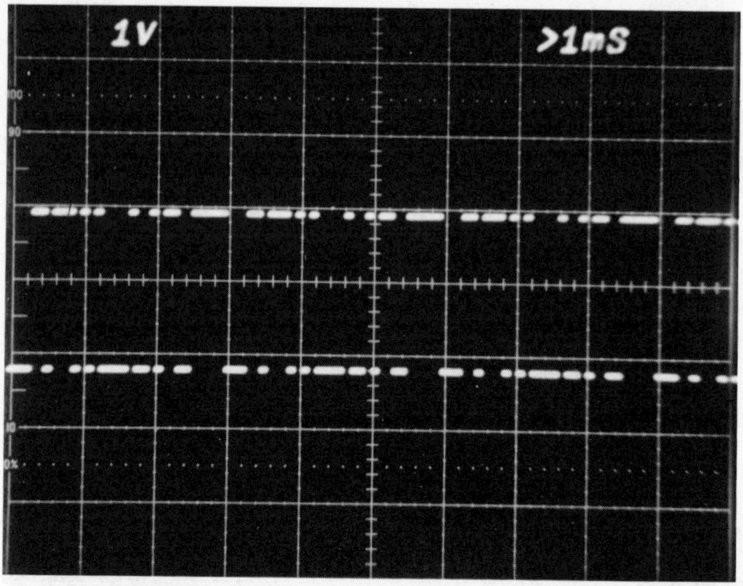

(c) Correct triggering by using variable sweep control (> indicates that timebase not calibrated)

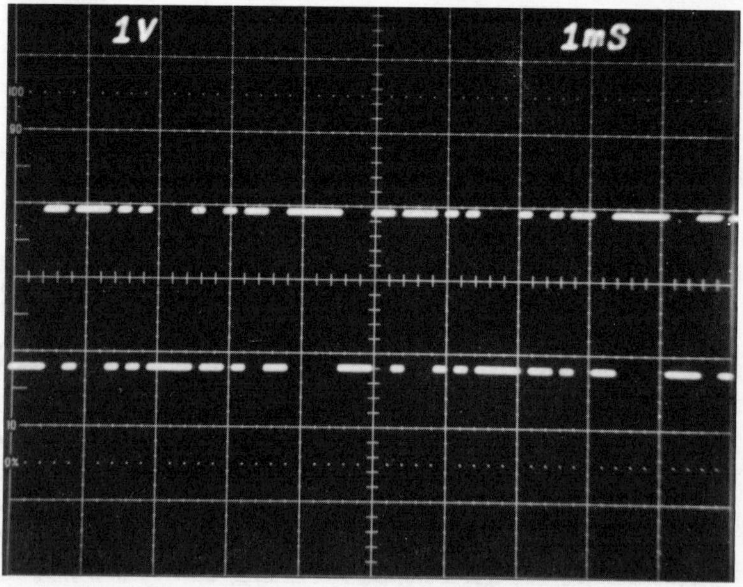

(d) Correct triggering by using variable hold-off control

FIG. 6.39(c) & (d). TRIGGERING FROM DIGITAL PULSE WAVEFORM

STABILITY CONTROL

In some oscilloscopes the sweep multivibrator (or sweep multivibrator and sweep generator) has a stability control. By varying this the triggering sensitivity of the timebase may be set:

(a) So that it cannot be triggered (low sensitivity).
(b) So that it can be triggered by pulses from the pulse generator (normal operation).
(c) So that it runs continuously at a frequency depending on the components of the sweep speed control. Under these conditions the timebase will not trigger correctly.

Such a control *must* be adjusted (with no signal input) until the timebase just does not run. Normally, once it is adjusted further adjustment is required only at infrequent intervals.

AUTOMATIC TRIGGERING

Up to the present 'normal triggering', as it is called, has been described. This means that in the absence of a signal there are no triggering pulses and the timebase does not operate. Even when there is a signal present the timebase will operate only if the trigger level control has been set to the correct value. Since the tube is blanked except when the timebase is operating nothing is seen on the screen. This can be confusing, particularly, for example, when the input signal decreases so much that the trigger level is not reached. Under this condition the trace disappears from the screen until the trigger level is operated. To prevent this it is desirable to have the timebase operating, even without an input, so that there is something visible on the screen and, when a suitable waveform is applied, to automatically lock or trigger.

There are two basic ways of doing this which will be described. Unfortunately, both are often called AUTO, but to distinguish them the second one will be called 'auto bright line'.

Auto

When the timebase is set to AUTO (often a switched position on the level control) the LEVEL CONTROL block (or the LEVEL CONTROL block and PULSE GENERATOR) is made into an oscillator (essentially an astable multivibrator). The frequency chosen is low, say 40 Hz, and this circuit will produce pulses continuously in the absence of a signal. These pulses trigger the sweep generator whatever the sweep speed setting. If the sweep speed is such that the complete sweep (*i.e.* sweep, flyback and hold-off) is slightly less than the period of oscillator (*i.e.* 1/40th second) the timebase will be triggered at 40 times per second. If the sweep speed is set to a low value so that the time of a complete sweep is greater than the period of the oscillator the timebase will be triggered by a next pulse after the release of the hold-off circuit, and will run at a low frequency. In the AUTO position the input coupling block is set to a.c. and approximately zero triggering level; the pulses from the level control block now trigger the timebase and a locked waveform should result under all conditions. However, it cannot be relied on for frequencies below about 40 Hz (the free-running frequency of the oscillator).

When the timebase has a stability control this must be set correctly. With no signal applied the oscilloscope should be switched to NORMAL and the stability control adjusted until the timebase *just does not run*, *i.e.* the trace disappears. This means it is set to condition (b). On switching to AUTO the trace should reappear. If the stability control setting is too low there is no trace when on AUTO. In practice, this system works well *provided* the stability control is properly set. There is, of course, no method of varying the trigger level, *i.e.*

the point on the waveform where triggering occurs; or altering the trigger circuit so that it triggers on a particular slope.

One disadvantage of this circuit is that at high sweep speeds of the timebase (*i.e.* small ms/div) the trace with no signal becomes very dim. This is due to the fact that the timebase is triggered only at the free-running frequency of the pulse generator (about 40 Hz), but the speed of the spot across the screen is high. There is therefore a long waiting time between traces. This is overcome in the bright line auto system.

Auto Bright Line or Auto Stability

With this arrangement the sweep generator is made to self-run in the absence of pulses. The basic arrangement is shown in figure 6.40. With the auto bright line OFF the circuit operates normally; the sweep multivibrator and sweep

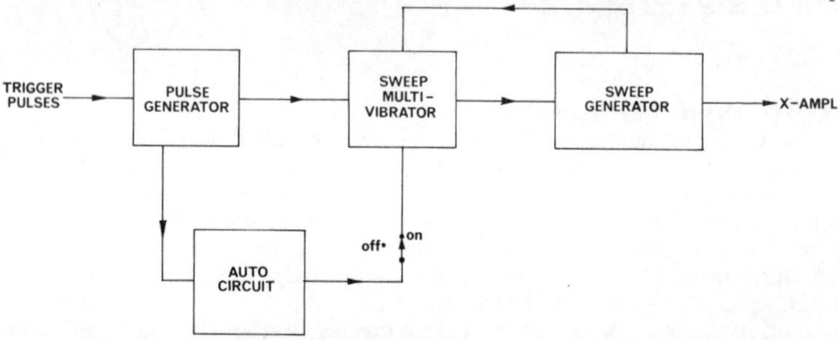

FIG. 6.40. BLOCK DIAGRAM OF AUTO BRIGHT LINE TRIGGER CIRCUIT

generator only operate when a pulse is fed to the sweep multivibrator. When the auto bright line circuit is switched to the ON position, provided that there are no pulses from the pulse generator, the auto circuit alters the conditions on the sweep multivibrator and sweep generator so that they free-run and produce repetitive sweeps. There are several ways in which the circuit can be made to free-run depending on the exact circuits used. When pulses come from the pulse generator they are fed to the auto bright line circuit, which now produces a voltage (*e.g.* by rectification and smoothing) so as to set the sweep multivibrator and sweep generator to the normal method of operation, *i.e.* operated only by trigger pulses. If the trigger pulses cease then the auto bright line circuit operates (in say 100 ms) to make the timebase free-running again. Thus there will always be a trace on the screen. If a waveform is applied to the Y input or external trigger input, but the trigger level control is set so that no trigger pulses are produced, then the waveform does not lock because the timebase free-runs. However, where the trigger level control is set correctly the waveform locks. It will be seen that when there are trigger pulses the circuit operates exactly as normal (the trigger level control and slope control thus having exactly the same effect as on normal setting). The only difference is that if no trigger pulses are available from the trigger pulse generator, a trace is still obtained but not locked. In other words, the stability control of the timebase is automatically set according to whether or not there are pulses – hence the name. Since the free-running frequency will be of the same order as when triggered the brightness remains approximately constant independent of the sweep speed setting. This circuit has the advantage that there is no stability control to set correctly.

There are some variations of this basic arrangement. When on AUTO the range of the level control may be limited so that the trace (unless very small) normally locks.

In one oscilloscope (Philips) the range of variation of the level control is automatically controlled so that it is no greater than the peak-to-peak value of the signal. This is also known as 'p-p auto' and is used by other manufacturers.

The automatic circuit of figure 6.40 must have a time delay so that it does not change over from triggered to free-run between trigger pulses. Thus there may be a short delay between the application of the signal and the circuit operating and the trace locking. For this reason AUTO should not be used below about 40 Hz.

It is useful to know when the timebase is being triggered (particularly if the trace is off screen); this may be indicated by a lamp display as on the Tektronix 7000 series. The lamp only lights when the sweep is actually being triggered, not when it is running free on AUTO.

SINGLE SHOT or SINGLE SWEEP OPERATION

This facility is not provided on some oscilloscopes and is mainly of use in examining transient phenomena. When the oscilloscope is placed in this position an additional circuit is used to prevent the timebase being triggered more than once. Details of a circuit will not be given but it may be a bistable multivibrator which will be called the lock-out bistable. Until the circuit is reset this bistable's state is such that it prevents pulses being applied to the sweep multivibrator (in a way similar to the hold-off circuit). To examine a transient a button is pressed which sets the lock-out bistable in the other condition so that trigger pulses can now pass to the sweep multivibrator. Usually a lamp is fitted to indicate that the circuit has been reset or is 'armed' ready for operation. The triggering arrangements are exactly as already described for repetitive traces and must be appropriately set. If a transient waveform is now applied, with the trigger arrangements correctly set, the timebase will be triggered and the sweep generator will operate producing a single trace. At the end of the trace the lock-out bistable is set to the lock-out position so preventing further triggering until the reset button is pressed. The speed of the sweep is, of course, settled by the SWEEP SPEED controls. This method of operation is mainly for photographing transients. If the transient is a fast one then photography (or a storage oscilloscope) is essential. Triggering can be from the Y-amplifier (internal) or external to the external trigger socket.

H.F. SWITCH

This is to be found mainly on the less expensive oscilloscopes. As the input frequency increases, triggering becomes more difficult, partly because of the problem of producing and maintaining sharp high frequency pulses. To overcome this at frequencies above about 1 MHz the h.f. switch can be operated. This converts the Schmitt trigger pulse generator into a high-frequency oscillator at, say, 500 kHz, which synchronizes with the incoming signal, and its output triggers the timebase. This results in a more stable trace than using the normal circuits. In some cases the frequency of oscillation of the pulse generator circuit is controlled by the level control, which should be adjusted for satisfactory locking.

Dual timebase and delayed sweep timebases are covered in Chapter 11.

7

Power Supplies, Cathode-Ray Tube Circuit and Z-modulation

THERE are a number of supplies required, as follows:

(1) A negative supply of, say, 1 to 2 kV for the cathode-ray tube.

(2) A high voltage positive supply of 5–12 kV for the electrode of a PDA tube where fitted.

(3) Several supplies to feed the amplifiers and timebase, etc., say, 20 to 150 volts, positive and/or negative.

The power supply arrangements will depend on whether the oscilloscope is to work off the normal 240 V, 50 Hz supply; or whether it is to be portable and operate off internal batteries or a low voltage supply of, say, 12 volts.

MAINS SUPPLY OSCILLOSCOPE

(1) In most oscilloscopes the negative supply voltage is obtained from a mains transformer using a voltage-doubler circuit with capacitance smoothing. The voltage may be stabilized by zener diodes. Variations of this voltage will alter the deflecting plate sensitivity and the calibration. One voltage-doubler circuit is shown in figure 7.1. When B is positive with respect to A, current flows in Re_2, and C_2 will become charged in the direction shown to

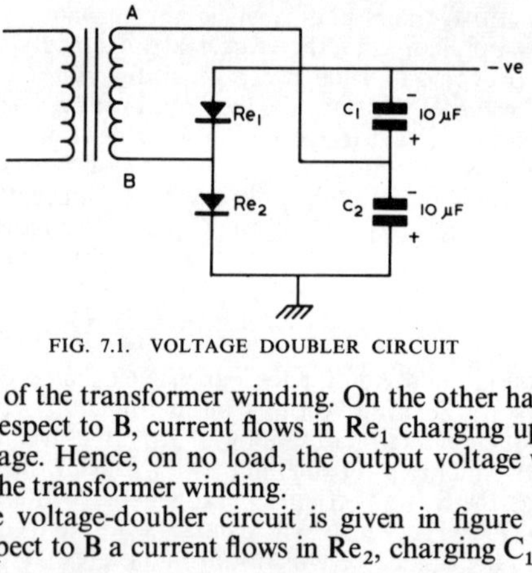

FIG. 7.1. VOLTAGE DOUBLER CIRCUIT

the peak voltage of the transformer winding. On the other half-cycle, when A is positive with respect to B, current flows in Re_1 charging up C_1 to the peak transformer voltage. Hence, on no load, the output voltage will be twice the peak voltage of the transformer winding.

An alternative voltage-doubler circuit is given in figure 7.2. When A is positive with respect to B a current flows in Re_2, charging C_1 in the direction

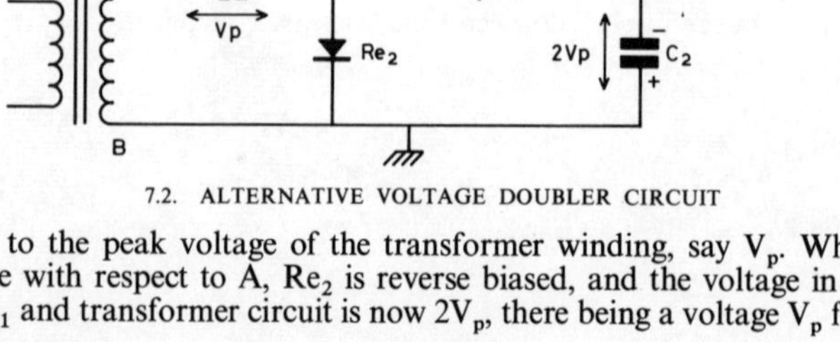

7.2. ALTERNATIVE VOLTAGE DOUBLER CIRCUIT

shown to the peak voltage of the transformer winding, say V_p. When B is positive with respect to A, Re_2 is reverse biased, and the voltage in the C_2, Re_1, C_1 and transformer circuit is now $2V_p$, there being a voltage V_p from the transformer and a voltage V_p from the capacitor C_1. Thus a current flows in Re_1 charging up C_2 to a voltage $2V_p$ in the direction shown.

A typical simple circuit for supplying the tube is given in figure 7.3. The voltage on A_2 must be made variable to give a focus control and the grid is

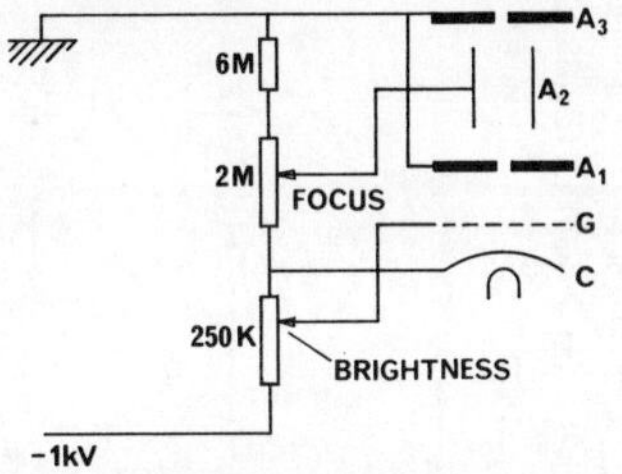

7.3. SIMPLE CATHODE-RAY TUBE CIRCUIT

given a variable negative voltage, with respect to the cathode, so that the brilliance of the trace can be varied.

In place of a mains transformer an e.h.t. oscillator may be used to provide the negative and possibly positive e.h.t. required.

A much simplified oscillator circuit (Advance Electronics) is shown overleaf in figure 7.4. The transformer T_1 has five windings, two being centre tapped. W_1 is the feedback winding to the bases of the transistors from the main collector winding W_2. Thus the circuit operates as a push-pull oscillator at 30 kHz. Winding W_3 and Re_1, together with smoothing circuit C_1, C_2 and R_1 provide the negative cathode supply and the focus voltage for A_2 from the potentiometer R_2. This focus potentiometer is in series with R_3 and R_4 to a stabilized +15 V supply. Under normal conditions the ratio of R_2 and R_3 to R_4 is such that the voltage on the base of Tr_4 is around zero volts. Tr_5 is a series control transistor fed from Tr_3. If the negative cathode voltage from Re_1 should rise then the voltage at the base of Tr_4 will go in the negative direction. This will reduce the current in Tr_4 and reduce the base-emitter voltage of Tr_3, so reducing its collector current and the base current of Tr_5. The emitter voltage of Tr_5 will drop, reducing the voltage on the oscillator until equilibrium is reached. In this way the negative cathode and focus supply are stabilized. This negative voltage can be set to the required value by R_4. Winding W_4, Re_2 and C_3 form a separate floating supply for the grid circuit. This forms part of the blanking and Z-modulation circuit. Windings W_3 and W_5 feed a tripler to supply the PDA electrode (see later). An overload protection circuit is included, but not shown in the figure for simplicity.

(2) The high positive voltage for the PDA electrode may be obtained from a mains transformer using a voltage multiplier rectifier, say a tripler to a septupler. A tripler circuit is shown in figure 7.5, which operates on the same principle as the doubler of figure 7.2. When B is positive with respect to A,

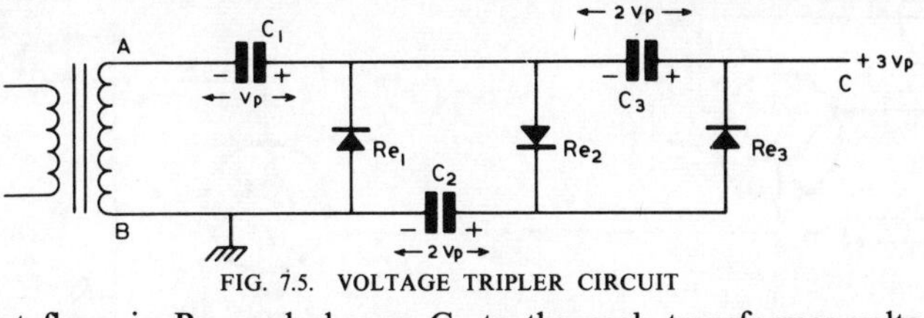

FIG. 7.5. VOLTAGE TRIPLER CIRCUIT

current flows in Re_1 and charges C_1 to the peak transformer voltage V_p in the direction shown. When A is positive with respect to B the transformer voltage and that across C_1 cause a current to flow in Re_2, charging up C_2 to

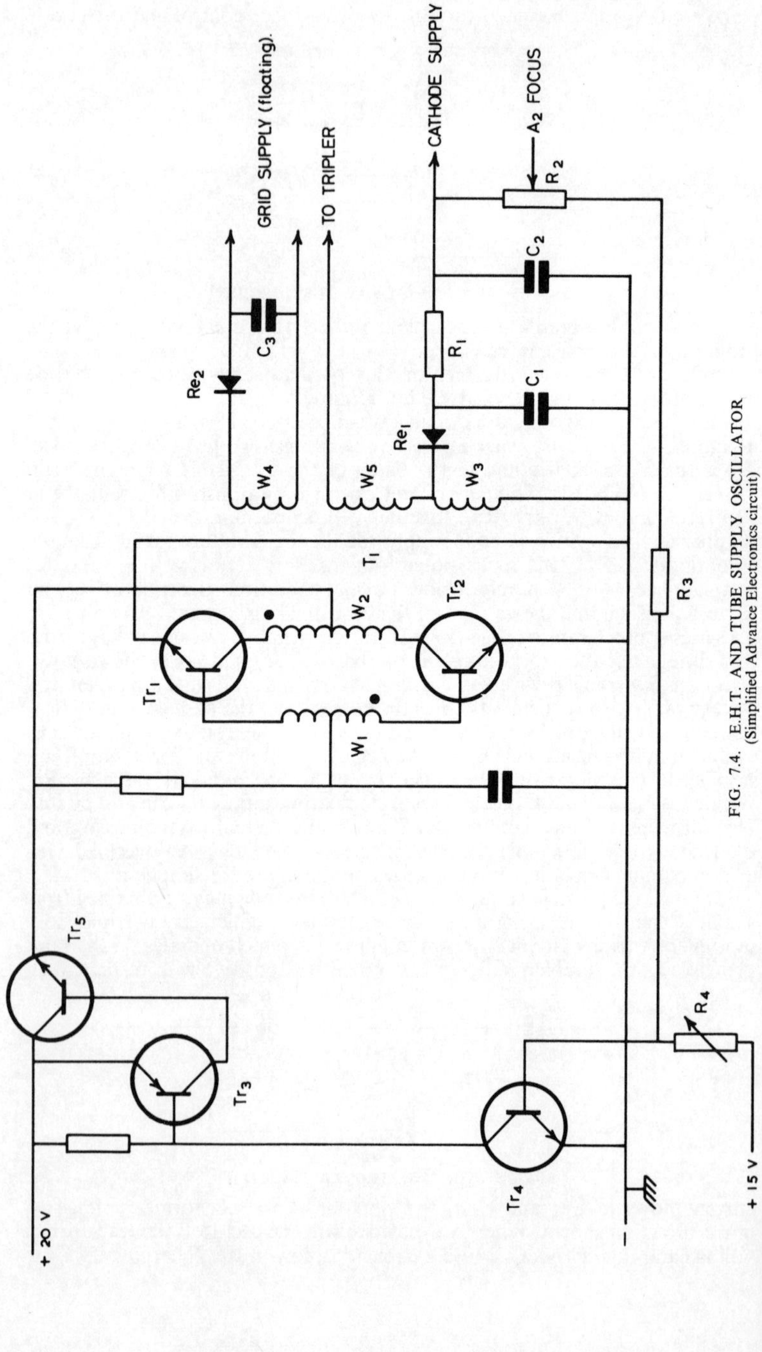

FIG. 7.4. E.H.T. AND TUBE SUPPLY OSCILLATOR
(Simplified Advance Electronics circuit)

$2V_p$ in the direction shown. Now consider B positive with respect to A and take the circuit C_2, Re_3, C_3 and C_1. In this circuit there is, in an anticlockwise direction, $+V_p$ from the transformer, $+2V_p$ from C_2 and $-V_p$ from C_1, a resultant of $2V_p$. Thus, a current flows in Re_3 charging up C_3 to a voltage $2V_p$ as shown. The voltage output at point C will be that across C_1 plus that across C_3, a total of $3V_p$. This voltage is applied, often through a limiting or protection resistor of, say, 1 mΩ, to the PDA electrode.

Alternatively, an oscillator may be used with a voltage multiplier as described earlier for the negative supply, the winding W_3 and W_5 of figure 7.4 feeding a tripler circuit similar to that just described. The tripler is followed by a simple R-C filter circuit.

(3) Amplifier and timebase supplies will be required from, say, 15 to 150 volts which may be positive or negative, depending on the oscilloscope, and generally will be more numerous with the complexity of the oscilloscope. Many possible stabilizer circuits are used. One is shown in figure 7.6 where Tr_4 is the series control transistor. Tr_1 and Tr_2 form a long-tailed pair, with the common-

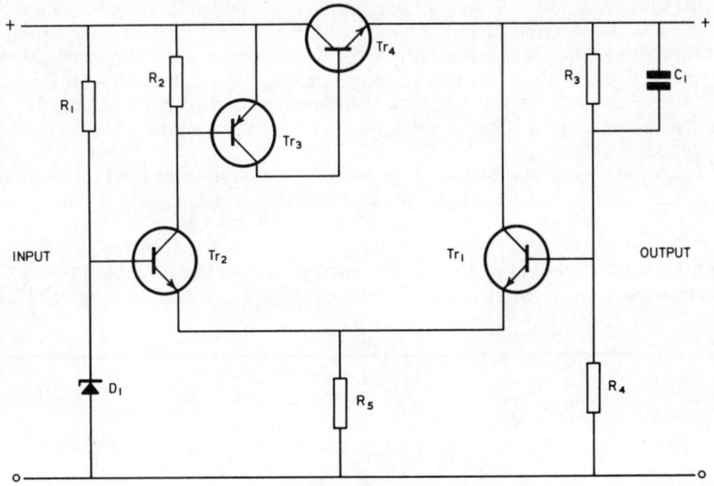

FIG. 7.6. VOLTAGE STABILIZER CIRCUIT

emitter resistor R_5. The base of Tr_2 is at a fixed potential, the same as the zener diode D_1, while the base of Tr_1 is fed with a fraction of the output voltage by the potential divider R_3R_4. Thus, if the output voltage rises, the base of Tr_1 goes more positive causing Tr_1 to conduct more, while transistor Tr_2 conducts less. This reduces the voltage across R_2 so decreasing the voltage on the base of Tr_3, which also conducts less. This reduces the base current of Tr_4 and so it tends to conduct less, the voltage rises across it so reducing the output voltage until equilibrium is reached. C_1 is used to feed the full ripple voltage to the base of Tr_1 so increasing the loop gain for a.c. and reducing the ripple in the output.

Advance Electronics in one of their oscilloscopes use a constant voltage transformer for stabilization.

PORTABLE OSCILLOSCOPES

This type of oscilloscope will not be dealt with in detail. It is only recently that such oscilloscopes have become available due to the use of transistors. Normally, the oscilloscope is made so that it can be operated off the mains, from, say, an external 12 volt supply, or off its own internal battery. Arrangements are usually incorporated to recharge the battery when the instrument is connected to the mains supply. When the instrument is on the external d.c. supply or on its own internal battery the voltage is stepped up by an oscillator or inverter to provide all the various voltages required. A standby position may be provided which just maintains a supply to the cathode-ray tube heater. This obviously reduces the loading on the battery or external supply, which is relatively large when the oscilloscope is in operation. On the mains, a suitable d.c. supply is obtained by rectification to feed the inverter. The basic oscilloscope is the same as non-portable types. Since the current consumption is relatively large the time of operation on rechargeable batteries may be quite short, say four hours. The time to recharge may be 14 hours. Replacement batteries are very expensive.

Z-MODULATION AND BLANKING

Blanking of the trace is often essential, for example, to prevent the display of the flyback. Blanking is an extreme case of Z-modulation, *i.e.* the beam is either fully on or fully off. As explained in Chapter 3, blanking can be achieved by the use of deflection blanking plates, but they cannot be used for Z-modulation (other than fully on or fully off). The advantage of using deflection blanking plates is that they are at approximately the final anode voltage A_3 (assuming the voltage on A_1 to be the same as on A_3). This is near earth potential and hence feeding the plates with a suitable potential, including d.c., is comparatively easy.

The other way of blanking is to use the grid, which must be used for normal Z-modulation so that the brightness can be varied continuously. Unfortunately, the grid is at a high negative potential to earth, say 1 to 2·5 kV. If only a.c. is to be fed to the grid, there is no great difficulty as it can be fed through a suitable capacitor, as in figure 7.7. R_1 is added to maintain a reasonable input impedance on the Z-input socket. R_2 prevents a static voltage being developed

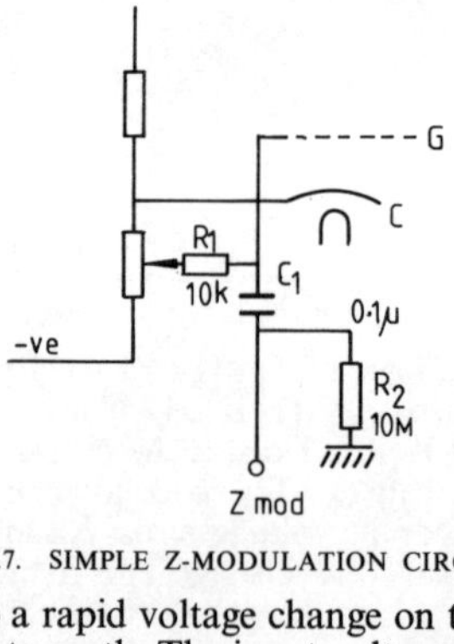

FIG. 7.7. SIMPLE Z-MODULATION CIRCUIT

on the Z socket, and also a rapid voltage change on the grid if the Z-input is suddenly connected, say, to earth. The input voltage must be limited so that the grid is not driven in a positive direction. This simple arrangement is all that is sometimes required.

Two improvements may be made:

(a) Using an amplifier between the Z-modulation input and the grid. This amplifier may also be used for other blanking waveforms and may need to have a good frequency response when switched beam oscilloscopes (see Chapter 10) and readouts are used (see Chapter 15).

(b) The use of a d.c. restorer on the grid side of C_1 so that the mean grid voltage does not change with variations of the mark-space ratio.

In many cases d.c. modulation is necessary or desirable. This is essential for blanking or unblanking when a low sweep speed timebase is used.

It may be assumed that the tube is normally ON and blanked when the trace is not required; or that the tube is normally OFF and the tube is unblanked or brightened (bright up) when a trace is required.

Many timebases go down to 5 sec/div or, say, 50 seconds per traverse; a.c. coupling at this low frequency is not practicable. As high frequency blanking and modulation are necessary the grid is commonly fed by two circuits. An a.c. circuit having a good frequency response; and a d.c. circuit of relatively low frequency response but, of course, going down to d.c. The a.c. circuit is as already described, *i.e.* capacitance coupling to the grid or, in some cases, to the cathode. The methods of getting essentially d.c. coupling vary.

In order to get d.c. coupling to the grid of the cathode-ray tube for blanking (or bright-up) Advance Electronics (in the OS 3000) use a separate floating grid voltage supply (see figure 7.4). A simplified circuit is shown in figure 7.8.

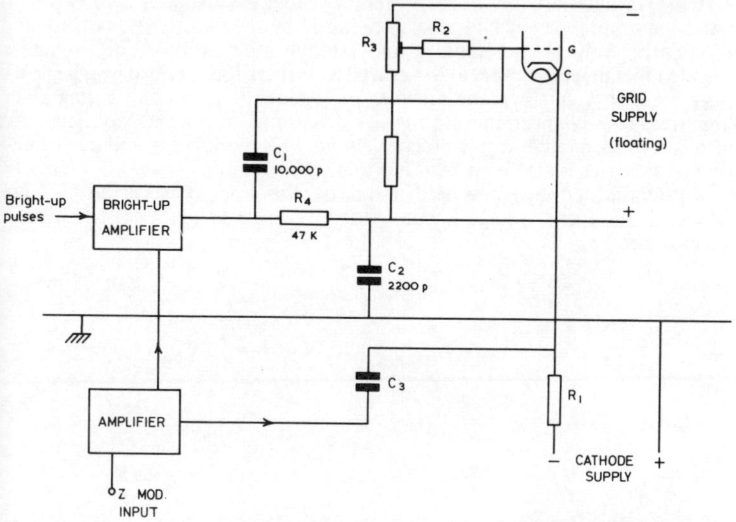

FIG. 7.8. Z-MODULATION AND BRIGHT-UP CIRCUIT (Simplified Advance Electronics)

The cathode is fed through R_1 from a separate supply. The grid is fed from the floating supply and the grid voltage can be varied relative to the cathode by the preset control R_3 to allow for variations in cut-off of different tubes. The positive side of the floating grid supply is fed through R_4 from the bright-up (or unblanking) amplifier. Thus, if the steady voltage of the bright-up amplifier varies, the floating supply moves in relation to earth, and hence cathode. Thus the grid voltage also changes relative to the cathode. To obtain rapid changes the grid is fed through C_1 so this forms the a.c. circuit. The time constant of C_1R_2 is much greater than that of C_2R_4, hence if a sudden voltage step takes place in the output of the bright-up amplifier, the grid changes rapidly by the voltage fed through C_1. Before this voltage has had time to decay the voltage across C_2 rises and changes the potential of the floating supply relative to earth, so maintaining the grid at the correct steady value. The BRIGHTNESS control varies the d.c. voltage on the bright-up amplifier, and therefore the voltage across C_2.

The Z-modulation input is fed to an amplifier and then to the bright-up amplifier. This Z-modulation amplifier also feeds the cathode of the cathode-ray tube, as regards a.c., through C_3. This is used so that the frequency response of the Z-modulation is not limited by the frequency response of the bright-up amplifier. The Z-modulation frequency response is d.c. to 40 MHz. This basic idea is also used by other manufacturers.

An interesting arrangement is used by Scopex (model 4D 25) for blanking. No Z-modulation facilities are provided. The bright-up pulses from the timebase are fed, as regards high frequency a.c., by a capacitor in the normal way. The steady component is obtained by the use of an optical coupler. During the timebase sweep, current is fed into the light-emitting diode of this coupler. The resulting light is fed to a light-sensitive transistor which feeds the grid through a d.c. amplifier. This is one way of overcoming the problem of the large d.c. voltage on the grid. Intertrace blanking (see Chapter 10) is by feeding pulses to the cathode through a capacitor. This principle is also used by Advance Electronics.

In some of their oscilloscopes, Tektronix basically use a modulated carrier system. An alternating voltage of relatively high frequency is used, which is varied in amplitude (by clipping) as required by the BRIGHTNESS control and modulating voltages (unblanking and external modulation). This voltage is then fed through capacitors to the grid circuit where it is rectified by a peak-to-peak rectifier circuit, and smoothed to provide the required bias voltage. The high frequency connection is through a capacitor in the normal way. The same idea can be used to control the focus voltage. This modulating and demodulating of a carrier is also used by other manufacturers.

Automatic focusing is now used in some oscilloscopes so that as the brilliance is varied, the focus voltage is automatically varied to maintain the tube in good focus.

8

Connection of Oscilloscope to Circuit under Test, Voltage and Current Probes

THE oscilloscope has to be connected to the circuit under test, which may sound simple but is not always the case. It is important that the connection should have the least possible effect on the circuit. The input resistance of the oscilloscope is usually 1 MΩ and its input capacitance may be, say, 25 pF (11 to 50 pF). The Y-input connection is normally a coaxial type socket (*e.g.* BNC), hence many operators use a length of coaxial cable. At low frequencies this is generally satisfactory, but a 1 metre length of coaxial cable will probably have a capacitance of 100 pF. No doubt the designer of the oscilloscope has tried to keep the input capacitance as low as possible, but is completely defeated by the use of the coaxial cable. The input capacitance will now be, say, 125 pF. One way of reducing the capacitance is to use a single live lead and a corresponding earth lead, but is not generally desirable owing to pick up on the lead and the possible radiation from it, which may cause oscillation and many misleading results. However, sometimes a more accurate waveform may be obtained, as shown in figure 8.1. At (a) a single short lead has been used, and at (b) about 0·6 metre of normal coaxial cable. At (a) the live lead has picked up interference (*e.g.* hum) which has blurred the waveform. However, the capacitance is less than that of the coaxial cable at (b), therefore the waveform (a square wave) is more correct in (a) than in (b).

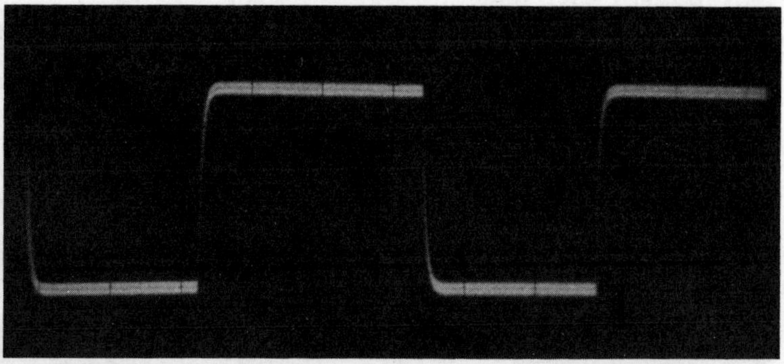

(a) Live lead and earth lead showing pick up of interference and some distortion of waveform

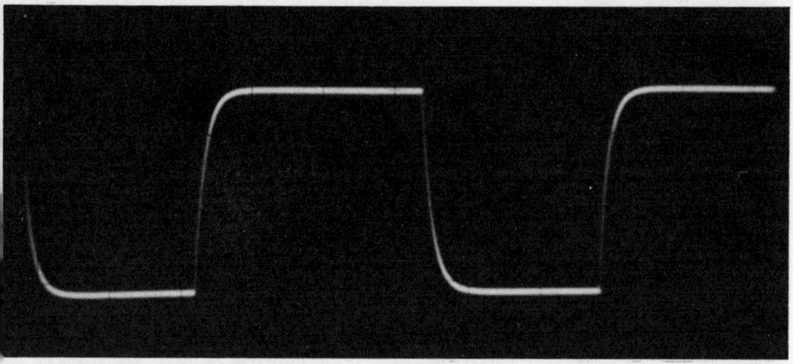

(b) 0·6 metre length of coaxial cable showing distortion of waveform due to large capacitance

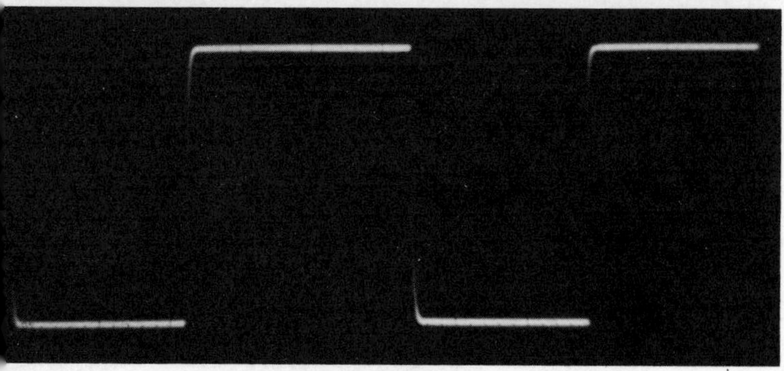

(c) × 10 probe showing improved rise time due to reduced capacitance

FIG. 8.1. EFFECT OF CONNECTION OF OSCILLOSCOPE TO CIRCUIT. IMPEDANCE OF CIRCUIT 470 kΩ AND FREQUENCY 1 kHz

The effect of this capacitance is obviously going to depend on the internal impedance of the circuit under test and on the frequency concerned. The device under test can be represented by the circuit of figure 8.2, where e is the source of e.m.f., R_i the internal resistance, and C_i the capacitance of the

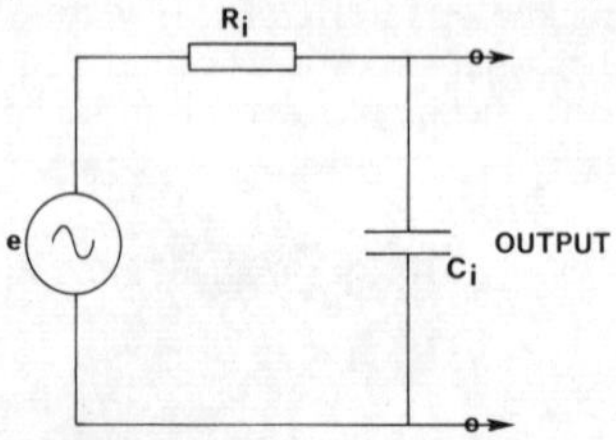

FIG. 8.2. EQUIVALENT CIRCUIT OF SOURCE OF SIGNAL

device itself. The input socket of the oscilloscope can be represented by the circuit of figure 8.3, where R_1 is the input resistance (nearly always 1 MΩ) and the input capacitance C_1, say 25 pF. When the two are connected the circuit is as figure 8.4. When a connecting cable is used C_i is increased by the

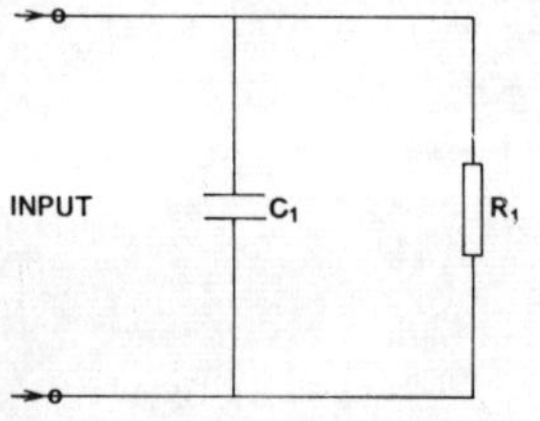

FIG. 8.3. EQUIVALENT CIRCUIT OF INPUT TO OSCILLOSCOPE

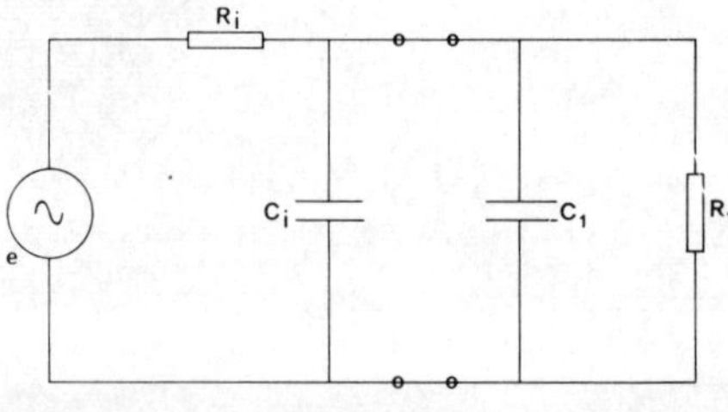

FIG. 8.4. COMBINATION OF THE CIRCUITS OF FIGURES 8.2 AND 8.3

capacitance of the cable. The circuit is a low-pass filter. For d.c. the voltage fed to the oscilloscope will be

$$e \times \frac{R_1}{R_1 + R_i}$$

In most applications R_1 will be large compared with R_i, and the input to the oscilloscope becomes equal to e. As the frequency is increased the reactance of $C_i + C_1$ becomes lower and, when comparable with the value of R_i, the results are upset. If a sine waveform is being examined the shape of the waveform will not be changed, but the amplitude fed to the oscilloscope will be reduced at high frequencies. When the reactance of $C_i + C_1$ is equal to R_i the voltage fed to the oscilloscope will be 70·7% of e (−3 dB). For example if $R_i = 320$ ohms and $C_i + C_1 = 50$ pF then the reactance of $C_i + C_1$ becomes 320 ohms at 10 MHz. Thus, the magnitude of the input to the oscilloscope will be incorrect at this frequency. Considering the circuit of figure 8.2 the output would, of course, drop even without the connection of the oscilloscope, but if $C_i = 10$ pF the voltage would not drop by 3 dB until a frequency of 50 MHz is reached.

To reduce the effect of the capacitance (where important) a correctly designed voltage probe should be used. There are two types of probe: the passive ones containing only R, C and L elements; and active probes using an active element such as an FET. The passive types are normally used.

PASSIVE VOLTAGE PROBES

Voltage probes called ×1 (or 1×) probes are available that produce no attenuation. They consist of a low capacitance coaxial cable which sometimes has the centre wire made of resistance alloy so as to reduce possible reflections. They are available in several lengths. A probe of 1 metre length has a capacitance of some 30 pF, so that when used with an oscilloscope having an input capacitance of 25 pF the probe input capacitance becomes 55 pF, which is still appreciable but less than with normal coaxial cable. The input resistance is near enough that of the oscilloscope's since the probe's resistance (which may include a resistor in the probe itself) is only a small fraction of 1 MΩ. They are not intended to operate at very high frequencies – say up to 20 MHz.

The probe more commonly used is one giving attenuation, usually attenuating by ten times and known therefore as an ×10 (10×) probe. This probe may be used for two reasons:

(a) To increase the voltage that can be applied to the oscilloscope.
(b) To reduce the effect that connecting the oscilloscope has on the circuit under test.

(a) Oscilloscopes vary in the minimum sensitivity, *i.e.* maximum VOLTS/DIVISION, and may be as low as 10 V/division. Assuming a 6 division maximum trace, this would mean a maximum a.c. input of 60 V p-p. With a 10-times probe this is increased to 600 V p-p, or 300 V peak, or 212 V r.m.s. Thus, even with the probe, the 240 V supply may overscan and an ×100 (100×) probe may be required. There are special probes with attenuations up to 1000 times for use with high voltages up to 20 kV.

(b) The basic circuit of a simple probe is given in figure 8.5, where R_1 and C_1 represent the input resistance (normally 1 MΩ) and input capacitance of the

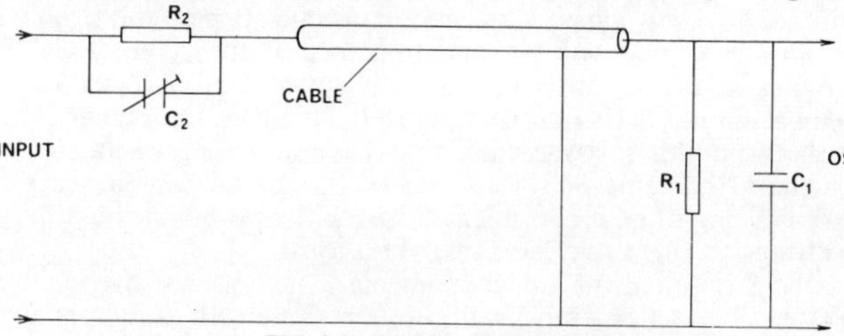

FIG. 8.5. SIMPLE PROBE CIRCUIT (×10)

oscilloscope. The probe itself has a resistor R_2 with a small preset capacitor in parallel. The coaxial cable should be of low capacitance and may have a resistance wire for the centre lead to reduce reflection from the incorrectly terminated coaxial cable. If R_1 is 1 MΩ and R_2 is made 9 MΩ then the attenuation for d.c. and low frequencies is 10. The input resistance is now 10 MΩ instead of 1 MΩ when a direct connection is made. Thus, using a probe has less effect on the circuit under test, particularly if it is one of high internal resistance. The capacitor C_2 must be added so that the attenuator becomes a capacitive divider at high frequencies, C_2 being adjusted to give an attenuation of 10 times. The capacitance of the cable must be added to C_1, so the value of C_2 will not be $\frac{1}{9}$th of that of C_1. However, the input capacitance will appreciably be less than that of the oscilloscope plus the cable, say 10 pF for an oscilloscope with an input capacitance of 20 pF (for a 1 metre length). Hence the use of this probe will reduce the capacitive loading effect on the circuit under test. This is shown in figure 8.1(c).

The probe is usually supplied with a number of different ends for use in different equipment, *e.g.* a clip-on probe, a fine probe for connection to integrated circuits.

Dual probes are also available that can be changed from $\times 1$ to $\times 10$ by changing the end or operating a button on the probe.

If a 100-times probe is used the input resistance is commonly 10 megohms (to avoid the use of very high resistors which tend to be unstable) but the input capacitance is reduced still further to, say, 3 pF for a 1 metre probe.

Since the cable is incorrectly terminated at both ends this type of probe is not suitable for frequencies above, say, 20 MHz to 50 MHz; or where very fast rise times are concerned. The probe rise time may be 5–10 ns.

Unless C_2 is adjusted to the proper value the frequency response of the probe will be incorrect. If too large the output will increase with frequency; if too small the output will decrease with frequency. If the frequency response is not uniform the probe will distort any non-sinusoidal waveform applied to it. The distortion can be serious, and is shown in figure 8.6 where (a) to (c) are taken using the 1 kHz square waveform. At (a) the value of C_2 has been correctly adjusted, and a good square wave is produced. At (b) the value is too small, and so the top of the waveform rises towards the right-hand side. At (c) the value of C_2 is too large and the top now slopes downwards to the right. Thus it is *most* important that C_2 is adjusted to the correct value, using a test waveform before use or *very* misleading results will be obtained.

If the square wave is of a different frequency slightly different results are obtained. The result is shown in figure 8.6 at (d), (e) and (f), when the test frequency is 5 kHz; (d) is correct; (e) C_2 too small, and (f) C_2 too large. The probe is most easily adjusted on a good square waveform of, say, 1 kHz to 5 kHz.

Many oscilloscopes have a test waveform available especially for setting up probes. A probe that has been in use on one oscilloscope must be checked before use on another oscilloscope because the value of C_2 depends on the input capacitance C_1. This varies with different oscilloscopes, and even with those of the same type there will be some tolerance on the value of C_1. The effect of changing the probe which was correctly adjusted on one oscilloscope to another may be similar to (b) or (c) of figure 8.6. This is the reason why, when dealing with the Y-amplifier, it was stated that the input resistance and capacitance must remain the same on all settings of the attenuator. However, if accurate work is being done it is prudent to check the probe on the different attenuator settings as there may be some variations.

When a probe is required for higher frequencies or short rise times a more complex type may be used. Compensation is now done with inductors at the oscilloscope end of the probe, and details of these probes are outside the scope of this book. However, it is just as important that the probe is adjusted correctly to match the oscilloscope. This type of probe may produce a rather different type of distortion, but the method of adjustment is the same. Probes of this type are made for use up to a frequency of at least 250 MHz.

Special probes are also available for oscilloscopes with a 50 Ω input impedance. The input resistance of the probe is now low, say 5 kΩ, but the capacitance is also low, say 2 pF.

It has been shown that the input capacitance of the probe (or lead) may cause changes in magnitude of high frequency sinusoidal signals and distortion of pulses or other non-sinusoidal signals. If one is interested in the phase of the voltage then, of course, this capacitance will cause a phase shift. The amount

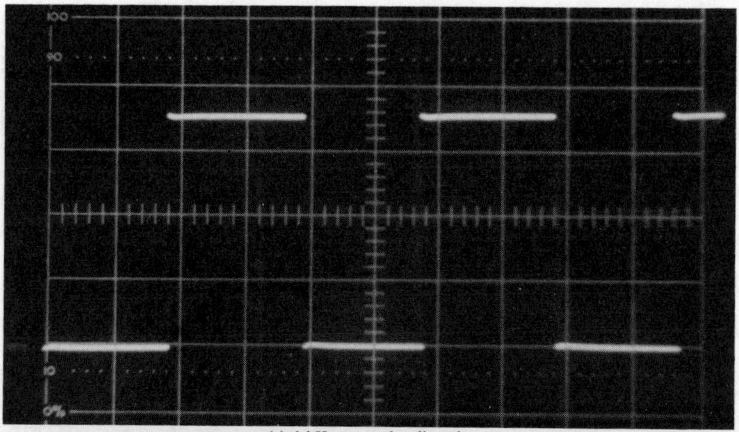

(a) 1 kHz, correctly adjusted

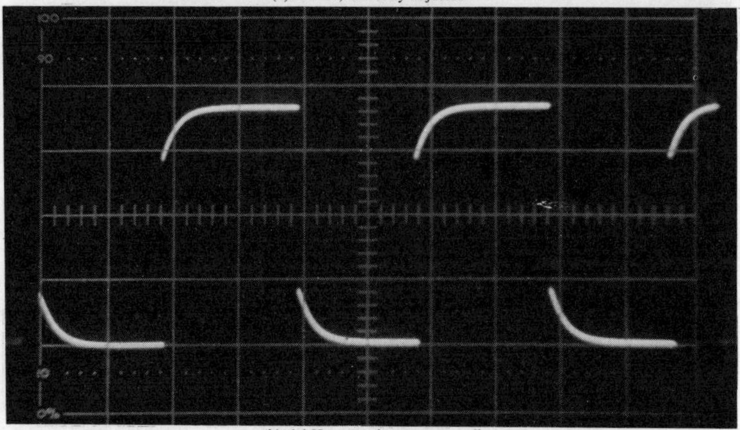

(b) 1 kHz, capacitance too small

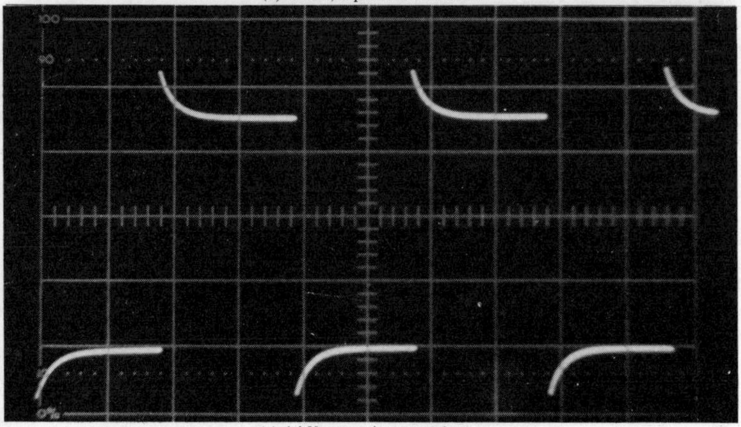

(c) 1 kHz, capacitance too large

FIG. 8.6(a)–(c). EFFECT OF ADJUSTMENT OF PROBE

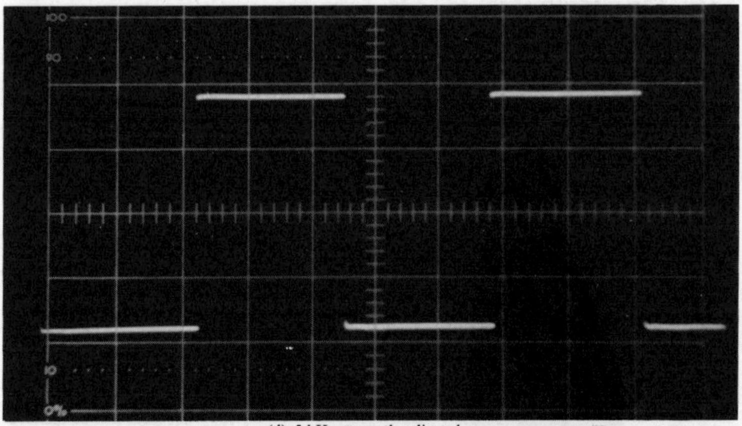

(d) 5 kHz, correctly adjusted

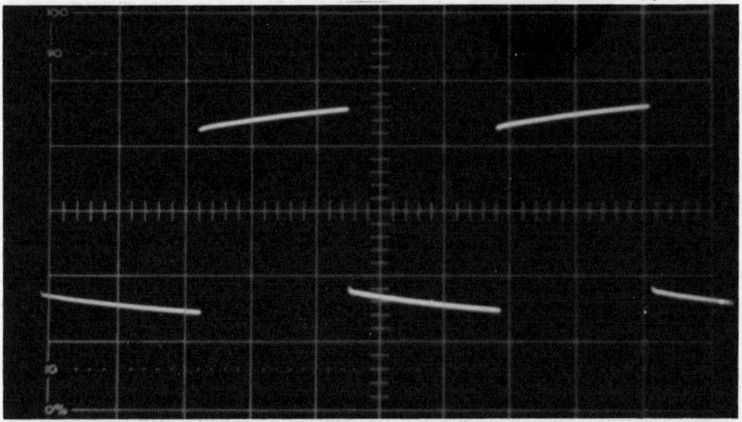

(e) 5 kHz, capacitance too small

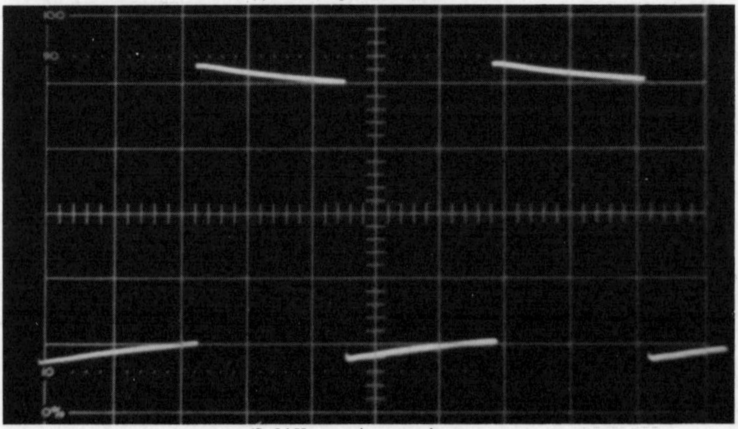

(f) 5 kHz, capacitance too large

FIG. 8.6(d)–(f). EFFECT OF ADJUSTMENT OF PROBE

increases with source impedance, input capacitance and frequency, and can be calculated in a particular case. Therefore, if the phase of two high frequency voltages are being examined care is necessary. If the frequency or source resistance is low the effect may be negligible, but if it is not negligible, then both circuits should be made identical so that the phase *difference* remains the same.

It is important to see the effect of the loading capacitance on the rise time. The rise time of a circuit is the time to rise from 10% to 90% of its maximum value. This is shown in figure 8.7, where t is the rise time. If the rise time of the voltage e in figure 8.1 is zero the rise time of the voltage across C_i is given by

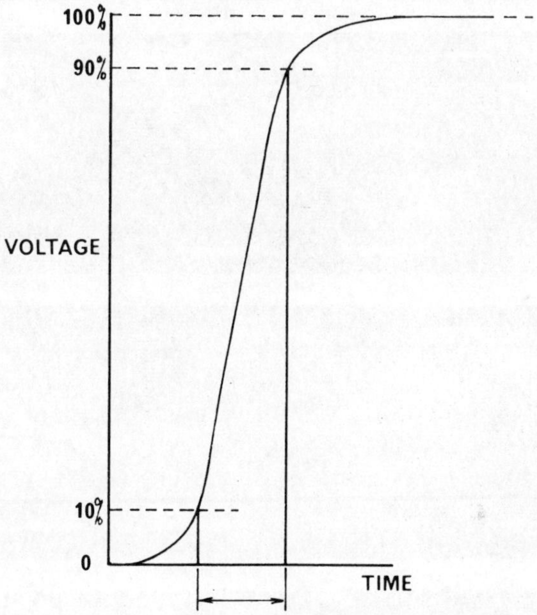

FIG. 8.7. RISE TIME OF WAVEFORM

$2{\cdot}2C_iR_i$. In the example quoted, if $R_i = 320\ \Omega$ and $C_i = 10$ pF, rise time t = $(2{\cdot}2 \times 320 \times 10 \times 10^{-12}) \times 10^9$ ns = 7 ns. If an oscilloscope with an input capacitance of 40 pF (oscilloscope plus probe) is connected across the circuit the effective C is now 10 + 40 = 50 pF. The rise time is $(2{\cdot}2 \times 320 \times 50 \times 10^{-12}) \times 10^9 = 35$ ns. Thus, if a rise-time measurement was made using the oscilloscope it would be 35 ns (neglecting any other effects which might increase it), whereas the circuit itself is only 7 ns. Thus a large error can result. The rise time of the oscilloscope and the probe should be small compared with that of the circuit under test, as mentioned in connection with Y-amplifiers.

The effect on a 1 MHz pulse of various connections will now be shown, the source resistance being about 500 ohms. These are shown in figure 8.8. At (a) is shown the waveforms using a short direct connection. While at (b) the effect of the capacitance of an × 1 probe. At (c) an × 10 probe (correctly adjusted) has been used, when it will be seen that the result is similar to (a). At (d) is shown the very misleading result obtained by using a coaxial cable about 0·6 metre long.

When looking at high-speed pulses the earth return is very important, and some results are shown in figure 8.9 using a 1 MHz pulse generator correctly

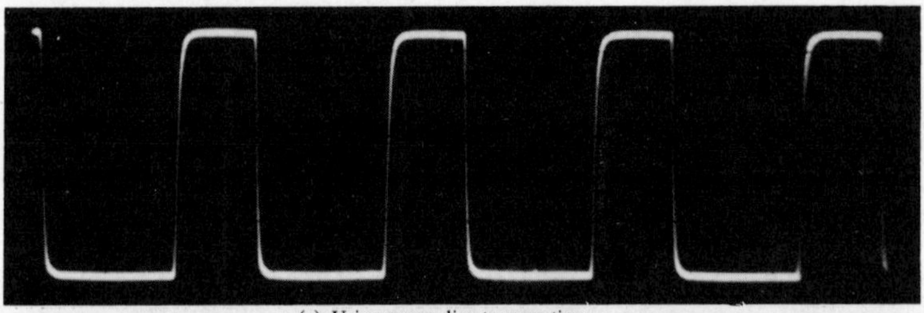

(a) Using SHORT direct connection

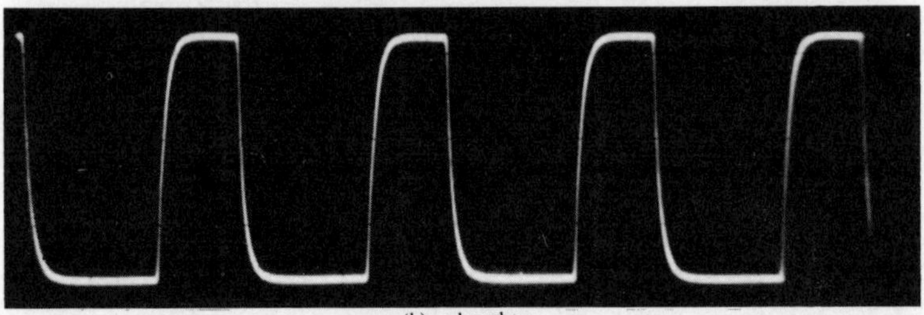

(b) × 1 probe

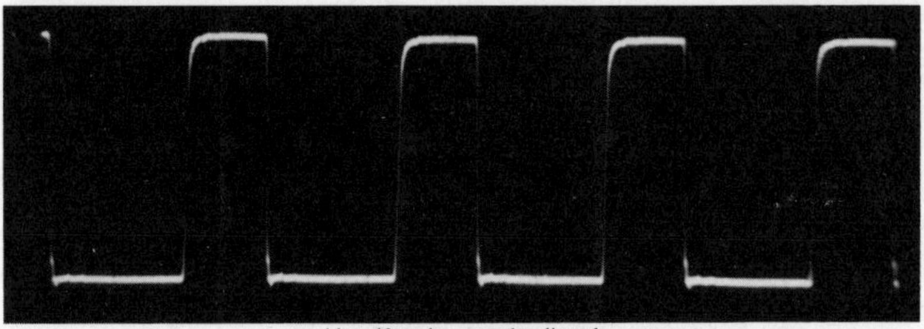

(c) × 10 probe correctly adjusted

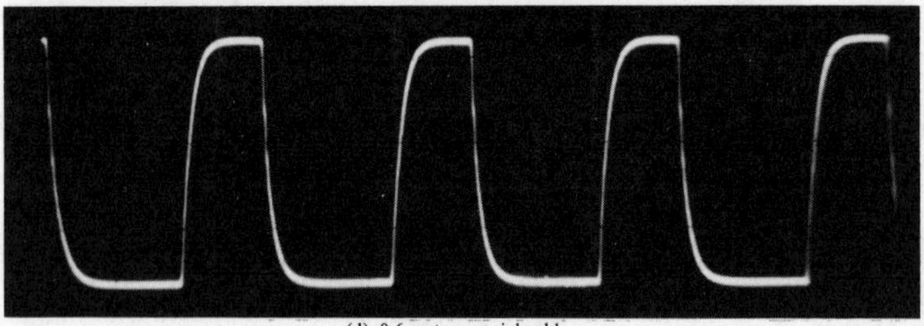

(d) 0·6 metre coaxial cable

FIG. 8.8. EFFECT OF CONNECTION OF OSCILLOSCOPE
1 MHz PULSES WITH SOURCE RESISTANCE 500 Ω

terminated with 50 ohm resistor. At (a) is shown the result of a *very short* lead between the terminating resistor and the oscilloscope input socket. At (b) is shown the result of using an earth return about 0·6 metre long between the terminating resistor and the oscilloscope (the live lead was still very short). Diagram (c) shows the same waveform using an × 10 probe between the terminating resistor and the oscilloscope, while (d) shows the same conditions but with an earth return of about 0·6 metre between the terminating resistor and the probe earth connection.

Thus it will be seen that when looking at high-speed pulses great care is

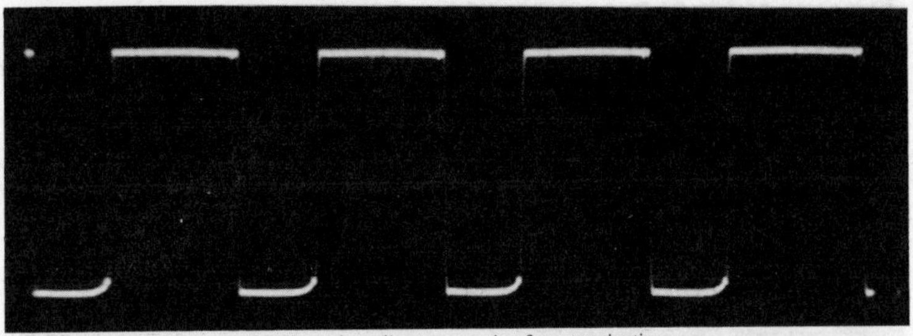

(a) VERY short direct connection from termination

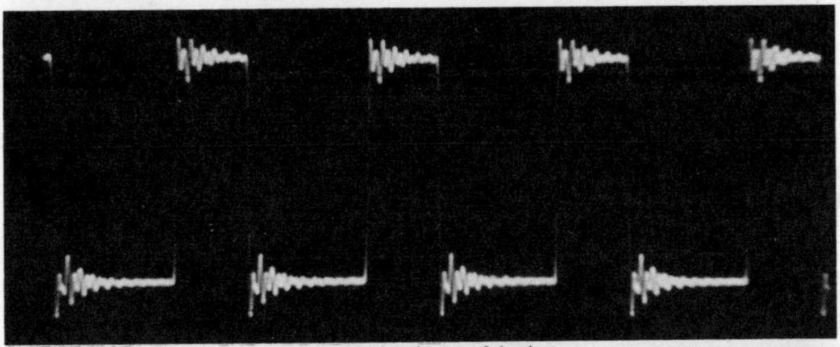

(b) As (a) but earth return 0·6 m long

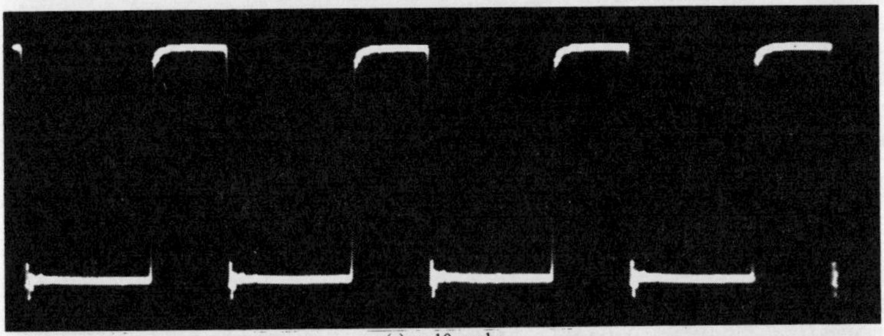

(c) × 10 probe

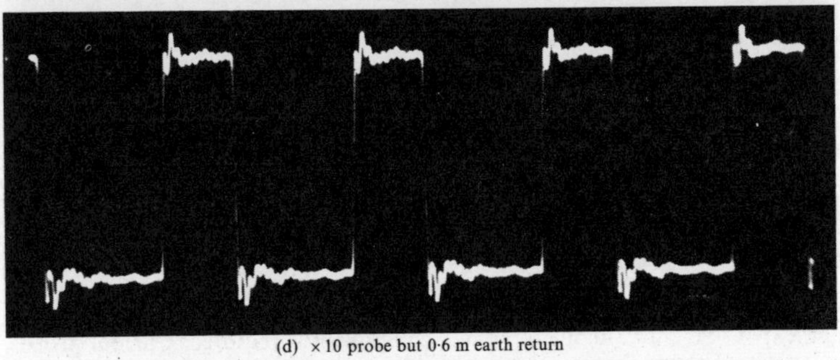

(d) × 10 probe but 0·6 m earth return

FIG. 8.9. EFFECT OF CONNECTING LEADS TO OSCILLOSCOPE. 1 MHz PULSES FROM PULSE GENERATOR CORRECTLY TERMINATED IN 50 Ω needed and short leads are essential. When the source of pulses should be matched to 50 ohms then more satisfactory results may be obtained by using an oscilloscope with a 50 Ω input connection, or a 50 Ω matching unit when using the 1 MΩ input.

The problems associated with probes for very high frequencies or very short rise times are involved and cannot be considered in this book, *e.g.* probes used with sampling oscilloscopes.

ACTIVE VOLTAGE PROBES

Sometimes a small input capacitance is required, but the loss of sensitivity by using, say, an × 10 probe cannot be tolerated. In such cases an active probe may be used. A simplified block diagram is shown in figure 8.10, where the first stage is an FET so that a high impedance is obtained. This may be 10 MΩ

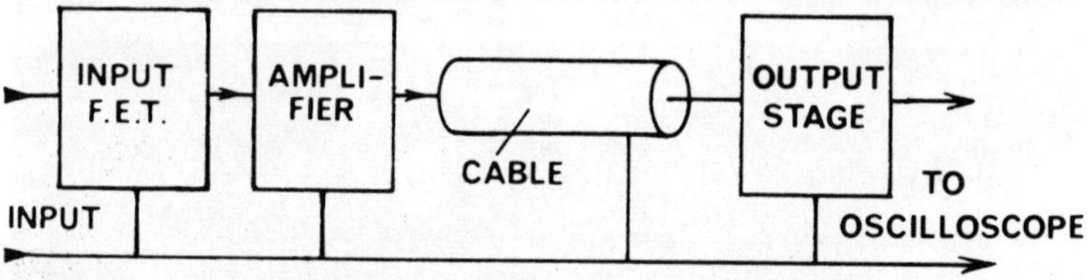

FIG. 8.10. BLOCK DIAGRAM OF ACTIVE VOLTAGE PROBE

and have an input capacitance of only 5 pF. The overall gain is unity and the bandwidth may be d.c. to 200 MHz. The input voltage that can be applied is limited to, say, 0·5 V although × 10 and × 100 attenuator probes may be used in front of the FET circuit.

Differential active probes are also available where the processing of the two signals takes place in the probe itself. Such devices may go up to 100 MHz.

CURRENT PROBES

It is sometimes necessary to measure or examine the current in a circuit, which must be done without upsetting the operating conditions of the circuit. In simple cases this can be done by using a low value series resistor, for example, as in the emitter circuit of a transistor, figure 8.11. The voltage across R is proportional to the current and will have the same waveform as the current at low frequencies; but capacitance effects will cause problems at high frequencies. One other problem which arises is that (unless a differential amplifier is used: see Chapter 9) one side of this added resistor must be earthy. However, these difficulties can be resolved by using a current probe. One passive type consists of a current transformer and requires breaking into the circuit, as shown in figure 8.12. Such instruments may have a sensitivity of 5 mV/mA, *i.e.* a voltage of 5 mV is produced for each milliampere flowing in the primary. The frequency response varies with the design, but may go up to 100 MHz or more. Probes are also made using the same idea where the lead is simply clipped in the probe, as in figure 8.13. This is a much more convenient arrangement. Being current transformers they give an output corresponding to the a.c. component of the current and often are only suitable for frequencies above 100 or perhaps 500 Hz.

An active current probe may be used incorporating a Hall effect device, which can be operated on d.c. and may go up to 50 MHz. This is more complex and expensive but valuable if the d.c. component is required. Current probes cover such a range that it is impossible to give typical figures and manufacturers' literature should be consulted.

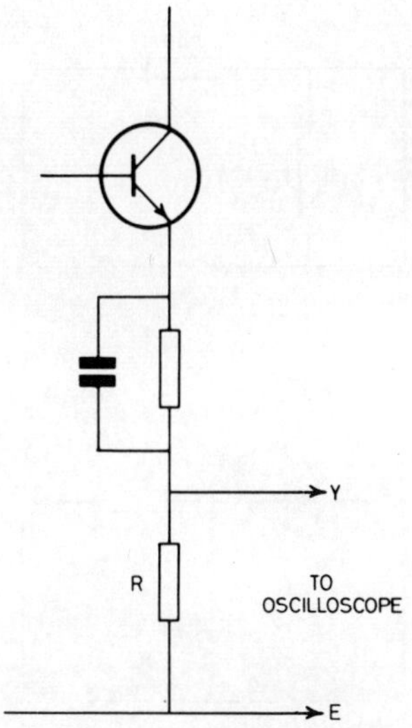

FIG. 8.11. MEASUREMENT OF CURRENT WAVEFORM BY USE OF SERIES RESISTOR

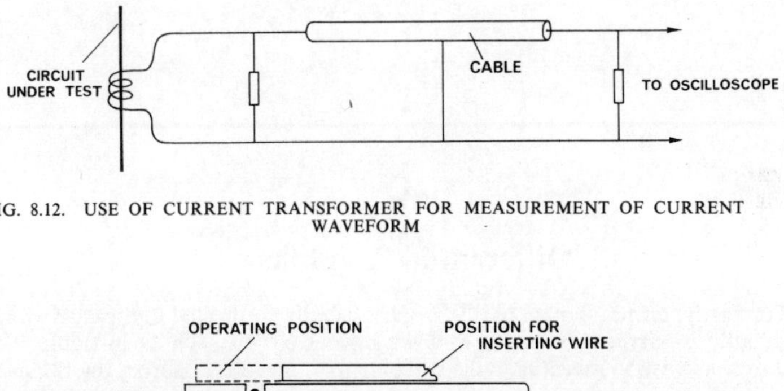

FIG. 8.12. USE OF CURRENT TRANSFORMER FOR MEASUREMENT OF CURRENT WAVEFORM

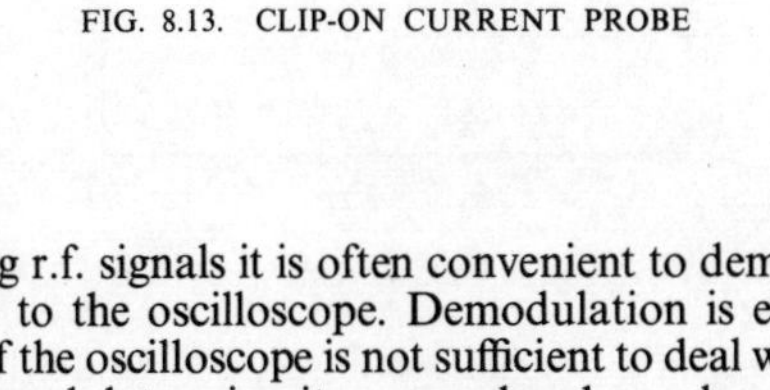

FIG. 8.13. CLIP-ON CURRENT PROBE

DIODE PROBE

When examining r.f. signals it is often convenient to demodulate the signal before applying it to the oscilloscope. Demodulation is essential if the frequency response of the oscilloscope is not sufficient to deal with the modulated signal. Normal demodulator circuits are used and may be a half-wave circuit, as in figure 8.14(a), or full-wave or voltage doubler as at (b) (overleaf).

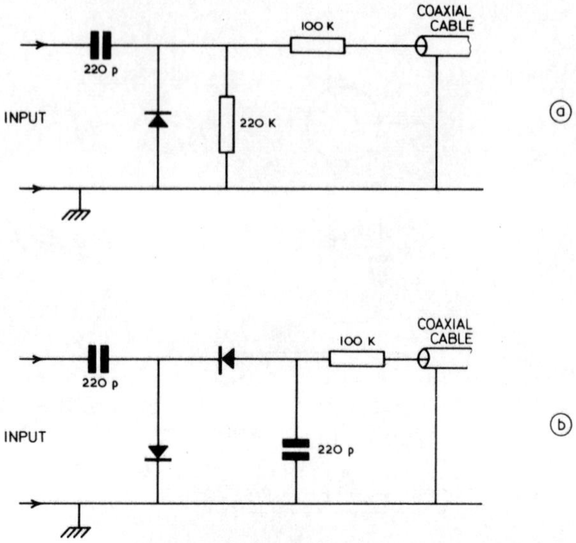

FIG. 8.14. DIODE PROBES
(a) Half-wave; (b) Full-wave

9

Differential Amplifiers

THE earthy terminal of an oscilloscope is usually connected to the case which invariably is earthed. Suppose that we have a circuit such as in figure 9.1, and that we wish to investigate the waveform of the voltage across the resistor.

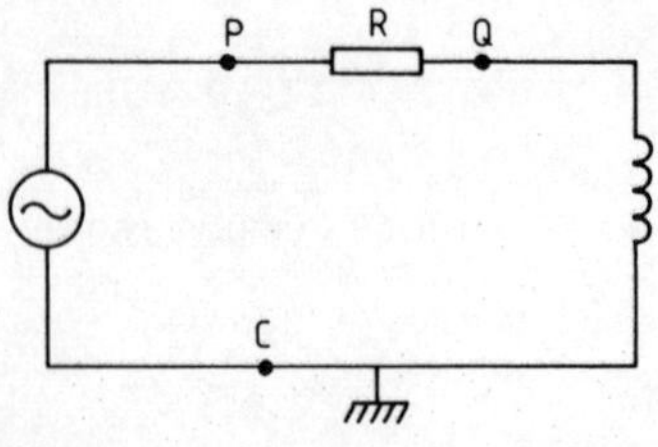

FIG. 9.1. CIRCUIT WHERE DIFFERENTIAL AMPLIFIER MAY BE USED

This is not possible with a normal oscilloscope amplifier because both sides of the resistor are at an a.c. potential to earth. It is sometimes possible to disconnect the earth from the oscilloscope and then connect the oscilloscope across R. This is a very dangerous practice if high voltages are involved as the whole case of the instrument becomes alive. Also, particularly if the impedance of the circuit or the frequency is high, results are misleading owing to the large capacitance of the oscilloscope's case to earth and the possibility of stray pick up.

A differential Y-amplifier is therefore used which has two inputs, A and B. If the same signal (called the common mode signal) is applied to two inputs of the amplifier there will be no trace on the screen. If different signals are applied then it is the *difference* between the two signals which appears on the screen. It is important to recognize that it is a 'difference' amplifier and nothing to do with differentiation. Sometimes the input is referred to as the 'differential input', which may be misleading since it is really the difference input. The term 'floating input' is sometimes used since the two inputs are free or floating with respect to earth. If the two inputs are a.c. the voltage seen on the screen at any instant is the difference between the instantaneous values at that instant.

A basic circuit is given in figure 9.2, where the two transistors Tr_1 and Tr_2 have a common-emitter resistor R_1. The value of R_1 should be high compared

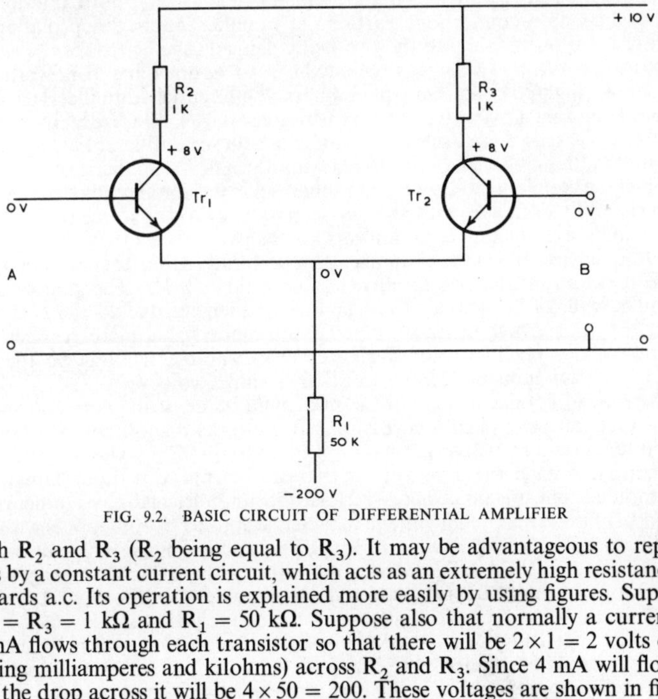

FIG. 9.2. BASIC CIRCUIT OF DIFFERENTIAL AMPLIFIER

with R_2 and R_3 (R_2 being equal to R_3). It may be advantageous to replace this by a constant current circuit, which acts as an extremely high resistance as regards a.c. Its operation is explained more easily by using figures. Suppose $R_2 = R_3 = 1$ kΩ and $R_1 = 50$ kΩ. Suppose also that normally a current of 2 mA flows through each transistor so that there will be $2 \times 1 = 2$ volts drop (using milliamperes and kilohms) across R_2 and R_3. Since 4 mA will flow in R_1 the drop across it will be $4 \times 50 = 200$. These voltages are shown in figure 9.2, the base-emitter voltage drop being neglected, *i.e.* assumed zero. Now suppose that both bases are raised by +10 volts (the common mode input). The drop across R_1 must now increase by 10 volts to a total of 210 volts, and hence the current $= 210/50 = 4{\cdot}2$ mA, therefore each transistor will pass a current of 2·1 mA. The drop across R_2 and R_3 will be $1 \times 2{\cdot}1 = 2{\cdot}1$ volts. Thus a 10 volt common mode signal results in an output of only 0·1 volt.

Suppose that the base of Tr_1 is raised to 0·1 volt and that of Tr_2 remains at zero – a difference-voltage of 0·1 volt. This will cause the current in Tr_1 to increase and so increase the voltage across R_1, but reduces the base emitter voltage of Tr_2. Hence the current in Tr_2 is reduced by almost the same amount since the current in R_1 must be practically constant. One may consider that the base-emitter voltage of Tr_1 is increased by 0·05 V and that of Tr_2 is reduced by 0·05 V. Assume that this causes the current in Tr_1 to increase by 0·5 mA (to 2·5 mA) and the current in Tr_2 to be reduced by the same amount (to 1·5 mA). The voltage across R_2 now increases to $1 \times 2{\cdot}5 = 2{\cdot}5$ V, an increase of 0·5 V. The voltage across R_3 will be reduced by 0·5 V. Thus a difference-input signal of 0·1 V produces an output across either transistor of 0·5 V, *i.e.* a gain of 5 times. However, the common mode signal of 10 volts only produces an output of 0·1 V of a gain of $0{\cdot}1/10 = 0{\cdot}01$ (actually a loss). The ratio of gain to a difference-signal to the gain with the common-mode signal is called the 'figure of merit of the amplifier'. In this example it is $5/0{\cdot}01 = 500$. It is also known as the 'common mode rejection ratio' (CMRR) and commonly given as a ratio, 500:1 in this case.

Bipolar transistors are shown in figure 9.2, but similar arrangements (*i.e.* field effect transistors) are used as in the Y-amplifiers described in Chapter 4.

Many oscilloscopes use plug-in amplifiers so that a differential amplifier can be used when required. The input facilities are as for a normal amplifier such as d.c. or a.c. connections, earthing of amplifier inputs. Stepped attenuators are fitted on each input, the two being ganged together to form a single VOLTS/DIV control. There may be facilities to control the bandwidth by switchable high-pass and low-pass filters. Differential amplifiers cover a wide range, some having very high sensitivities of, say, 10 μV/div. In general the higher the sensitivity the less the bandwidth. In some differential amplifiers the bandwidth may be small compared with normal Y-amplifiers, say 1 MHz.

When the coupling mode of the amplifier is d.c. the common mode rejection ratio generally decreases with increase in frequency owing to the difficulty of maintaining exactly similar conditions on the two halves of the amplifier at high frequencies. When the coupling to the amplifiers is a.c. then the common mode rejection falls at low frequencies, say below 50 Hz. The gain of each amplifier will, of course, also fall at low frequencies due to the coupling capacitor. As the magnitude of the common mode signal increases then the common mode rejection ratio decreases. For example, it might be 100,000 with a 1 volt common mode input and 10,000 with 20 volts.

There is, of course, a limit to the maximum value of the common mode signal that can be applied, but it is important to distinguish between (a) the maximum non-destructive input voltage; and (b) the maximum common mode input voltage. (a) is the voltage that can be applied without damage to the amplifier, but this input may overload the amplifier and give an incorrect display. (b) is the maximum voltage that can be applied to both inputs without overloading. This will vary with the range in use; it might be ± 10 volts on the 1 mV/div to 50 mV/div ranges, and ± 400 volts on the 100 mV/div ranges and above. A DIFFERENTIAL BALANCE control is sometimes fitted so that the amplifier can be set for maximum common mode rejection ratio.

If a differential amplifier is used to display the voltage across R in figure 9.1, one input would be connected to P and the other to Q. The earth connection of the oscilloscope would be connected to the earth or common lead C of the circuit. The difference between the two inputs which is the voltage across R would then be displayed.

An OFFSET voltage control might be provided. This offset voltage is used to cancel any d.c. voltage applied to one of the inputs so that the a.c.-only component may be examined.

PRECAUTIONS IN USING DIFFERENTIAL AMPLIFIERS

It must be noted that any differential amplifier is not perfect and that there will be *some* output due to the common mode signal. To take an example, suppose that the common mode rejection ratio is 10,000 and that a common mode signal of magnitude 10 volts is applied on the 1 mV/div range. Suppose now that the difference signal is 3 mV, *i.e.* the difference between the two signals applied. On the 1 mV/div range this will produce a deflection of 3 divisions.

The common mode signal of 10 volts is equivalent to a difference-signal of

$$\frac{10 \text{ volts}}{\text{CMRR}} = \frac{10}{10{,}000} = 1 \text{ mV}.$$

This will produce a deflection of 1 division and the total waveform on the screen will be the sum of the two. The common mode signal of 1 mV compared with 3 mV of wanted or difference-signal is not negligible and obviously will upset the result. If the common mode rejection ratio had been 100,000 then it would have only been 0·1 divisions compared with 3 divisions of wanted signal, which could be neglected for most purposes.

The common mode rejection ratio quoted for a differential amplifier is that of the amplifier itself, and is true for *equal* signal applied to both inputs. There are two reasons why in practice these two inputs might not be equal, although connections are made to two points having the same voltages.

It is common practice to use probes (say × 10) to connect between the equipment and the oscilloscope. For a normal amplifier this generally produces more accurate results as both the resistance loading and the capacitance loading are reduced. In the case of differential amplifiers the result may be unsatisfactory. If the two probes are identical they will not introduce any error; but, of course, probes are not identical and there will be tolerances on the resistance and capacitance values. Assume one probe to be perfect with a ratio of exactly 10/1; and the other to be 1% out with a ratio of 10·1/1. If equal voltages of, say, 10 volts are applied to the probes then the inputs to the two amplifiers will be

$$\frac{10}{10} = 1 \text{ volt, while the other will be } \frac{10}{10{\cdot}1} = 0{\cdot}990 \text{ V}.$$

This is equivalent to applying a difference voltage of $1 - 0{\cdot}990 = 0{\cdot}01$ volt. A true difference-signal applied to the probes to give an equivalent deflection on the screen would be 0·1 volt (due to the 10/1 ratio of the probes). The 10 volts common mode gives the same deflection as 0·1 difference-signal, or the common mode rejection ratio is now only 100/1 even if the amplifier itself has an infinite common mode rejection ratio.

The second reason for an apparent reduction in the common mode rejection ratio is the internal impedances of the sources of voltage. If these are the same then no error will result, but if they are different, which is quite likely since they are at different parts of the circuit, then a reduction in the CMRR will occur. Consider figure 9.3 where circuit (1) has an e.m.f. e and an internal resistance of 10 kΩ; and circuit (2) has the same e.m.f. e but internal resistance 20 kΩ. It is assumed that the two circuits are connected directly to a differential amplifier of input resistance 1 MΩ (assuming both inputs to have exactly the same input resistance). The actual voltage fed to input (1) will be

$$e \times \frac{1{,}000{,}000}{1{,}010{,}000} = 0{\cdot}9900 \text{ e}$$

and that fed to input (2) will be

$$e \times \frac{1{,}000{,}000}{1{,}020{,}000} = 0{\cdot}9804 \text{ e}.$$

This is equivalent to a difference-voltage of $0{\cdot}9900 - 0{\cdot}9804 \text{ e} = 0{\cdot}0096$ e volt. Thus, an apparent common mode input of e volts produces a deflection

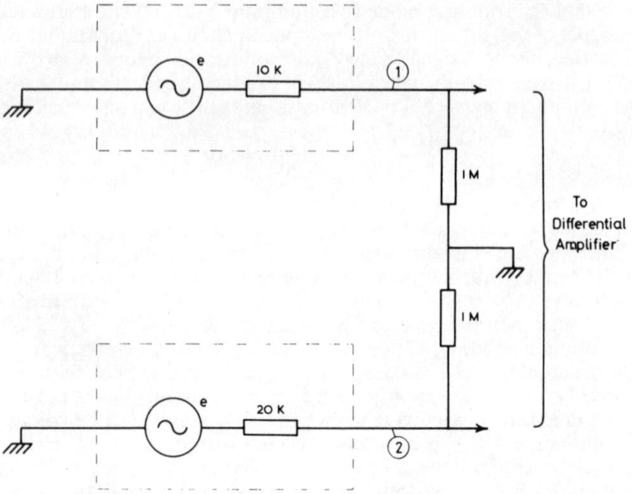

FIG. 9.3. EFFECT OF INTERNAL RESISTANCE OF SOURCE ON CMRR

corresponding to a difference-signal of 0·0096 e. The apparent common mode rejection ratio is

$$\frac{e}{0{\cdot}0096\ e} = 104:1$$

assuming, of course, that the differential amplifier itself has an infinite common mode rejection ratio. This effect can be reduced by (a) taking the voltage from points of low internal resistance; and (b) increasing the input resistance of the differential amplifier. Some amplifiers have this facility.

The above calculation has been done on the resistive elements only, but similar or worse errors occur at high frequencies due to differing source impedance and capacitances.

Care is needed as regards the earth connections, particularly when high sensitivity differential amplifiers are in use. The screens of the probes should be connected together and to the amplifier earth at the amplifier, as in figure 9.4. They should *not be connected to the earth of the apparatus under test.* A single wire should connect the oscilloscope earth to the apparatus earth. This prevents the possibility of circulating currents in the probe screens, which can induce unwanted signals into the amplifier.

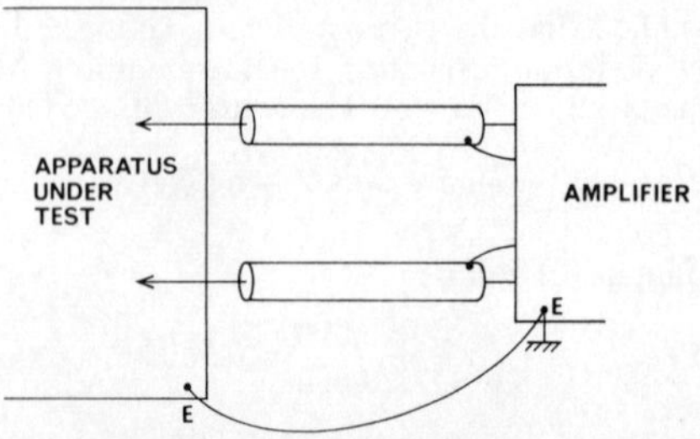

FIG. 9.4. EARTH CONNECTIONS WHEN USING DIFFERENTIAL AMPLIFIER

DIFFERENTIAL COMPARATOR AMPLIFIERS

Basically this is a differential amplifier with an additional facility; it may be used as a differential amplifier in the way described. However, a variable, and accurately known d.c. voltage source is available that can be applied to one or the other of the inputs. The d.c. source goes in both positive and negative directions and has a voltage equal to the maximum voltage that can be applied to one input. The idea is to use what is known as the 'slide back' technique using the oscilloscope as the null indicator. Suppose that a voltage of approximately +10 volts is applied to input A and that the amplifier is set to a sensitivity of 10 mV/div. The spot will go off the screen but can be brought back by the application of +10 volts to input B, this second voltage coming from the variable source. Thus, if the variable voltage is adjusted until the trace is in its central position (assuming it to be in the central position with no voltage applied to either input) then this is a measure of the unknown voltage applied to input A. If a deflection of 0·1 division can be read, the input voltage can be measured to an accuracy of $0{\cdot}1 \times 10\,\text{mV} = 1\,\text{mV}$, or 1 mV in 10 volts, or 0·01 %. This is assuming that the source of variable voltage is exactly known and that the common mode rejection ratio is infinite. Both these factors will cause errors, but the possible overall accuracy can be calculated if the accuracy of the variable voltage is known (say 0·1 %) and the common mode rejection ratio is known.

This method is not limited to d.c. voltage measurements. Suppose we have a pulse, or any other waveform, superimposed on a relatively large d.c. voltage. The variable voltage is now varied until the top of the pulse corresponds to the zero line and the value of voltage source noted, say V_1. The variable voltage is now reduced until the bottom of the pulse corresponds to the zero line and the voltage noted, say V_2. The magnitude of the pulse is now $V_1 - V_2$. In a similar way the peak-to-peak value of an a.c. voltage can be measured. The variable voltage is varied in, say, the positive direction until the top of the waveform is on the zero line and the voltage noted, say V_3. The variable voltage is now made negative and adjusted until the bottom of the waveform corresponds to the zero line and the new voltage noted, say V_4. The peak-to-peak value of the voltage is now $V_3 + V_4$. In both cases the accuracy of measurement can be much higher than using the direct measurement off the oscilloscope screen. The differential amplifier must be designed to recover quickly after being overdriven. The variable voltage may have a digital readout or a dial readout (multiturn potentiometer).

10

Multitrace Oscilloscopes

THERE are many applications requiring more than one trace on the screen at the same time so that waveforms at two points in a circuit may be examined. This could be done by two oscilloscopes, but apart from the obvious expense it is not easy to compare waveforms at particular instants in time. But, for example, if two traces are produced on a single tube using a common horizontal deflection then comparison of the two waveforms is easy at any instant. In addition, difference measurements (say) can be made. There are three basic ways of obtaining multiple traces, using monoaccelerator or PDA tubes:

(1) By a single-gun dual-trace or split-beam tube.
(2) By a multigun tube.
(3) By beam switching using a single-beam tube.

(1) DUAL-TRACE or SPLIT-BEAM TUBES

This was the earliest method of obtaining two traces on a single screen and was used by Cossor in the 1930s. The idea is to use a single gun with means of splitting the beam so that each portion passes through different Y-plates. The beams then pass through common X-deflecting plates. The basic idea is shown in figure 10.1. The splitter plate is a thin plate placed in the centre of the two Y-deflecting plates, so splitting the beam into two. This plate is connected to

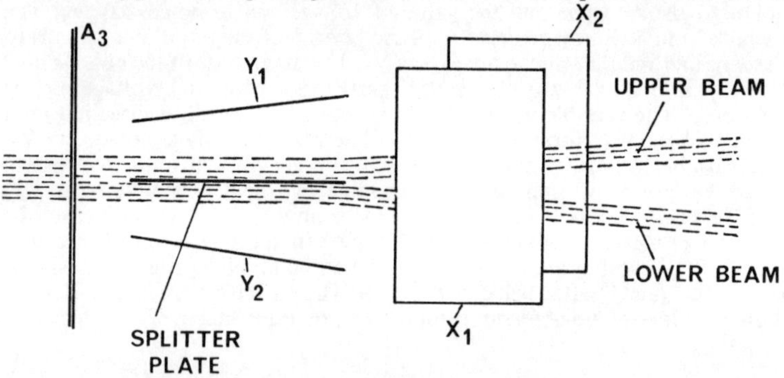

FIG. 10.1. SPLIT-BEAM TUBE

the final anode A_3. A positive voltage applied to Y_1 relative to A_3 will deflect the upper part of the beam upwards; a positive voltage applied to the Y_2 plate will deflect the lower part of the beam downwards. Therefore, for the same signal applied to the two Y-plates the two traces are 180° out of phase, or in opposite directions. This can be confusing. An alternative is to use two holes in the anode and two separate sets of Y-deflecting plates. This allows symmetrical deflection to be used and avoids the 180° phase difference on the two beams.

The split-beam arrangement has the following advantages:

(a) Tube construction simplicity in comparison with two-gun tubes.
(b) It only requires one brightness control and one focus control, and therefore reduces the cost.
(c) Since the two traces are deflected by the same X-plates the time-scale is the same for both traces. This makes precise comparison easy. Also, there is no possibility of phase error measurement (provided one allows

for the possible 180° phase difference on the two Y-plates where this occurs) which can happen in a switched-beam oscilloscope if operated incorrectly [see (3) BEAM SWITCHING].

(d) There is only one heater to fail and only one heater-cathode insulation to break down.

(e) There are no problems on transient displays [see (3)].

The split beam arrangement has the following disadvantages:

(a) The brightness of each trace is half of that obtained by the equivalent normal tube.

(b) The brightnesses of the traces cannot normally be controlled separately, but is possible in some tubes.

(c) Only asymmetrical deflection can be used with the arrangement shown in figure 10.1.

(d) There is some interaction between the plates. (Some compensating arrangements can reduce this).

(e) The upper part of the beam may cut off towards the bottom of the screen, and the bottom part cut off at the top of the screen.

(f) If Y delay lines are required then two must be provided, one for each trace.

Very few oscilloscopes now use this type of tube.

(2) MULTIGUN TUBES

These tubes use completely separate guns with separate Y-deflecting plates. The X-plates may be separate, one pair for each gun, but it is more usual for them to be common to all guns. Only two guns are used as a rule (although four-gun tubes have been made) and the explanations that follow will assume two-gun tubes, for simplicity.

In comparison with the dual or split-beam tube the arrangement has the following advantages.

(a) The brilliance and focus of each trace can be controlled separately.

(b) Other things being equal, the traces should be twice as bright.

(c) Symmetrical deflection can be used.

(d) There is no problem of 180° phase difference on the two traces provided the plates are correctly connected.

(e) If separate X-plates are used then two traces can be produced having different time-scales.

(f) As in the split beam there is no possibility of false phase measurements [see (3)], when common X-plates are used.

(g) As in the split-beam there are no problems in displaying two transients [see (3)]:

The arrangement has the following disadvantages.

(a) The tube is much more complex.

(b) Separate brightness and focus controls must be provided, which may have advantages but increase the cost.

(c) Due to tolerances in the positioning of the two guns the beams may not be in the same position along a horizontal line. Assuming common X-plates it is essential that they fall in the same position if accurate phase comparisons and measurements are to be made. To overcome this a small separate deflecting system may be used, commonly between the A_1 and A_2 anodes, to correct for the manufacturing tolerances. It is essential that when deflected horizontally the two traces are parallel; again, some correction may be necessary for this.

(d) Since there are two heaters there is a greater possibility of heater failure or heater-cathode shorts.
(e) As with the split-beam arrangement if Y-delay lines are used then two are required.

Multigun tubes are rarely used in oscilloscopes at present.

(3) BEAM SWITCHING

This system makes use of a normal single-beam tube. It has the advantages of a simple tube, only one brightness control (normally) and one focus control, with no difficulties in the alignment of two separate beams. This method of obtaining multiple traces is used almost universally at the present time.

The basic idea is to time-share the same beam between the two traces; it may be considered to be a time division multiplex system. For simplicity it will be assumed that only two traces are required. There are two basic ways of doing the time sharing.

(a) ALTERNATE TRACES. Signal A is displayed for one traverse of the spot across the screen, then signal B is displayed for one traverse, and so on. The switching is now done at half the timebase repetition rate.
(b) CHOPPED TRACES. The switching is now done at a high frequency, say 50kHz to 1 MHz, as shown in figure 10.2. On any particular horizontal traverse of the beam each trace will consist of short sections and only half of the actual waveform will be completed. However, if the switching frequency is not locked to the timebase, the position of the short sections on each waveform will move and a continuous trace will be visible.

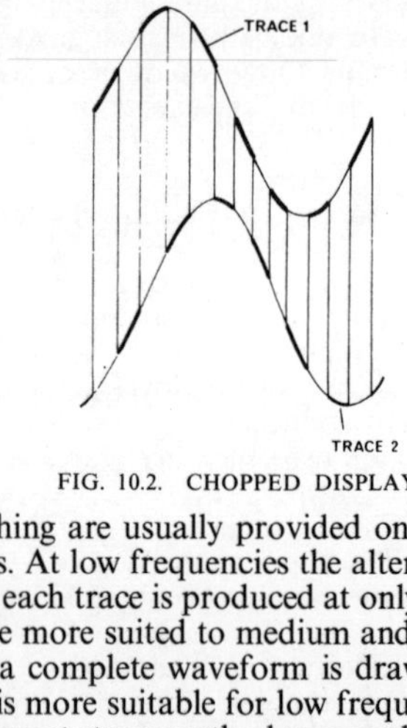

FIG. 10.2. CHOPPED DISPLAY

Both types of switching are usually provided on the oscilloscope because each has its limitations. At low frequencies the alternate trace method causes bad flickering because each trace is produced at only half the timebase repetition rate. It is therefore more suited to medium and high frequency displays; the advantage is that a complete waveform is drawn on each traverse. The chopped trace system is more suitable for low frequencies and overcomes the disadvantage of the alternate trace method as regards flicker. As the frequency is increased the portions of the waveform drawn on each trace become longer as do the missing portions, hence the effect of the beam switching becomes more visible. The frequency of the waveform being examined must normally be considerably less than the chopping frequency. In early oscilloscopes the chopping frequency was low, say 10–20 kHz, but much higher frequencies are now used of 100 kHz to 1 MHz. In the chopped system the beam must be cut off during the movement of the spot from one trace to the other. This can be done by feeding a suitable switching waveform to the grid or blanking electrode of the cathode-ray tube from the chopping waveform generator.

In most oscilloscopes it is possible to switch from alternate trace to chopped trace, the user selecting the mode depending on the frequency and type of waveform under examination. In some oscilloscopes when the timebase sweep speed is low it is set to the CHOPPED mode, and at high sweep speeds (high repetition rates) is automatically changed to ALTERNATE mode.

In general it is preferred that the chopping waveform generator is not a locked multiple of the timebase repetition rate so that the portions of the display waveform 'run round' the trace and give the appearance of a continuous trace, often making it difficult to know that it is a chopped trace. Unfortunately, in many cases the chopping waveform generator locks into step at a multiple of the timebase repetition frequency and only sections of the waveform are drawn. This is shown in figure 10.3(a). At (b) is shown the effect when this locking does not occur, and (c) shows the result of running the oscilloscope in the alternate

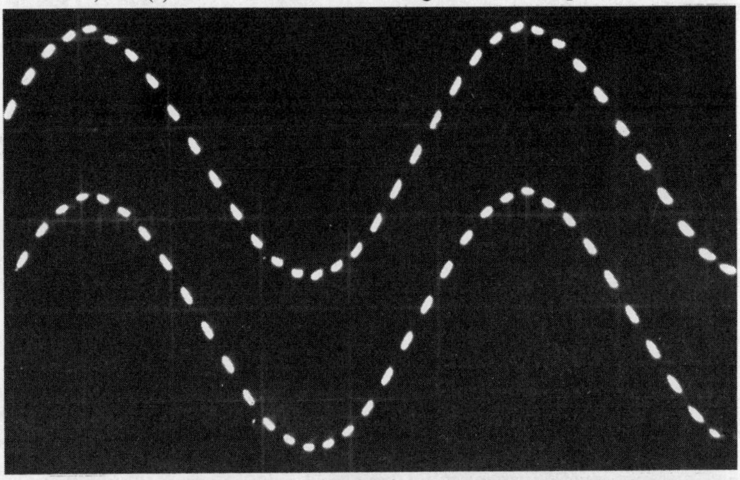

FIG. 10.3(a). CHOPPED TRACE: CHOPPING FREQUENCY LOCKED TO SWEEP GENERATOR

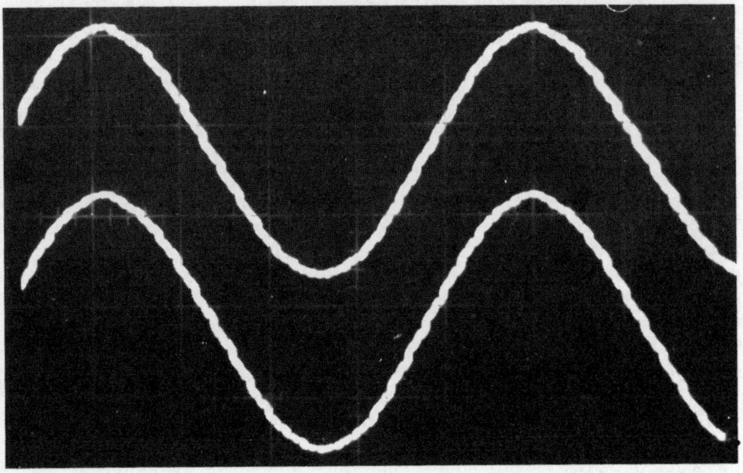

FIG. 10.3(b). CHOPPED TRACE: CHOPPING FREQUENCY NOT LOCKED TO SWEEP GENERATOR

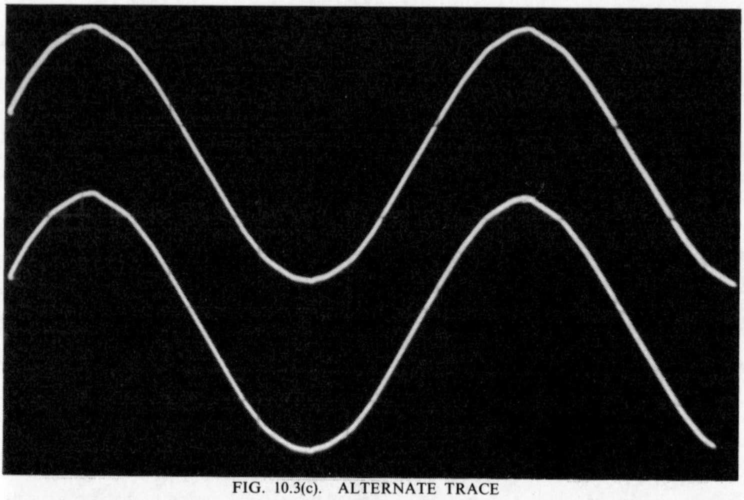

FIG. 10.3(c). ALTERNATE TRACE

trace mode which overcomes, of course, any distortion of the waveform due to chopping. This locking was more common when valves were used; it depends on how well the circuits are isolated. In some oscilloscopes it is almost impossible to see this chopping even at high frequencies.

In both the alternate and chopped modes it is possible to shift each beam separately in the Y or vertical direction. The X-shift control, of course, moves both traces by the same amount.

TRIGGERING

When using alternate traces the method of triggering is very important as false phase relationships can be displayed. If the two signals are of the same frequency or are frequency related and the phase relationship is required, then the timebase *must be triggered by one signal only*. Suppose the timebase is triggered by signal A only. When tracing out signal A it will be triggered at a certain instant depending on the setting of the level control, say at X in figure 10.4(a). When the beam is tracing out signal B the timebase is still triggered by signal A and will be triggered *at the same instant* X (assuming correct triggering), as before. Thus signal B will be traced out with the correct time (and phase) relationship to signal A. Alternatively, the trace can be triggered by signal B at all times. The timebase must NOT be triggered by signal A when displaying A and by signal B when displaying B, *i.e.* ALTERNATE triggering. If this is done then the timebase will be triggered at instant X when displaying signal A, and it will be triggered by signal B *when it has the same voltage as* A had at X, and displayed as shown in figure 10.4(d). Thus the two waveforms will appear approximate in phase at all times. The phase relationship actually displayed will depend on the magnitude of the traces and on the level of triggering. This is illustrated in figure 10.5. At (a) the display is ALTERNATE but the triggering is from the larger trace (channel A). This is therefore the correct phase relationship. At (b) the display is in the CHOPPED mode, which confirms the correct phase relationship given by (a). At (c) (d) and (e) ALTERNATE displays are shown with ALTERNATE triggering. The difference between them is that the trigger level is altered (reduced) between photographs. It is seen that the phase relationship changes greatly, and hence ALTERNATE trigger MUST not be used when correct phase relationships are important. This is illustrated again in figure 10.6, where one of the waveforms is a square waveform. The correct phase relationship is

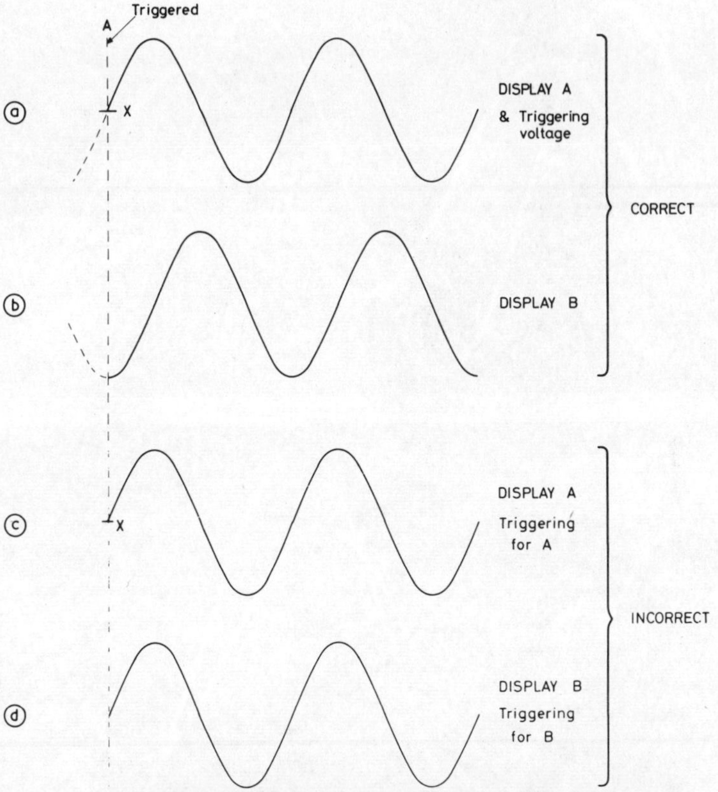

FIG. 10.4. TRIGGERING ON ALTERNATE TRACES
(a) and (b) Correct. Off signal A.
(c) and (d) Incorrect. Off A on one trace and B on other trace.

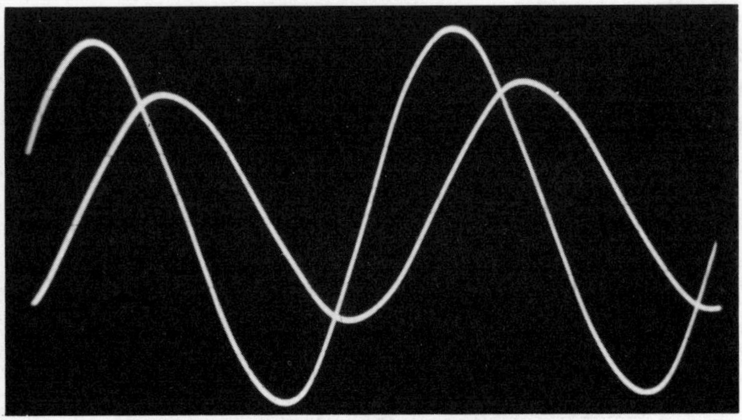

(a) Alternate display triggered off larger trace (correct phase relationship)

FIG. 10.5(a). METHODS OF TRIGGERING

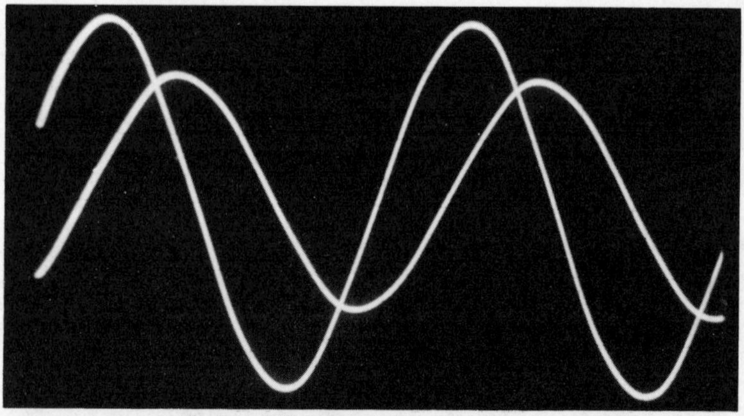

(b) Chopped display (correct phase relationship)

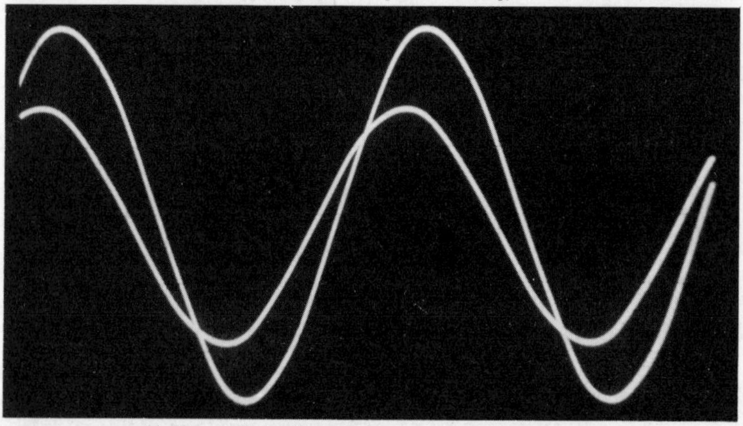

(c) Alternate display, alternate triggering (incorrect phase relationship)

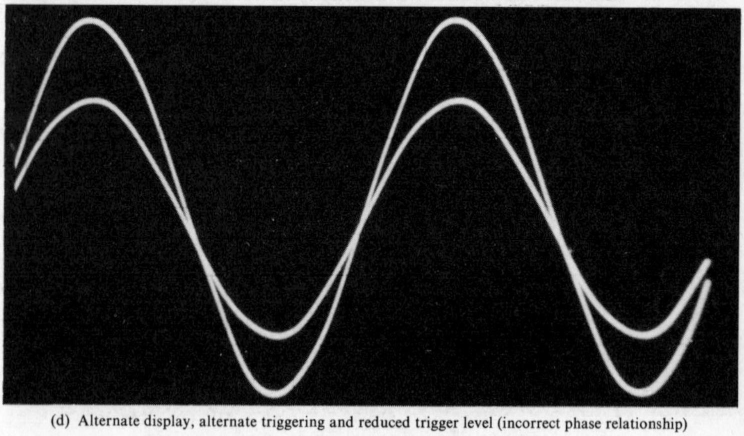

(d) Alternate display, alternate triggering and reduced trigger level (incorrect phase relationship)

FIG. 10.5(b)–(d). METHODS OF TRIGGERING

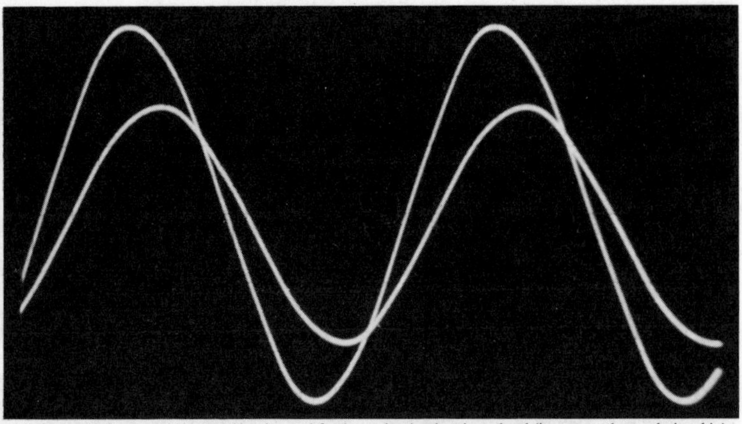

(e) Alternate display, alternate triggering and further reduction in trigger level (incorrect phase relationship)

FIG. 10.5(e). METHODS OF TRIGGERING

given at (a), when a CHOPPED display was used. At (b) (c) and (d) are ALTERNATE displays and ALTERNATE triggering, with the triggering level changed between photographs. Again, it is seen that almost any phase relationship display is

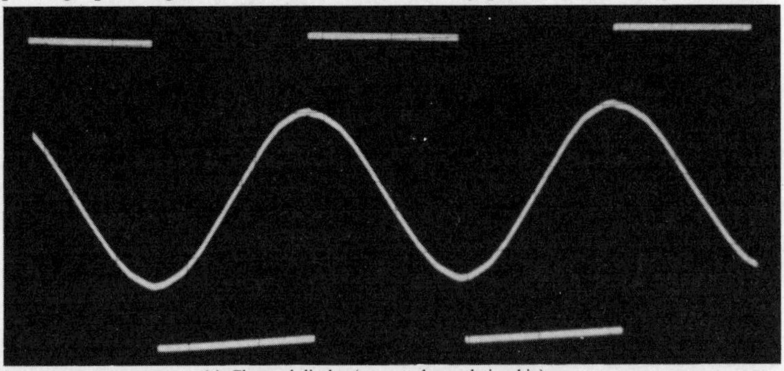

(a) Chopped display (correct phase relationship)

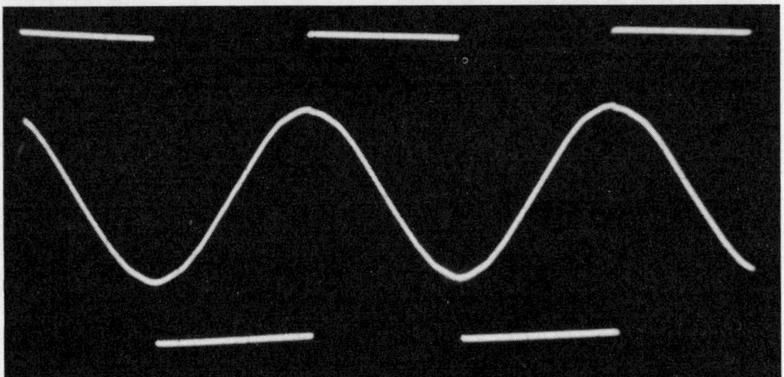

(b) Alternate display, alternate triggering (correct phase relationship by chance)

FIG. 10.6(a)–(b). METHODS OF TRIGGERING

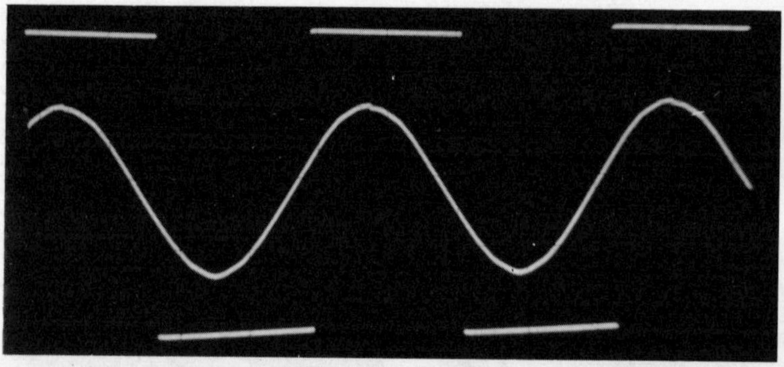

(c) Alternate display, alternate triggering and different trigger level (incorrect phase relationship)

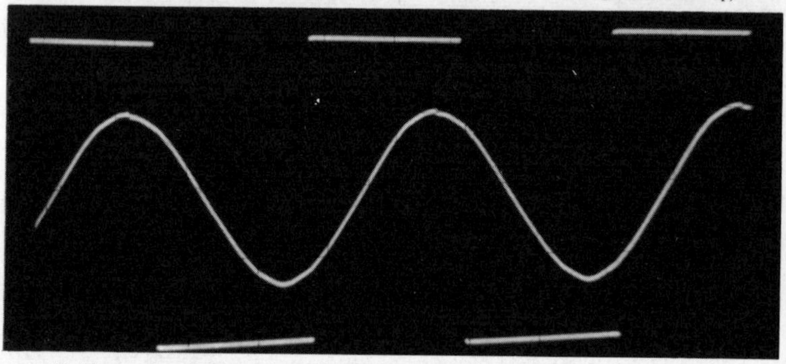

(d) Alternate display, alternate triggering and different trigger level (incorrect phase relationship)

FIG. 10.6(c)–(d). METHODS OF TRIGGERING

possible. Correct phase relationship is obtained if the triggering is from one (either) signal only.

In the case of two waveforms with a sharp leading edge, such as two-square waveforms, the waveforms will always appear in phase. Many oscilloscopes now do not allow one to trigger the timebase in this way. If such a facility is available great care is necessary to avoid very misleading results. It is a mode of operation that should be used only for special purposes. One oscilloscope has this method of triggering but only by the use of an external connection so that one is well aware that it is being used. Unfortunately, one manufacturer calls this mode 'normal'.

When in the CHOPPED mode the triggering is not important as it cannot alter the phase angle displayed because both traces are being drawn during a single traverse of the timebase. Triggering may be from either signal or from the sum of the two signals.

It is possible to show waveforms of different and non-related frequencies on the screen. If the timebase is triggered by signal A for both traces then signal A will be stationary; but signal B will not be stationary and a moving (unlocked) trace is obtained, as in figure 10.7(a). If the timebase is triggered by signal B then (if the triggering is adjusted correctly) signal B will be a locked trace and signal A will be moving.

If signal A is used for triggering when being displayed and signal B is used when being displayed, *i.e.* ALTERNATE triggering, both signals will be locked and appear stationary on the screen, as in figure 10.7(b). *The sweep speed will, of course, be the same for both traces* as this is determined by the timebase sweep

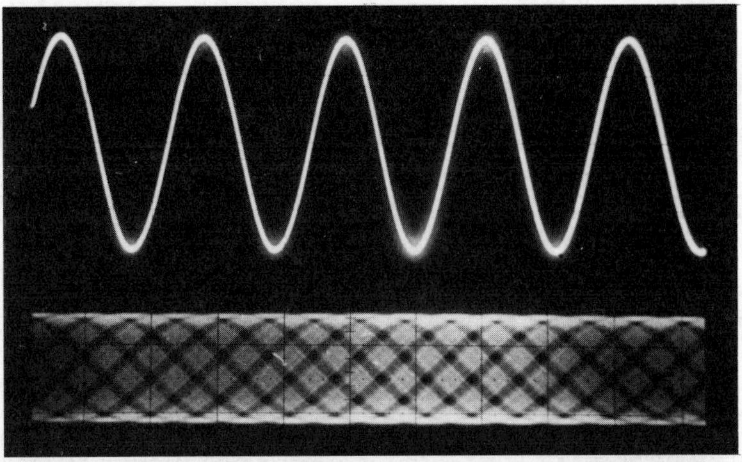

(a) Alternate display, triggering to upper display

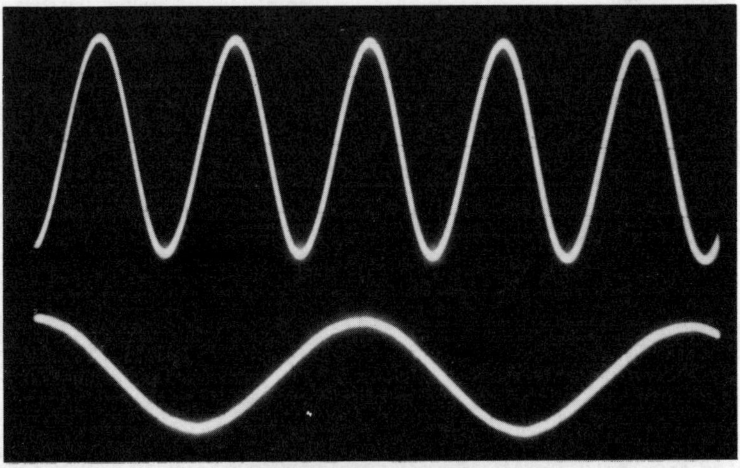

(b) Alternate display, alternate triggering

FIG. 10.7. TWO WAVEFORMS OF DIFFERENT AND UNRELATED FREQUENCIES

setting. If two such locked traces are obtained on the screen it *must not be inferred* that they are of the same frequency or frequency related. This can be checked by switching so that the timebase is triggered by one signal only on both traces. If they are not frequency related, one will not be locked and will move or produce multiple traces.

It is possible to make approximate frequency comparisons using this display since the time-scale is the same for both traces. For example, in figure 10.7(b) there are approximately 5 cycles of the upper waveform and approximately 2 cycles of the lower waveform. Thus the frequency relationship is approximately 5 to 2. This can be used to get, say, one frequency at approximately the correct value relative to the other. Then the triggering can be changed to, say, signal A and the frequency of signal B adjusted until the display of B is stationary with respect to the display of signal A. Both will have a certain relationship between complete cycles. Easier, and possibly more reliable methods, are given in Chapter 13 on uses of the oscilloscope.

There are facilities to add or subtract the two signals, which provide the user with basically a differential amplifier. If the oscilloscope is connected as in figure 10.8, and the amplifier set to the difference of Y_1 and Y_2, the waveform seen on the screen will be that across the resistor R. In a similar way, if

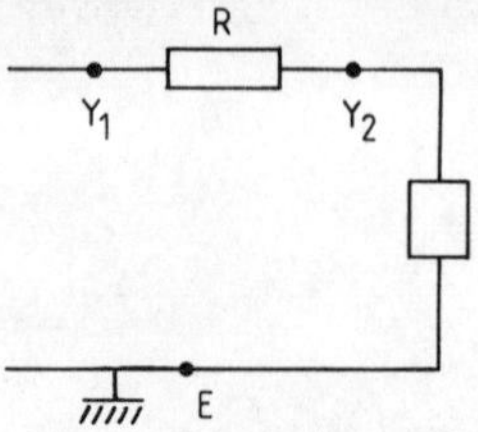

FIG. 10.8. USE OF DIFFERENTIAL AMPLIFIER

the two signals are nearly alike but in antiphase, then the difference between them can be displayed by setting the amplifier to the sum position. This arrangement does not have as large a common mode rejection ratio as a properly designed differential amplifier and, of course, there are limits to the magnitude of the common mode signal that can be applied. Figure 10.9 shows two waveforms. At (a) the two waveforms are in the correct phase and polarity. At (b) the upper trace has been inverted. Both these are taken in the ALTERNATE display mode but triggered off one trace only. At (c) is shown the same display but in the CHOPPED mode. At (d) is shown the sum waveforms of those shown at (a), while at (e) is shown the difference waveform, *i.e.* by setting the oscilloscope to the ADD position and switching the amplifier for the upper trace to INVERT.

In some oscilloscopes it is possible to switch one of the Y-amplifiers to feed

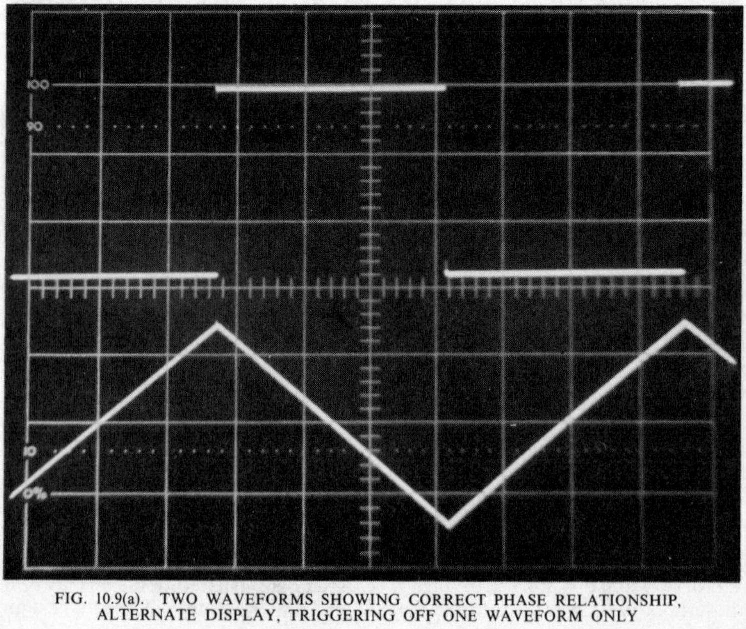

FIG. 10.9(a). TWO WAVEFORMS SHOWING CORRECT PHASE RELATIONSHIP, ALTERNATE DISPLAY, TRIGGERING OFF ONE WAVEFORM ONLY

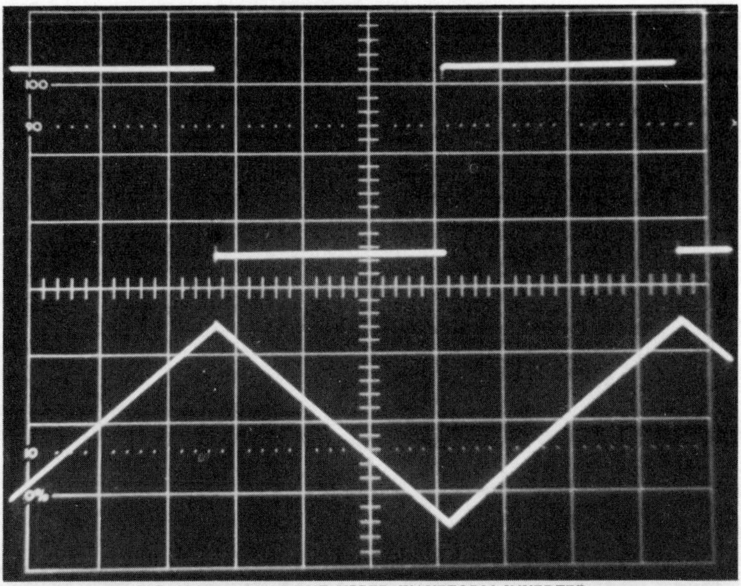

FIG. 10.9(b). AS (a) BUT UPPER WAVEFORM INVERTED

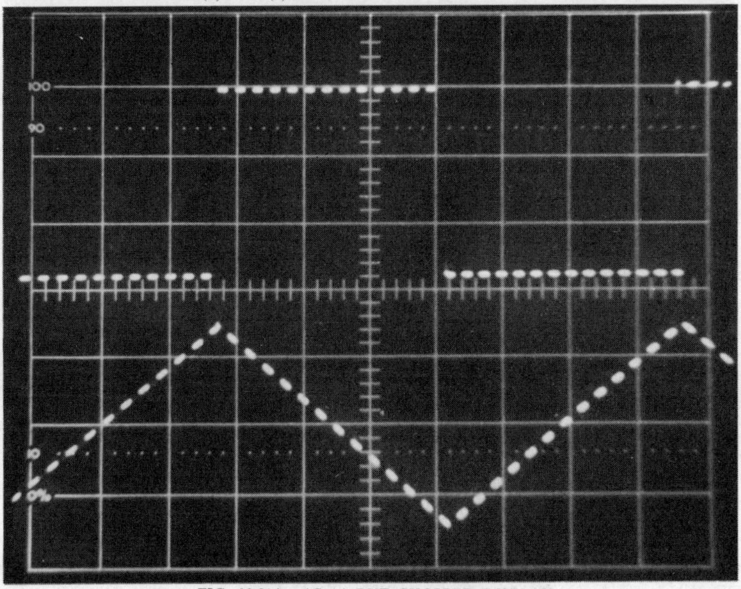

FIG. 10.9(c). AS (a) BUT CHOPPED DISPLAY

the X-plates so that an X-Y display is obtained. The advantage is that identical amplifiers are used in both X and Y directions and the normal Y facilities are available for both directions of display.

In one of their oscilloscopes Philips have facilities to display the product of the two waveforms (*i.e.* one multiplied by the other). This multiplier operates

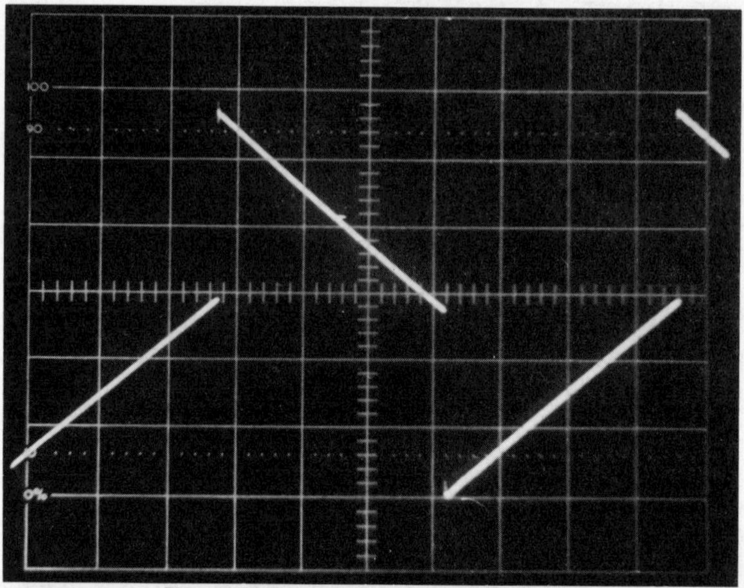

FIG. 10.9(d). SUM OF WAVEFORMS SHOWN AT (a)

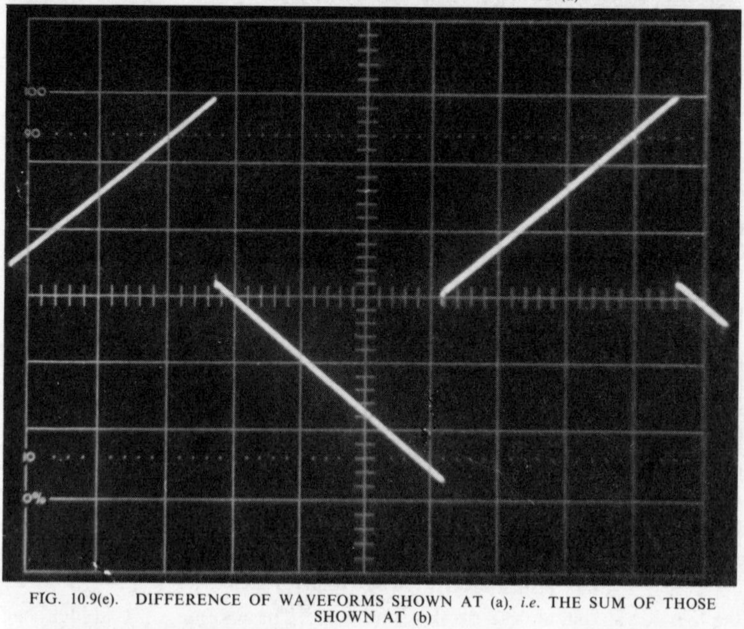

FIG. 10.9(e). DIFFERENCE OF WAVEFORMS SHOWN AT (a), *i.e.* THE SUM OF THOSE SHOWN AT (b)

up to 100 MHz and enables one to display a waveform corresponding to the instantaneous power in the circuit.

Problems arise when single-shot operation is used with switched beam oscilloscopes. If the oscilloscope is set to ALTERNATE then only one trace will be obtained as obviously only one trace can be produced on a single trace or traverse across the screen. When the CHOPPED mode is used, provided the trace time is much longer than the period of the chopping frequency, a double trace will be obtained, but only half of each waveform will be drawn. If the transient waveform is not too detailed and chopping takes place many times during the single traverse, the result is satisfactory. If, however, the speed of trace is increased the result is not satisfactory as a rule. Eventually, of course, when the time of trace corresponds to that of a half-cycle of chopping frequency only a single trace results.

We will now compare switched beam operation with dual trace and multiple gun oscilloscopes. The arrangement has the advantages:

(a) A simple tube is used and much of the cost of an oscilloscope is in the tube.
(b) There is only one brilliance and focus control, which reduces cost and complexity.
(c) Symmetrical deflection is used with reduced distortion – not always possible with split-beam tubes.
(d) No problem of 180° phase difference as in some split-beam tubes.
(e) The two traces must be in the same position along the X-axis at any instant, and no adjustment is required as in a multigun tube.
(f) The traces will be parallel.
(g) The tube can be used on one beam only with full brilliance and all facilities.
(h) There is only one heater to go open circuit and one heater-cathode insulation to fail.
(i) Where a Y-delay line is used only one is required.
(j) Two modes of display, alternate and chopped.
(k) Relatively cheap circuitry to do the switching.
(l) It is easy to extend the idea to more traces – say four.

It has the disadvantages of:

(a) Less brilliance than a two-gun tube, other factors being equal.
(b) Possibility of false phase information if not correctly used with some oscilloscopes (*i.e.* alternate triggering).
(c) Difficulty on single traces, *i.e.* transients.
(d) Generally, no provision to vary brightness of one trace relative to the other, but this is possible.

MULTIPLE TRACES

More than two traces may be displayed using this switching technique. The maximum is usually four since the brightness is reduced by multiple traces, and it becomes difficult to interpret more than four traces on the screen. Four traces may be obtained by using a double-trace oscilloscope frame (*i.e.* case) and two double-trace amplifiers, as shown in figure 10.10. This is the arrangement used on the Tektronix 7000 Series, and is quoted as an example.

This oscilloscope consists of a main frame (available in various forms) with places for three or four plug-in units, which may be amplifiers or timebase units. Single plug-in or dual plug-in amplifiers are available, and it will be assumed that two dual amplifiers have been fitted. Many other special plug-in units are available, and some examples are given in Chapter 15.

The correct setting of the mode switches and triggering is important. The main frame MODE switch can be set to L.H., R.H., ALT., CHOP or ADD. When set to L.H. the left-hand amplifier plug-in is used, which may give a single trace or dual trace depending on the setting of the amplifier controls. When set to R.H. the same conditions apply to the right-hand amplifier plug-in. When set to ALT, beam switching takes place on the alternate basis between right-hand and

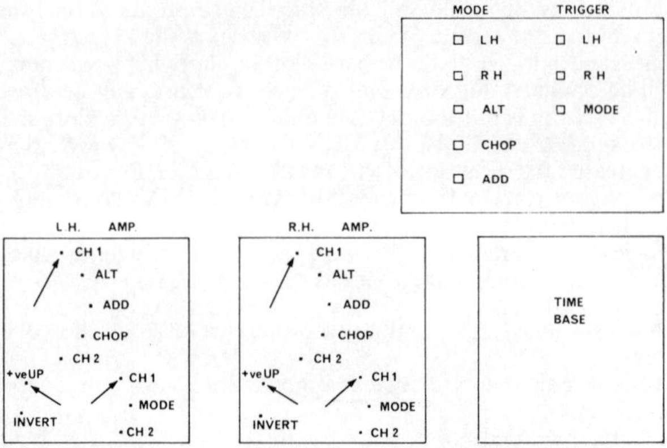

FIG. 10.10. ARRANGEMENT FOR DISPLAYING FOUR TRACES
(Tektronix 7000 series)

left-hand amplifiers. Each amplifier may be set to SINGLE trace, ALT trace or CHOPPED trace, and so 2, 3 or 4 traces can be obtained. With the main frame control set to CHOP, beam switching between amplifiers occurs on a chopped basis. In the ADD position the outputs of the two amplifiers are added.

Each amplifier has a MODE switch giving channel 1 on its own, channel 2 on its own, the sum (ADD) of the two inputs and also the difference since one channel can be reversed. Each amplifier can be set to ALT or CHOP between its two inputs.

Turning to triggering, the main frame triggering control can be switched to left-hand or right-hand amplifiers, and to a position called MODE. This means that the triggering is connected to the same signal as the setting of the MODE switch, *i.e.* L.H., R.H., ALT, CHOP and ADD. In the CHOP position the triggering signal is obtained from the addition (sum) of the two amplifier inputs as in the ADD position. On each amplifier the triggering can be selected to signal 1 or 2 and MODE. Again, the MODE position means that the triggering is obtained from the source indicated on the mode switch, excepting that when in the CHOP position the triggering is from the sum of the two inputs. If, therefore, the trigger control is placed in the MODE position and the MODE switch is in the ALT position, the triggering is in the alternate position, trace 1 being triggered by signal 1 and trace 2 being triggered by input 2. As already explained, this is a dangerous position and can produce phase errors; but it can be used to display two signals of unrelated frequency. In fact, by also using the frame control in the same way four unrelated frequencies can be displayed. Figure 10.11 shows four traces using switched beams. This display is using CHOPPED mode between all traces (*i.e.* CHOP setting on main frame and both amplifiers).

SWITCHING CIRCUITS

There are two parts to the switching system. First, the circuit that switches the cathode-ray tube from one input to the other; and secondly, the device, usually a multivibrator, producing the switching signal.

A basic block diagram is shown in figure 10.12.

The device must be fast, particularly in the chop mode or a considerable percentage of the total time will be lost in the spot moving from one trace to the other. One method uses switching diodes; the basic arrangement is in figure 10.13. Many manufacturers use this method.

Each channel has shunt diodes D_3D_4 and D_6D_7 and series diodes D_1D_2 and D_5D_8. The junction of D_3D_4 is fed with a switching square wave of

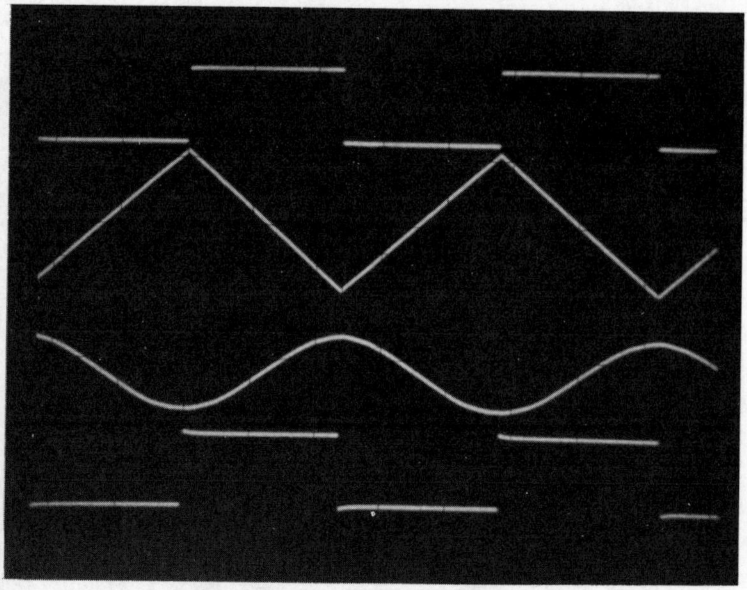

FIG. 10.11. FOUR-TRACE DISPLAY (CHOPPED MODE)

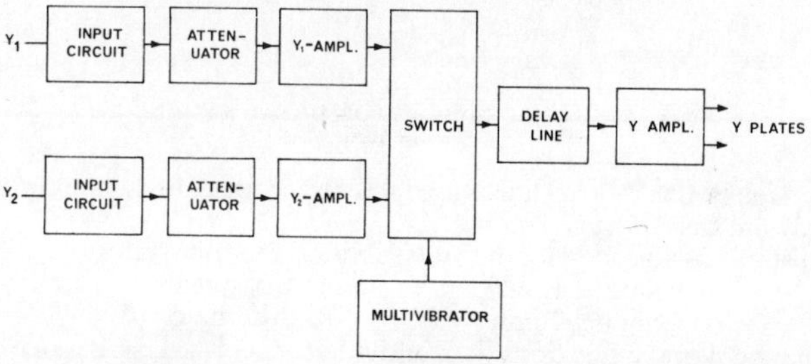

FIG. 10.12. BLOCK DIAGRAM OF SWITCHED TRACE OSCILLOSCOPE

opposite polarity to that on the junction of D_6D_7. If the junction of the shunt diodes D_3D_4 is at zero potential, D_3 and D_4 are reversed biased by the current flowing through D_1 and D_2 from the + 30 V supply. This supply also feeds the amplifier stage through D_1 and D_2, and a signal is therefore fed from input 1 to the main Y-amplifier. Hence this channel is operating. On the next half-cycle the junction is given a positive potential, so making D_3 and D_4 conduct, but reverse biasing D_1 and D_2 so that no signal passes to the Y-amplifier. The other channel works in the same way, but in antiphase.

Shunt and/or series FETs may be used as switches, the gates being fed with suitable waveforms. Tektronix use an integrated circuit in some of their oscilloscopes.

So that each trace can be positioned separately the shift control for each channel must be placed before the switching circuit. From the point of view of the switching circuit the position or shift voltage is dealt with in the same way

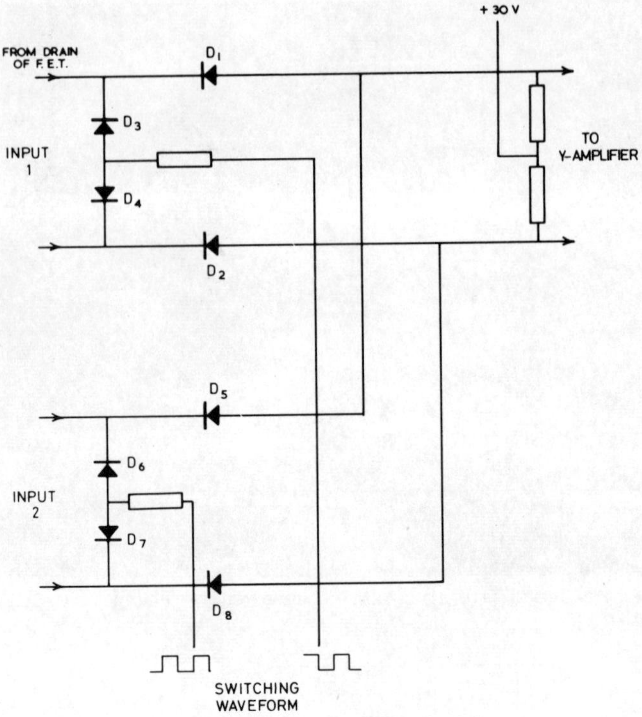

FIG. 10.13. USE OF DIODES FOR SIGNAL SWITCHING
(Scopex Instruments)

as a d.c. signal. If a Y-delay line is used it is placed after the switching circuit so that only one delay line is required.

It must be possible to place the multivibrator into four states:

(a) Channel 1 only (c) Chopped
(b) Channel 2 only (d) Alternate.

In channel 1 and 2 positions the multivibrator is biased so that one or other of the two transistors is fully conducting. In mode (c) it must be set to run free (*i.e.* as an astable multivibrator) at an appropriate frequency, say 100kHz to to 1 MHz. In mode (d) it is set as a bistable and fed with pulses from the time-base so that it switches over at the end of each scan of the timebase.

This multivibrator must also supply suitable pulses to blank the trace during the transition period between the two traces. A suitable waveform is fed to the grid of the cathode-ray tube or deflection blanking plates so as to cut the beam off as it moves from one trace to the other.

11

Delayed Timebase or Dual-timebases (Strobe Timebase)

When examining some waveforms, such as a television waveform, it is desirable to see a small section in detail. For example, if the timebase is run at picture frequency, the result of integration of the synchronizing pulses is as in figure 11.1, and no detail is visible of the field synchronizing pulses. The timebase cannot be run at a higher frequency because the waveform only repeats

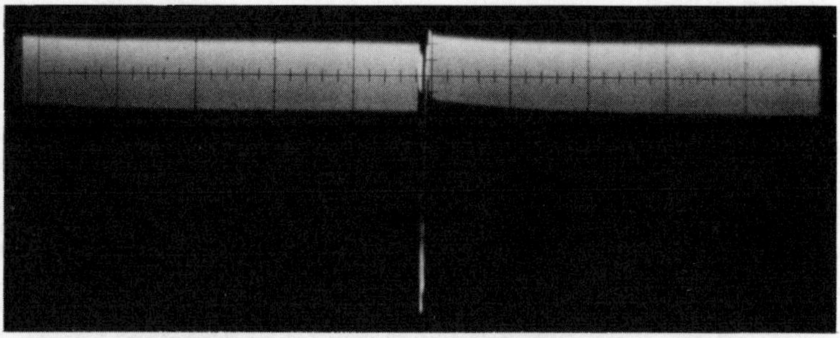

FIG. 11.1. INTEGRATION OF TELEVISION SYNCHRONIZING PULSES (625 LINES) WHEN TIMEBASE OPERATED AT REPETITION RATE EQUAL TO PICTURE FREQUENCY *i.e.* 25 Hz

itself at picture frequency. Some improvement is possible where there is an × 10 horizontal expansion. The result of × 10 expansion is given in figure 11.2 and and often is adequate. However, for some purposes it does not give sufficient

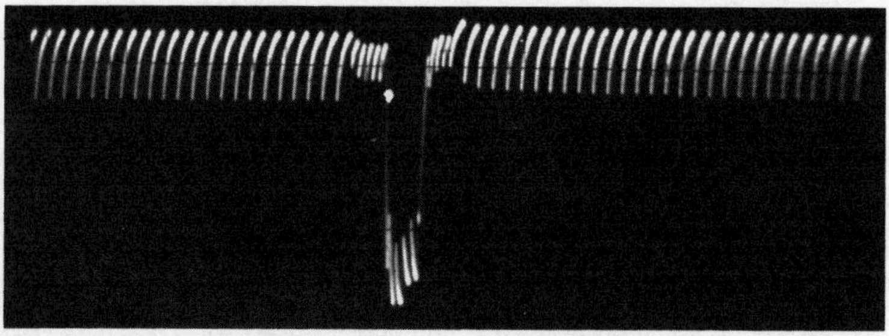

FIG. 11.2. AS FIGURE 11.1 BUT WITH × 10 HORIZONTAL (X) EXPANSION

resolution of the waveform and the delayed timebase idea was developed. This uses two timebases, and hence sometimes known as a 'dual-timebase'. This will be explained in terms of a television waveform, but as will be seen later it can be used in other applications.

A timebase with a high trace speed is required corresponding, say, to 2 lines of the picture which is triggered once per picture, *i.e.* 25 times per second. In order that the correct 2 lines are displayed the instant of triggering of this timebase must be variable. One timebase is used to produce the variable delay time and the other the fast sweep. There are three variations of the idea, which will be called (a) delayed sweep; (b) delayed triggered sweep; and (c) mixed sweep.

(a) Delayed Sweep

This is also called 'delayed presentation of sweep'. The basic idea is shown in figure 11.3. This is drawn using a simple television type of waveform as it is impossible to draw a complete 625-line waveform. It is shown as a non-

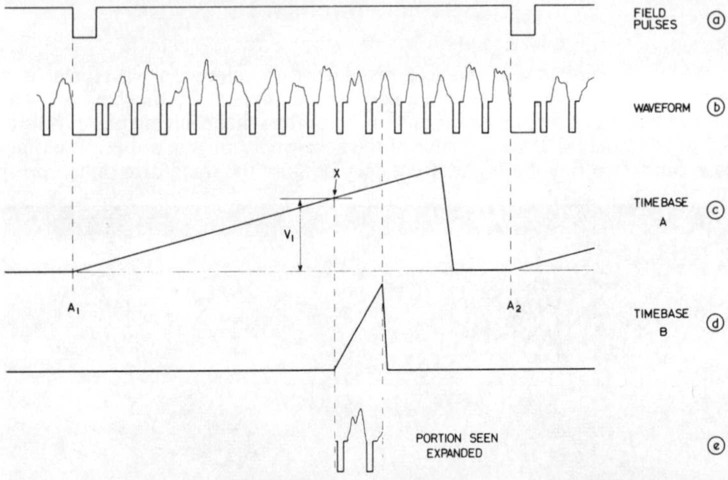

FIG. 11.3. OPERATION OF DELAYED SWEEP

interlaced system with only 15 lines per picture ('picture' and 'field' now have the same meaning), and a single broad (or long) pulse for the field pulse. At (b) is shown the television waveform, and it will be assumed that we wish to look at one particular line in detail. It will also be assumed that field pulses as shown at (a) are available for triggering the oscilloscope through the external triggering socket. At (c) is shown the normal, or main timebase, which is triggered at instants A_1 A_2, etc., by the field synchronizing pulses shown at (a). Thus, this timebase runs at picture frequency, *i.e.* it is triggered once per picture. In an actual television waveform this timebase would be triggered at 25 times/second, or a submultiple. With this timebase operated normally on the waveform of figure 11.3, about 11 lines would be displayed on the screen (*i.e.* those lines corresponding to the sweep) and it would be difficult to see the detail of any one line. If the timebase were run at line frequency all the lines would be superimposed and, because they are all different, would result in a blurred video portion.

There is some confusion about naming the two timebases. The author will use the term A for the main timebase. This is the one normally used when the delay feature is not in operation and produces the delay when delayed sweep is used. The second timebase will be called B. This produces the trace after the delay. Some manufacturers use the two terms in reverse. In some oscilloscopes either timebase can be used on its own, but more usually only A can be used as a normal timebase.

When delay sweep is in use timebase A does not produce any X-deflection but operates only as a variable delay circuit and feeds a comparator circuit. The comparator circuit is also fed with a variable voltage V, commonly obtained from a multiturn potentiometer. This control is called the DELAY TIME MULTIPLIER. When the amplitude of the voltage of timebase A reaches that of the variable voltage source, say V_1, the comparator gives out a pulse which triggers the timebase B at point X. The sweep speed of timebase B is greater than that of A, shown at (d), and it is B which gives the X-sweep. The portion of the waveform displayed corresponds to the sweep of timebase B and is shown at (e). This section, of course, is expanded to the full screen width. Hence we now see approximately one line only. No further sweep across the screen takes place until timebase B is again triggered by A, reaching voltage V_1 on its next sweep. Thus timebase B is triggered only once per picture. Since the sweep speed is high and yet operating only once per picture, when changing from normal timebase A to delayed sweep by timebase B, the brilliance of the trace will decrease considerably. The AMOUNT of waveform displayed depends only on the sweep speed of timebase B, which can be varied in the normal way. Thus a half-line can be displayed by doubling the sweep speed of timebase B. (Timebase B must always be set to a sweep speed higher than that of A).

The POSITION of the displayed section is determined by the position of point X, and hence on the voltage V_1. Thus the position of the display can be varied by varying V_1 using the DELAY TIME MULTIPLIER control. To assist in obtaining the correct display there is an intermediate condition known as 'A brightened (or intensified) by B'. The procedure is as follows:

(i) The oscilloscope is run in the normal way using timebase A which is triggered by the field synchronizing pulses using external triggering. This timebase must be run at the repetition rate of the waveform or a submultiple of it. In a normal television waveform this is 25 Hz or a submultiple. This display is shown in figure 11.4(a) for integrated synchronizing pulses on a 625-line system.

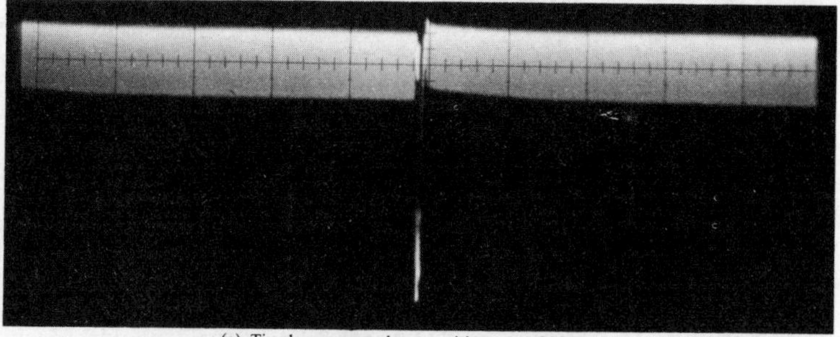

(a) Timebase operated at repetition rate of 25 Hz

FIG. 11.4. INTEGRATED SYNCHRONIZING PULSES

(ii) The timebase is now switched to 'A intensified by B'. In this position timebase A still produces the X-deflection but it now also triggers timebase B, from the comparator as already described. However, timebase B is not used for the X-deflection but it produces a brightening pulse equal in length to its sweep time. This pulse is applied to the grid of the cathode-ray tube so that a part of the waveform is brightened as at (b). The *position* of the brightened part can be varied by the DELAY TIME MULTIPLIER (the multiturn potentiometer) which varies voltage V_1. The LENGTH of the brightened portion is determined by the sweep speed of the timebase B which is varied by the TIME/DIV knob of timebase B. When the position and length of the brightened portion correspond to that part of the waveform to be examined in more detail, the timebase is switched to 'delayed sweep' and the selected part is now displayed on the screen as at (c).

The time-scale is now settled by the sweep speed of timebase B, and can be varied as required by altering the TIME/DIV setting of timebase B. If sufficient expansion cannot be obtained when timebase B is running at maximum sweep speed, the X-expansion ($\times$ 10 or $\times$ 5) control can be used which will give further expansion but, of course, allowance must be made for this when using the time-scale. The portion of the waveform seen can be varied by using the DELAY

(b) Sweep A intensified by sweep B

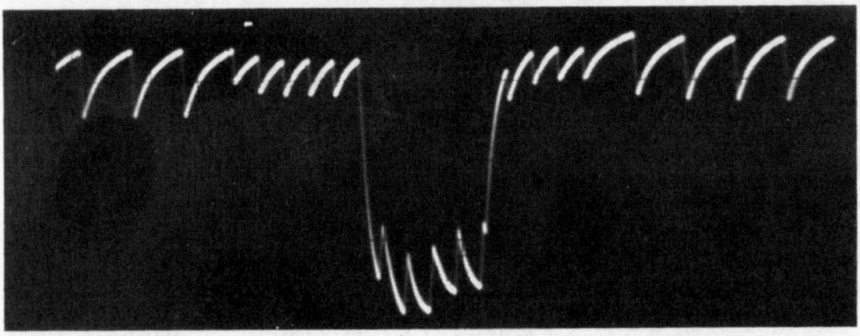

(c) Sweep B (delayed expanded sweep)

FIG. 11.4(b)&(c). INTEGRATED SYNCHRONIZING PULSES

TIME MULTIPLIER. Operating this control varies V_1 continuously, hence the selected part moves steadily across the waveform and any particular portion of it can be selected as required. As already mentioned, this display is not very bright and the faster the sweep speed of timebase B the dimmer it becomes. With modern oscilloscopes using PDA tubes there is usually no difficulty in getting a trace easily visible. It is worth mentioning that if the system is used to display one line of a 625-line television waveform, this corresponds to an X-expansion of at least 625 times. In practice it is possible to display only a part of a line if required. There are other uses of the DELAY TIME MULTIPLIER which will be explained later.

A drawback to this arrangement is that with large expansions there is likely to be some jitter on the display. There are three reasons for this:

(i) There may be slight variations in the triggering of the timebase A which will cause jitter of the display. The amount depends on many things, but it will be reduced if there is a sharp leading edge to the waveform that triggers the timebase. For a television waveform it is generally desirable to use external triggering from a related pulse.

(ii) The timebase B is triggered when the voltage of timebase A reaches a certain critical value, which is settled by the comparator. Due to noise, etc., there will be some small variation in the instant that the comparator triggers timebase B. This will be reduced in a good oscilloscope; by careful design it can be kept small, but will always be present to some extent.

(iii) There may be some jitter in the waveform under examination.

Suppose that we have a waveform as in figure 11.5(a), that this triggers timebase A at point X, and that we are examining the portion between the dotted

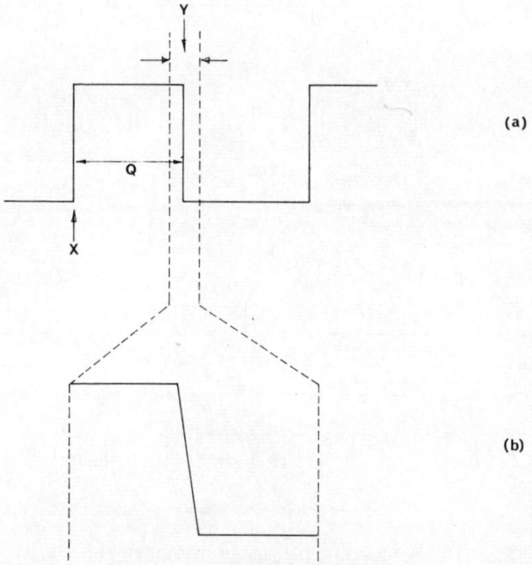

FIG. 11.5. EFFECT OF JITTER
(a) Waveform to be examined; (b) Expanded sweep

lines shown expanded at (b). If there is any variation in the time Q, jitter will occur however good the oscilloscope.

The advantage of this arrangement is that the portion to be displayed can be varied continuously by means of the DELAY TIME MULTIPLIER, also only one triggering signal is required, *i.e.* to trigger timebase A.

(b) Delayed Triggered Sweep

Also called 'delayed generation of sweep' and 'delayed gate trigger mode'. One problem of delayed sweep is that any variation in the delay time in triggering timebase B, or any variation in the waveform being viewed, the resulting delayed trace will jitter. As a *very large expansion* of the waveform is possible by this method, any small timing errors are greatly exaggerated and cause bad jitter. When on delayed triggered sweep, instead of the comparator triggering timebase B, it opens a gate so that timebase B can be triggered by a suitable signal. This is explained by reference to figure 11.6, which again uses the simplified television type waveform. At (b) is shown the waveform, and at (a) the field synchronizing pulses. Timebase A is set up as previously, preferably triggered externally by the field pulses at (a). The oscilloscope is first switched to 'A intensified by B.' Under these conditions the sweep is again from timebase A. However, when the magnitude of the timebase reaches the value V_1 it opens a gate [as at (d)] allowing triggering pulses to pass to timebase B. In this case it is assumed that the external trigger input for timebase B consists of composite pulses as shown at (f). Line pulses would be equally suitable. Now the gate is opened at X_1, but timebase B is not triggered until point X_2 (it being assumed that the timebase is set to trigger on the negative-going edge of the line pulses). Thus, timebase B intensifies the trace and indicates the portion selected. The result appears as in figure 11.4(b). The oscilloscope is now switched to delayed triggered sweep and then timebase B as shown at (e) is used to produce the X-sweep. A single line is displayed as at (g), expanded to the full screen width. The display is like 11.4(c) (but probably with less jitter).

Slight variations in the operation of the comparator will have no effect (except under the conditions where its operation corresponds to the leading

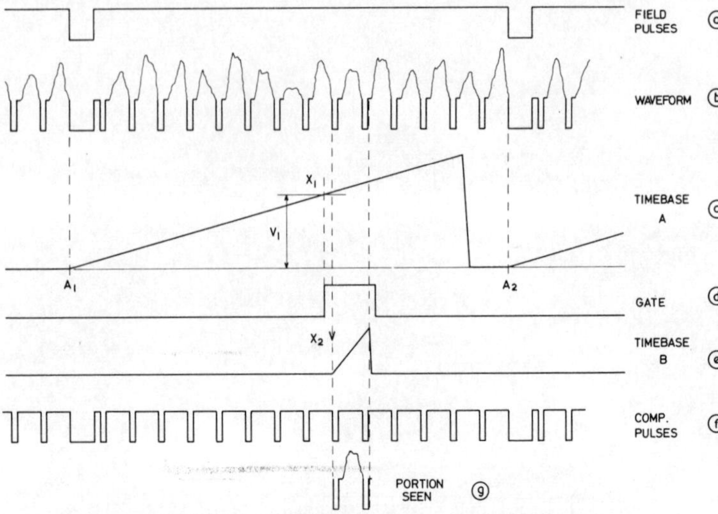

FIG. 11.6. OPERATION OF DELAYED TRIGGERED SWEEP

edge of a line pulse) because it is only opening the gate for timebase B. Hence, jitter due to this is eliminated. Similarly, any jitter on the waveform itself will have a negligible effect because the line being displayed is actually triggering the timebase. One may consider that timebase B is set to display one line, but it is only allowed to do this once per picture because trigger pulses can reach timebase B only when the gate at (d) is opened.

There is one disadvantage, however, of this arrangement. Timebase B, in the example shown, can only be triggered at the start of a line (on one or other edge of a line synchronizing pulse). Thus as the voltage V_1 is decreased nothing happens until the gate is opened before the next pulse. This is shown in figure 11.7, where line pulses are shown at (a). At (b) timebase B will trigger on line Q even though the voltage V is varied from V_2 to V_1. When it is reduced to V_3,

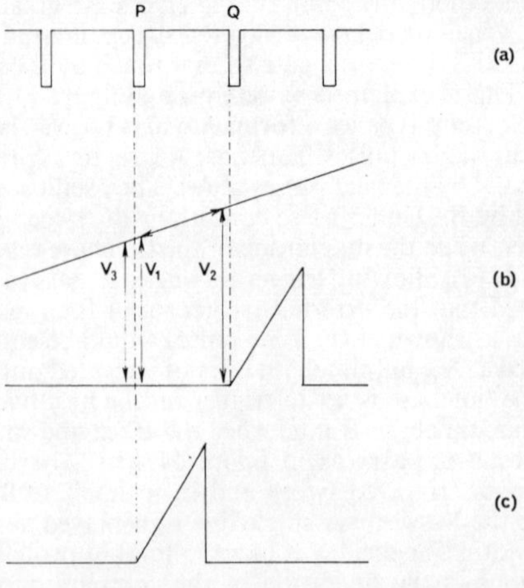

FIG. 11.7. EFFECT OF VARYING DELAY TIME MULTIPLIER WHEN USING DELAYED TRIGGERED SWEEP

timebase B triggers on line P, as shown at (c). (There will be some uncertainty between voltage V_1 and V_3). There are sudden jumps from one line to the next as the DELAY LINE MULTIPLIER is operated. The timebase B need not be triggered by external line pulses; internal triggering from the displayed waveform could be used by suitable adjustment of the trigger level control of timebase B. If the waveform of figure 11.5 were being displayed in this way, with the trigger slope control set to negative and at the correct level then, provided the gate opened before the negative edge of the pulse, the timebase would be triggered at point Y, but part of the negative edge may be missing. Varying the DELAY LINE MULTIPLIER would have no effect in one direction (reducing V); it would only open the gate earlier, but the timebase would still only be triggered at instant Y. Moving it in the other direction would eventually open the gate after Y, and timebase B would not be triggered at all and there would be no trace.

As before, if the maximum speed of timebase B is not sufficient then the trace expansion (say × 10) can be used, and in this case it is usually more successful due to the reduced jitter.

Figure 11.8 shows the effect of delayed sweep and delayed triggered sweep. At (a) is shown the complete waveform as displayed on timebase A only. At (b) is shown the same waveform with a portion brightened by the B timebase, the oscilloscope being set to the 'A intensified by B' position and the DELAY LINE MULTIPLIER set to cover the positive-going edge. At (c) is shown this section expanded, *i.e.* set to B delayed by A, so that a much clearer display of the leading edge is obtained. There will be some slight jitter, which does not show on the photograph. At (d) is the result when set to delayed triggered sweep, the triggering of timebase B being internal (*i.e.* from the input waveform), the trigger slope being set to positive and the level to a low value so as to include most of the leading edge. Any jitter is now almost completely

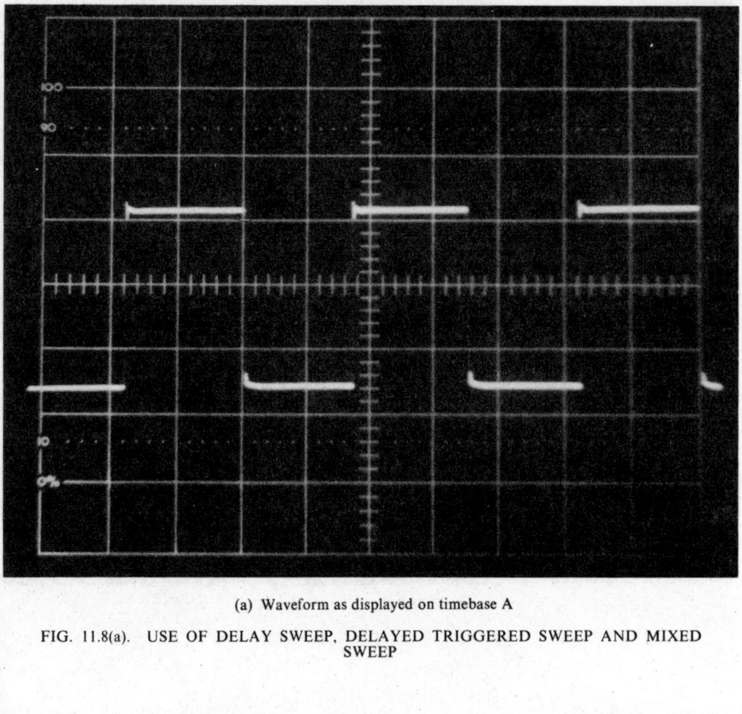

(a) Waveform as displayed on timebase A

FIG. 11.8(a). USE OF DELAY SWEEP, DELAYED TRIGGERED SWEEP AND MIXED SWEEP removed. At (e) is shown the same waveform, but with maximum Y-expansion also in use; much more detail can now be seen that at (a).

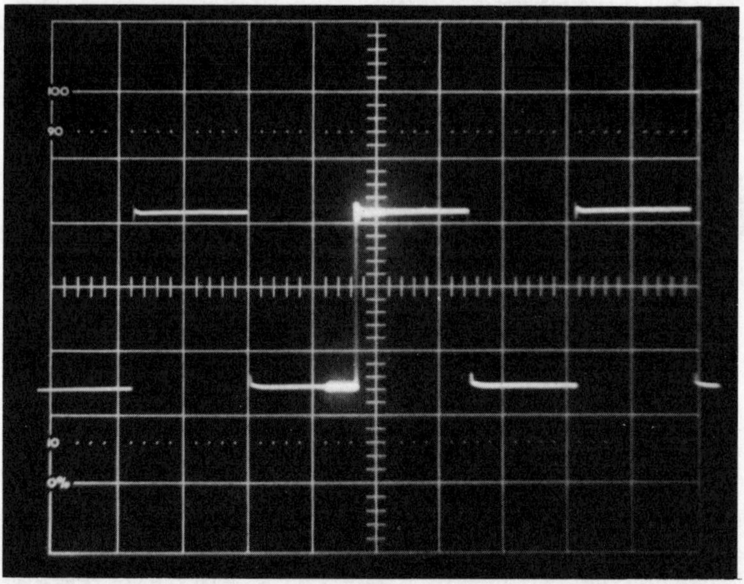

(b) Sweep A intensified by sweep B

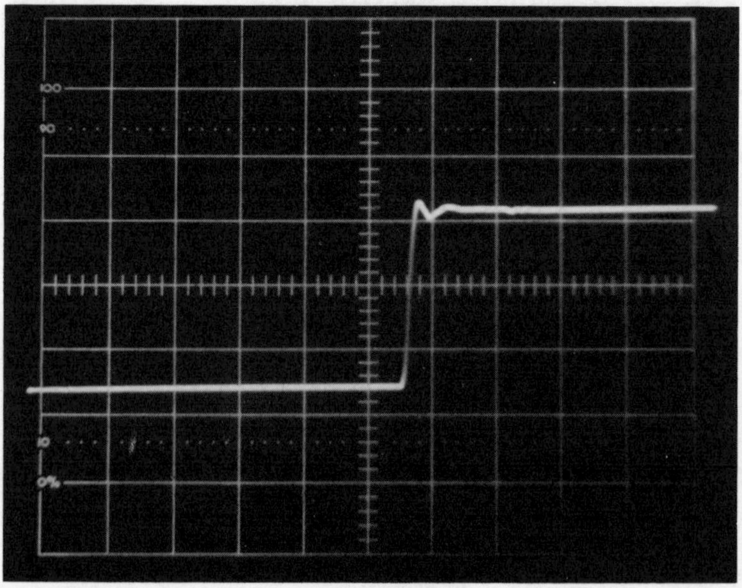

(c) Delayed sweep display

FIG. 11.8(b)–(c). USE OF DELAY SWEEP, DELAYED TRIGGERED SWEEP AND MIXED SWEEP

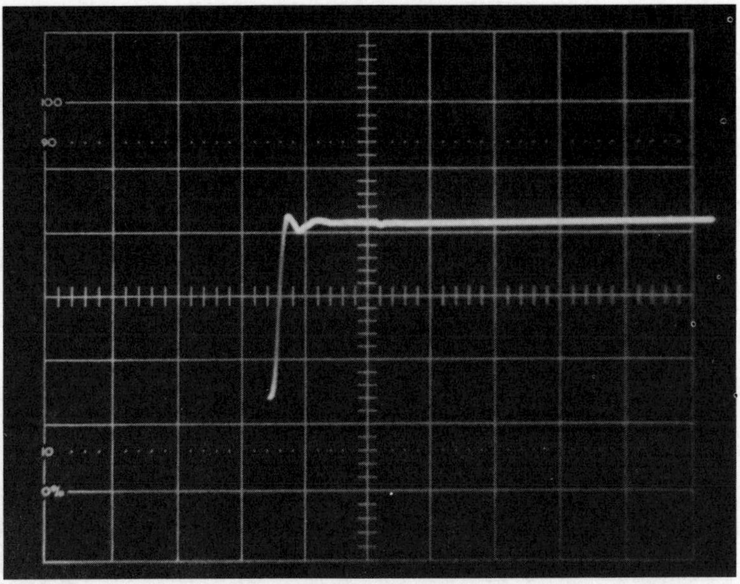

(d) Delayed triggered sweep display

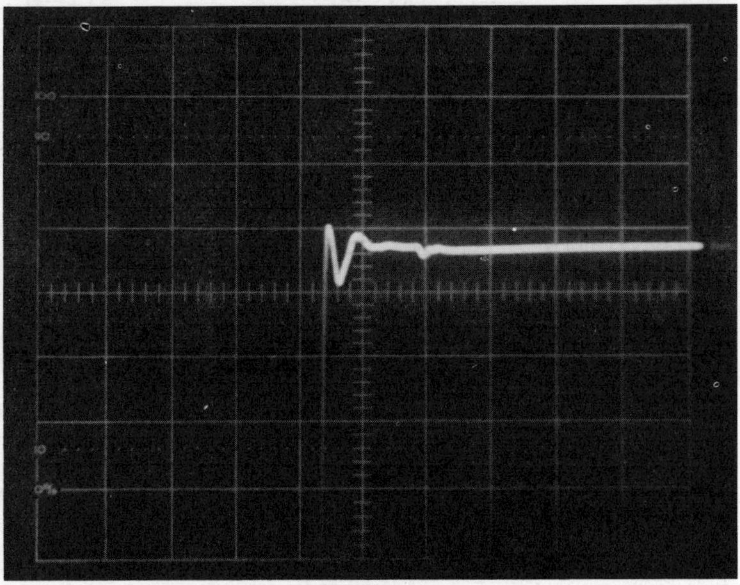

(e) As (d) but increased Y deflection to show more detail

FIG. 11.8(d)–(e). USE OF DELAY SWEEP, DELAYED TRIGGERED SWEEP AND MIXED SWEEP

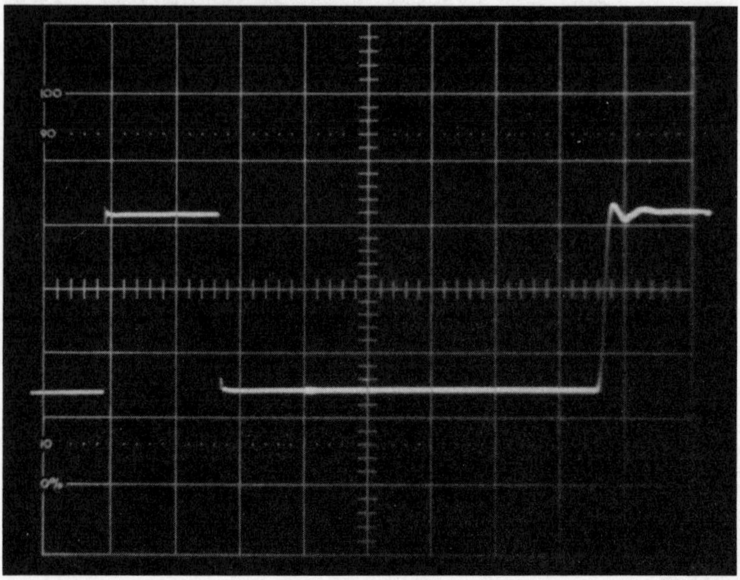

(f) Mixed sweep display

FIG. 11.8(f). USE OF DELAY SWEEP, DELAYED TRIGGERED SWEEP AND MIXED SWEEP

(c) Mixed Sweep

In this mode one again starts with timebase A on its own, with suitable triggering and sweep speed. The oscilloscope is then set to 'A intensified by B' as previously, and either the delayed sweep or the delayed triggered sweep is used. On now switching to MIXED SWEEP, timebase A operates and produces the sweep up to the point when it reaches the delay voltage V_1. At this point the horizontal deflection changes to timebase B, giving an expanded trace. This is shown for the same waveform as previously in figure 11.8(f), where the first positive pulse is that obtained by timebase A and the next positive-going edge is expanded in the same way as the other traces.

The principle is shown in figure 11.9. Timebase A is triggered as shown at A_1 A_2 etc., and runs until point X, when (assuming delayed sweep) timebase B is triggered and produces the sweep at an increased speed.

The operation is perhaps made more clear in figure 11.10. At (a) and (b) delayed sweep is used and the difference between the two is that the DELAY TIME MULTIPLIER setting has been changed so that at (b) the faster sweep starts later. The instant of the start of the faster sweep can be varied *continuously*

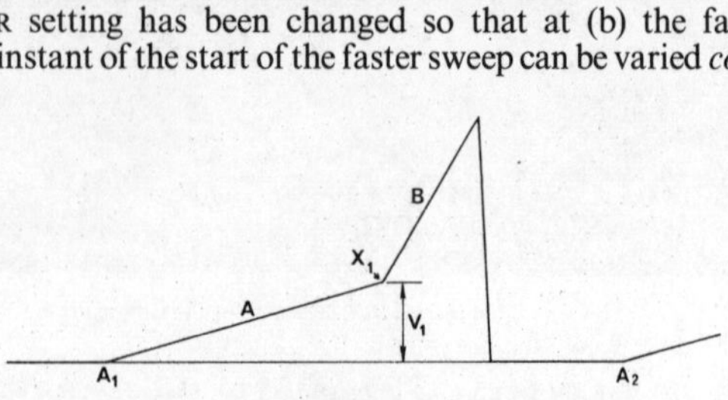

FIG. 11.9. MIXED SWEEP OPERATION

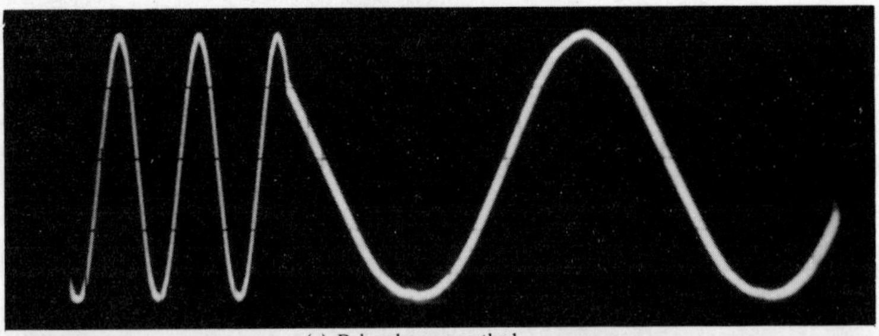

(a) Delayed sweep method

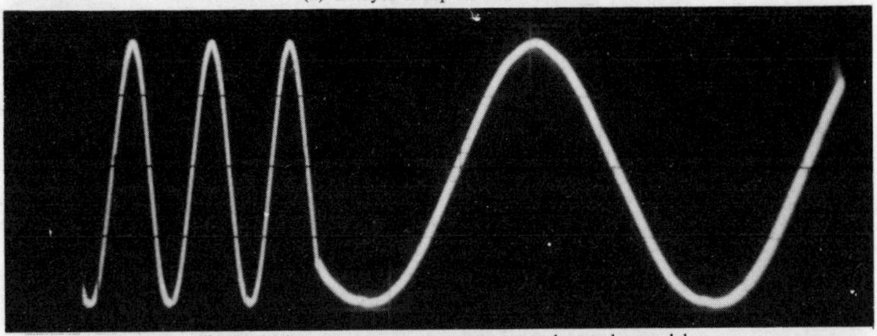

(b) As (a) but DELAY TIME MULTIPLIER control set to longer delay

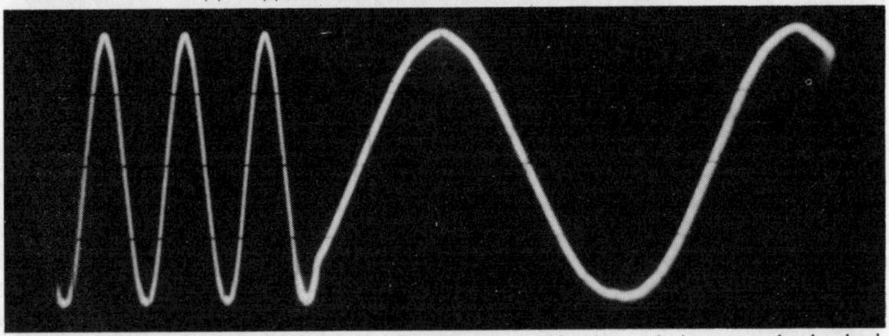

(c) Delayed triggered sweep method. Timebase B trigger set to positive slope and trigger control to low level

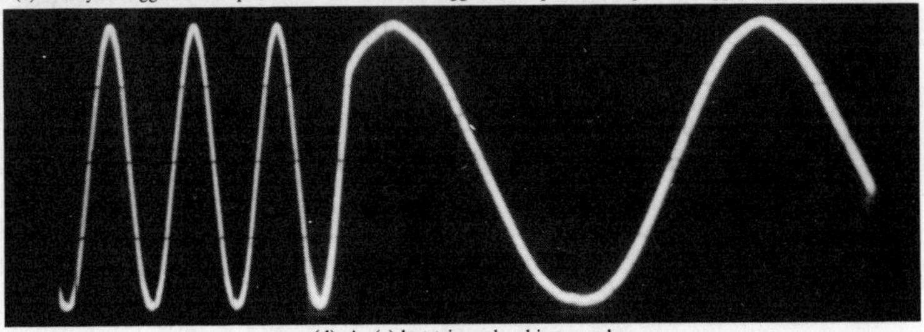

(d) As (c) but trigger level increased

FIG. 11.10(a)–(d). MIXED SWEEP DISPLAYS

by the DELAY TIME MULTIPLIER. Figures (c) to (f) are using delayed triggered sweep and internal triggering of timebase B. The instant of start of this fast sweep depends not only on the setting of the DELAY TIME MULTIPLIER but also on the level setting and slope polarity of the B timebase. At (c) and (d) the slope setting is positive. At (c) the level has been set low so that when the gate opens, timebase B will trigger as soon as the waveform goes in a positive direction.

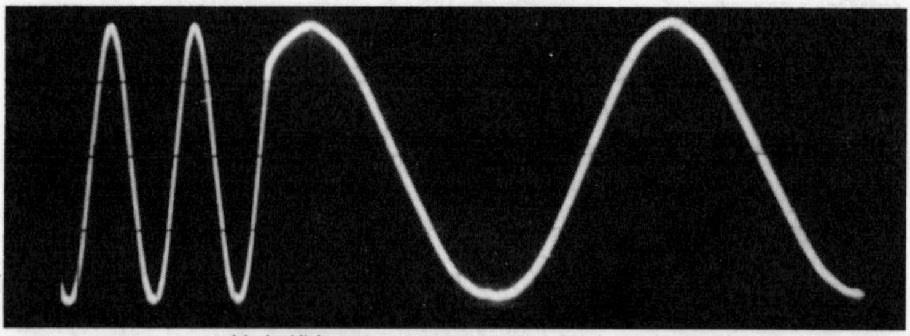

(e) As (d) but DELAY TIME MULTIPLIER reduced

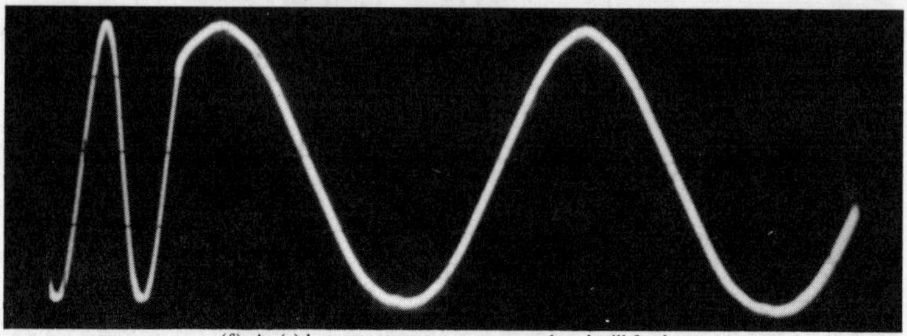

(f) As (e) but DELAY TIME MULTIPLIER reduced still further

FIG. 11.10(e)–(f). MIXED SWEEP DISPLAYS

At (d) the level has been set higher, hence the higher sweep speed starts later. At (e) and (f) the DELAY TIME MULTIPLIER has been reduced, other factors being the same as at (d). It will now be seen that triggering takes place *at the same part of a cycle* but now occurs on earlier cycles. As one turns the DELAY TIME MULTIPLIER the point at which timebase B operates jumps from cycle to cycle.

USE OF DELAY TIME MULTIPLIER FOR TIME MEASUREMENT

This is not just a variable control but an accurate timing device and it can be used for precise measurements of time. Suppose that the waveform of figure 11.11 is displayed on timebase A as at (a), and that the length of the pulse is required. The oscilloscope is switched to ‘A intensified by B’ and the

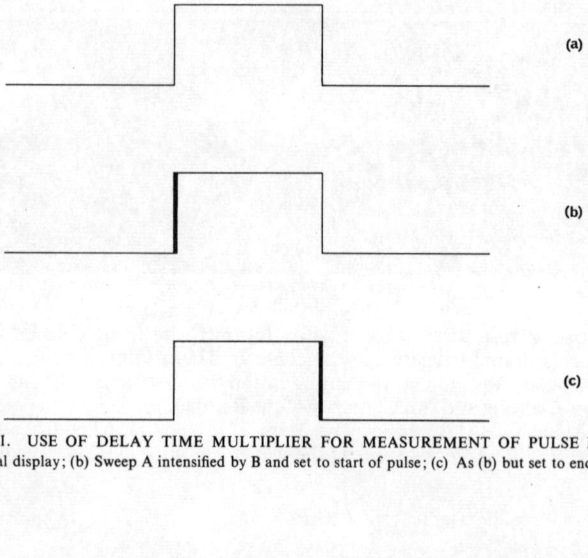

FIG. 11.11. USE OF DELAY TIME MULTIPLIER FOR MEASUREMENT OF PULSE LENGTH
(a) Normal display; (b) Sweep A intensified by B and set to start of pulse; (c) As (b) but set to end of pulse

B sweep speed set high so that only a small section is brightened. This section is moved by the DELAY TIME MULTIPLIER until it coincides with the leading edge, as shown at (b). The reading of the DELAY TIME MULTIPLIER is noted. It is then moved until the brightened portion corresponds to the trailing edge as at (c). Again the reading is noted. The difference between the two readings is multiplied by the A timebase sweep setting and the result is the time of the pulse. For example, if the setting for (b) is 2·7, for (c) 6·3, and the A sweep speed setting is 20 μs/div, the width of the pulse is (6·3–2·7) × 20 μs = 72 μs. This is an accurate measurement if the circuit is suitably designed as the resolution of the DELAY TIME MULTIPLIER is good, being a 10-turn potentiometer, and there are no parallax errors. This method of time measurement is shown in figure 11.12.

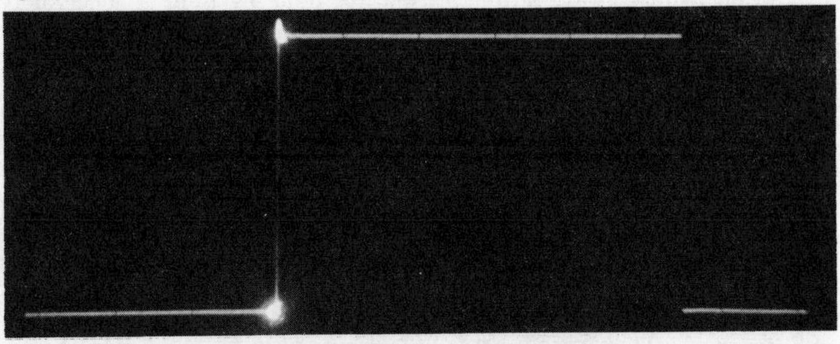

(a) Intensified portion set to positive leading edge

(b) Intensified portion set to negative trailing edge

FIG. 11.12. USE OF DELAY TIME MULTIPLIER TO MEASURE PULSE LENGTH. PULSE FREQUENCY 10 kHz

12

Calibrators

SOME form of calibrator is provided on many oscilloscopes. The calibration of the vertical amplifiers and the sweep speed of the timebase should be checked from time to time. The methods used will depend on the type of oscilloscope and will be more involved on the more expensive instruments. The calibration output is often used to check the setting of the probes, while in others a separate output is provided for setting up the probes.

One simple arrangement is to use the mains supply frequency and generate a low voltage square waveform. This may be done by using a zener diode together with a potential divider. A simple circuit is shown in figure 12.1, the zener diode D_1 being fed from a suitable a.c. voltage through R_1, If D_1 is a

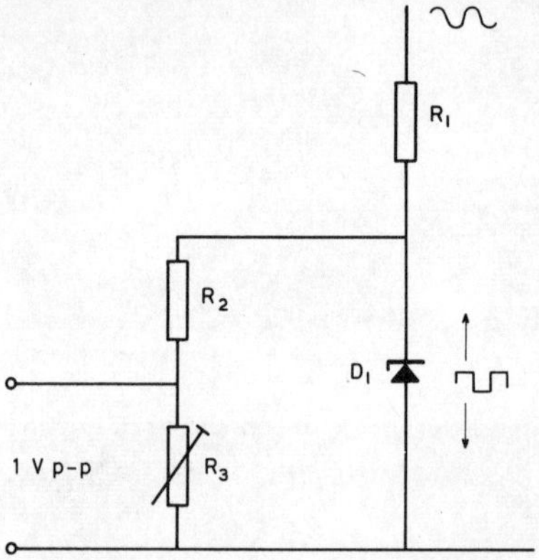

FIG. 12.1. SIMPLE CALIBRATOR CIRCUIT

10 V zener diode then an attenuator R_2 R_3 may be used to produce a 1 V p-p output. In the cheaper oscilloscopes only one voltage is provided; if the Y-amplitude is correct or corrected at one setting of the Y-attenuator, it should be correct at all settings since the accuracy of the attenuators should not change. Generally, a preset gain control is fitted on the Y-amplifier(s) so that the sensitivity can be adjusted if required. The same calibration voltage can be used to check the timebase sweep speed. The check is only approximate (since the mains frequency varies) but probably is accurate enough. The sweep speed can only be checked for two or three settings. It does not follow that the others will be correct because other settings use different C and R values on the sweep generator. This low frequency waveform is not suitable for checking a probe and another output ('probe test') may be provided. This may be a square wave output from the timebase which is set, say, to 1 kHz. The probe is adjusted for a level trace on the screen (the timebase output is a pulse corresponding to the trace, and so only a line is obtained on the screen corresponding to the top of the pulse – *i.e.* it is the unblanking signal).

Expensive oscilloscopes may have a more complex calibrator. Usually a square wave oscillator of about 1 kHz is used with some means of obtaining a stable output, and an attenuator may be used so that several voltages are available. This frequency is high enough for checking the adjustment of probes. D.C. calibration voltages may also be provided.

A calibration current source might be provided so that current probes can be calibrated. This may consist of a loop carrying a known current; or the voltage calibration points may be designed to give an accurately known current when short circuited.

When more extensive calibration and servicing are involved special calibration units are made which give the facilities to accurately calibrate and set up oscilloscopes.

PLUG-IN UNITS

Calibration is rather more difficult with plug-in units. For example, part of the Y-amplification is in the plug-in and part in the main frame. An error in the Y-gain may be in either unit. It may be corrected by adjusting the gain of the plug-in, which will be satisfactory when the same plug-in is used in the same frame. However, if the plug-in is changed to another frame then the calibration will be incorrect if the original error was due to incorrect gain of the first frame. To overcome this the actual gain of the frame and the actual gain of the amplifier must be checked. Special equipment may be needed and reference must be made to the manufacturer's handbook for details.

When plug-ins are changed between frames a check should therefore be made of the calibration if the calibrated features are to be used.

13

Some Uses of an Oscilloscope

THE number of uses of an oscilloscope is extremely large and only a selection can be described.

THE OSCILLOSCOPE AS A D.C. VOLTMETER

As modern oscilloscopes have calibrated d.c. amplifiers the instrument can be used as a d.c. voltmeter. It is only necessary to determine the movement of the trace on the screen when the voltage is applied. It is best to use the oscilloscope with the timebase in operation as this prevents a bright spot with the possibility of burning the screen. The timebase should be set to a speed sufficiently high to avoid flicker, say several 100 Hz (1–10 ms/div), and if it is set to AUTO, a trace will always be visible. With no voltage applied and the input shorted (or, where applicable, the amplifier grounded by the appropriate switch) the trace is moved to lie on a convenient major mark on the graticule. The voltage is applied, the new position noted, and the vertical movement determined. The applied voltage is then the vertical movement in divisions multiplied by the vertical sensitivity setting (in volts/div or mV/div). It is important that if there is a continuous vertical sensitivity control (*i.e.* control of amplifier gain), it MUST be placed in the CALIBRATE position. Normally, an upward movement corresponds to a positive applied voltage and downward to a negative applied voltage. In some oscilloscopes the trace can be inverted by a switch. The range covered depends on the oscilloscope, but may be from, say, 10 mV/div to 50 volts/div direct, and, say, to 500 V/div with 10/1 attenuator. Without the attenuator the input impedance is reasonably high compared with a voltmeter, being normally 1 MΩ on all ranges and 10 MΩ with 10/1 attenuator. The accuracy obtainable will not be very high, say 5 %, depending on the oscilloscope and whether an external or internal graticule is used. One advantage of using the oscilloscope in this way is that one knows exactly what is being measured. If there is any ripple or high frequency superimposed on the d.c., this will be visible (and may have to be removed before an accurate measurement can be made); whereas when using a d.c. voltmeter there is no such indication and the reading obtained may not be accurate because one is unaware of some ripple, etc.

Figure 13.1 shows (a) the first position, and (b) the second position. The timebase was set to 1 ms/div as indicated by the top right-hand figure. (These figures are produced automatically on this particular oscilloscope: see DIGITAL READOUT in Chapter 15). The change from one position to the other is 3·6 divisions. The volts/div is 50 as indicated by the top left-hand figure. The voltage is therefore $3{\cdot}6 \times 50 = 180$ V.

THE OSCILLOSCOPE AS AN A.C. VOLTMETER

Although not essential it is better to use a timebase to guard against the possibility of burning the screen by a stationary spot. Again, the AUTO setting of the timebase is most convenient. When an a.c. voltage is applied a trace will be obtained, and its magnitude is measured on the graticule. A timebase speed producing a large number of cycles is often convenient as it is easier to read the magnitude. The magnitude is most easily read if the Y-shift control is adjusted to place the bottom of the trace on a major division of the graticule. The total deflection on the screen is the peak-to-peak amplitude when multiplied by the setting of the sensitivity control (volts/div). The peak amplitude is half of this and the r.m.s. value is the peak value divided by $\sqrt{2}$ or 1·414 (*i.e.* the peak-to-peak value divided by 2·828) *provided the waveform is sinusoidal.*

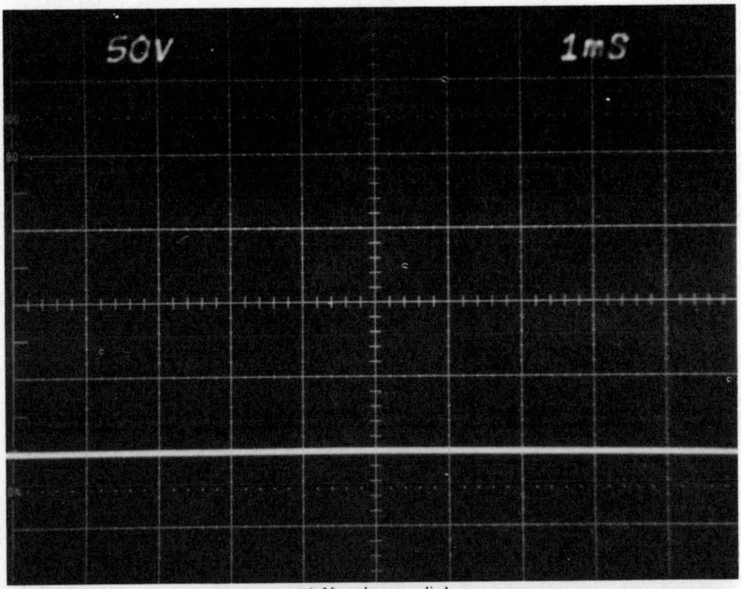

(a) No voltage applied

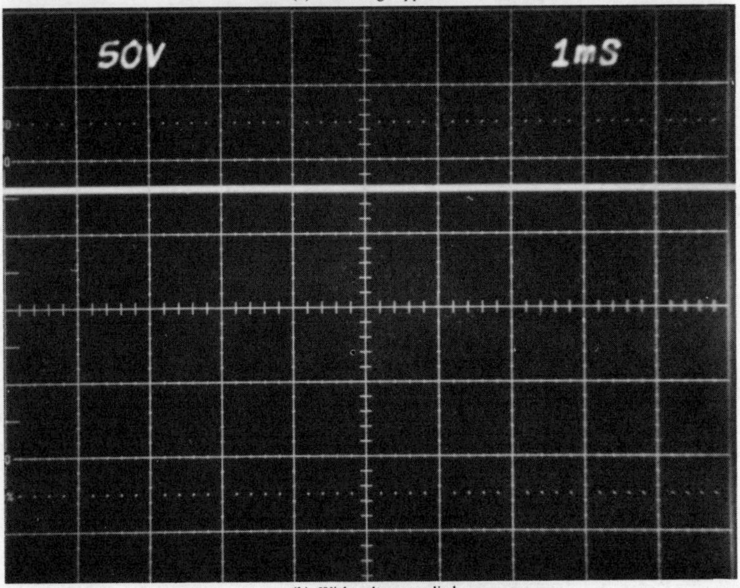

(b) With voltage applied

FIG. 13.1. USE OF OSCILLOSCOPE AS A D.C. VOLTMETER

If there is a variable gain control it MUST be set to the CALIBRATE position, and the frequency being measured must be well within the bandwidth of the oscilloscope. At the limit of the bandwidth the sensitivity will only be 70·7% of that at lower frequencies. If the a.c. input position is used, care is necessary

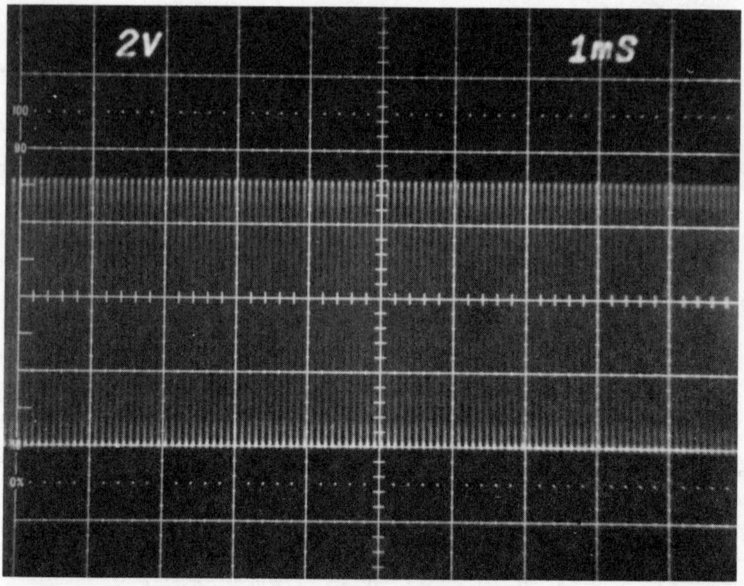

(a) Waveform for measurement of amplitude

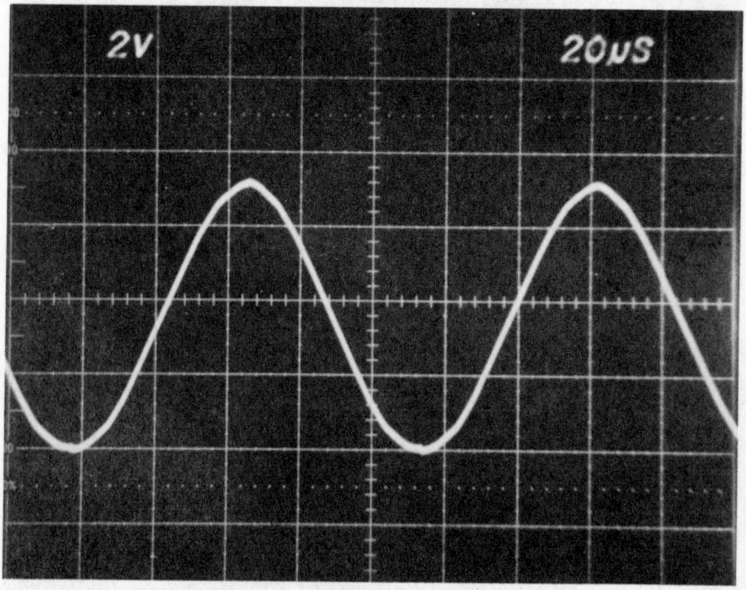

(b) Display of waveform

FIG. 13.2. USE OF OSCILLOSCOPE AS AN A.C. VOLTMETER

at low frequencies because the sensitivity will be reduced at these frequencies. Unless there is also d.c. present it is best to use the d.c. input position.

The measurement of voltage is shown in figure 13.2. At (a) is shown the trace for measuring voltage; the sweep speed is 1 ms/div and the volts/div is 2. The number of divisions covered by the trace is 3·6, hence the peak-to-peak value of the voltage is 7·2 volts. The peak value is 3·6 volts and the r.m.s. value is

$$\frac{3{\cdot}6}{1{\cdot}414} = 2{\cdot}55 \text{ volts.}$$

At (b) the sweep speed has been increased to 20 μs/div to show the actual waveform, which is approximately sinusoidal. From (b) the frequency is easily determined. One cycle covers 4·8 divisions, hence the time of a cycle is $4{\cdot}8 \times 20 = 96$ μs. The frequency is therefore

$$\frac{1{,}000{,}000}{96} = 10{,}416 \text{ Hz.}$$

Using the oscilloscope in this way enables the operator to see what is being measured (by looking at the waveform) and can often prevent false readings. An a.c. voltmeter usually measures mean voltage, but is calibrated in r.m.s. for a SINE WAVEFORM. If the waveform is not sinusoidal large errors can result. This is a most useful feature of using an oscilloscope as a voltmeter, as most misleading results are possible under certain conditions by only using an a.c. voltmeter or electronic voltmeter. If higher accuracy is required a voltmeter can be used in parallel with the oscilloscope. It is interesting to note that Grundig produce an electronic multimeter with a small built-in oscilloscope to solve just this problem.

THE OSCILLOSCOPE AS A D.C. OR A.C. AMMETER

One method is to use a current probe as described in Chapter 8. Determining the current is similar to that already described for voltage, but using the appropriate sensitivity range of the probe.

If a probe is not available, the current can be measured by using a series resistor and connecting the oscilloscope across the resistor. The resistor should be kept to a minimum value to avoid upsetting the circuit conditions. Its value will depend on the circuit, but must be such that the voltage drop across it is sufficient to give at least a reasonable deflection on the oscilloscope on the most sensitive range. The voltage across the resistor is determined as already explained and the current then calculated by Ohm's law. The value of the resistor must be known, of course. If a normal vertical amplifier is being used, it is essential that one end of the resistor is at earth potential, often possible by rearrangement of the circuit. For example, in figure 13.3 one may wish to measure the a.c. collector current, but if the resistor is placed at point P, then neither side of the resistor is at earth potential. One possibility is to place it at Q in the emitter circuit. Since the emitter current is usually almost the same as the collector current this may be satisfactory. Another possibility is to place the resistor at point R and connect the Y-input of the oscilloscope between the lower end of the resistor and the zero or earth line. There should be no a.c. voltage between the positive line and earth (which can easily be checked). If there is, it can be removed by placing a large capacitor, C_1, from the positive line to earth, as shown. In this way the a.c. voltage across the resistor is fed to the oscilloscope together with the supply voltage, the latter being easily removed by switching the oscilloscope to the a.c. input position. If it is not possible to arrange for one end of the resistor to be at earth potential then a differential amplifier may be used, as explained in Chapter 9.

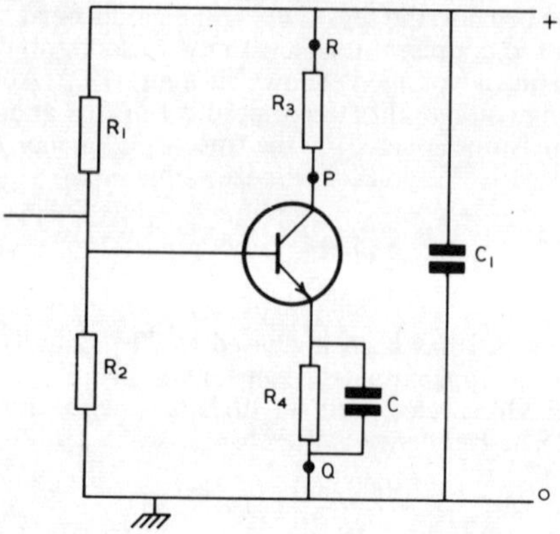

FIG. 13.3. TRANSISTOR CIRCUIT IN WHICH IT IS REQUIRED TO MEASURE THE COLLECTOR ALTERNATING COMPONENT OF CURRENT

WAVEFORMS

The most common use of the oscilloscope is to display waveforms, *i.e.* voltage against time. In this case the linear timebase is used as already described. Where a current waveform is required a current probe may be used or a series resistor, as explained in the previous section.

It is not the intention of the author to explain how the oscilloscope can be used in specific applications, *e.g.* fault-finding in an audio frequency amplifier, as the applications are so numerous. Only general points will be described. It is assumed that the reader is familiar with the equipment being tested and that he can interpret the traces obtained. In some cases the d.c. component is important, and the oscilloscope must be switched to the d.c. input position. In others it is desirable to remove the d.c. component so that a greater amplification can be used to obtain a more detailed trace of the a.c. component.

If the waveforms to be observed are at low frequencies, say up to 50 kHz, there is usually no difficulty, a probe or a direct lead may be used. The effect of capacitance is only likely to be important if the circuit is of high impedance and square waves are being considered. The difficulties become greater at the higher frequencies. If correct traces are to be obtained a probe becomes essential; it MUST be correctly adjusted and designed for the frequency under investigation. In rise-time measurements the rise time of the probe becomes important as well as that of the oscilloscope. The rise times should be small compared with that of the waveform under investigation. The graticule may have 10% and 90% marks so that if the trace is made of suitable amplitude, possibly by using the Y-amplifier variable gain control, the rise time can be read off directly. The use of the expanded X-trace or delayed timebase assists in accurate measurement of the rise time.

Some examples of waveform measurements will now be given.

(1) Pulse waveform. The waveform is a pulse at 2 MHz and it is intended to show how the trace can be expanded. Figure 13.4 shows at (a) the waveform at 0·1 μs/div sweep speed and 2 v/div vertical sensitivity. At (b) the sweep speed has been increased to 0·05 μs/div, the maximum speed of the oscilloscope. At (c) the sweep speed control is at 0·1 μs/div but the × 10 expansion is used corresponding to a sweep speed of 0·01 μs/div.

(2) This is an example showing how it is possible to obtain a misleading result. The actual waveform is the sum of two sinewaves of different frequencies and might occur with hum superimposed on a high frequency sine wave-

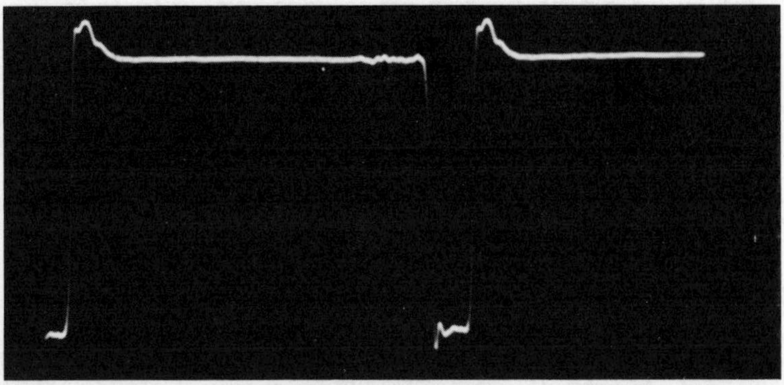

(a) 0·1 μs/div sweep speed

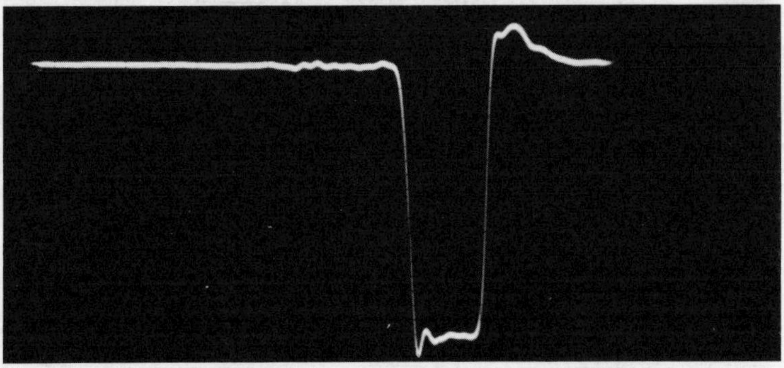

(b) 0·05 μs/div sweep speed

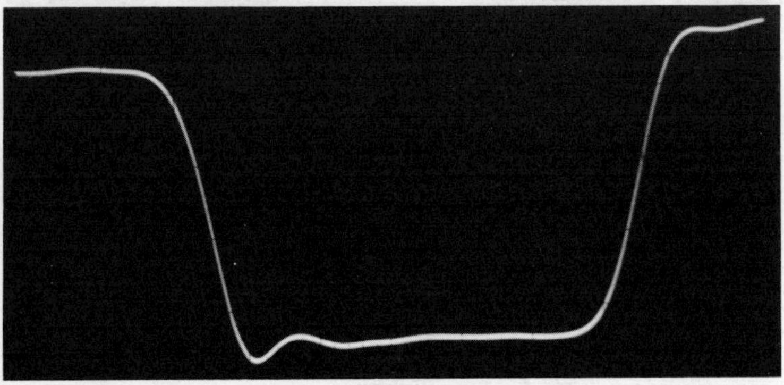

(c) 0·1 μs/div sweep speed setting but × 10 horizontal expansion used resulting in an effective sweep speed of 0·01 μs/div

FIG. 13.4. PULSE WAVEFORMS

form. The results are shown in figure 13.5. At (a) is shown the waveform at a low sweep speed so that the low frequency can be seen. The triggering was external from the low frequency source. The waveform at (b) is taken at a higher sweep speed and shows two cycles of the higher frequency. The trigger-

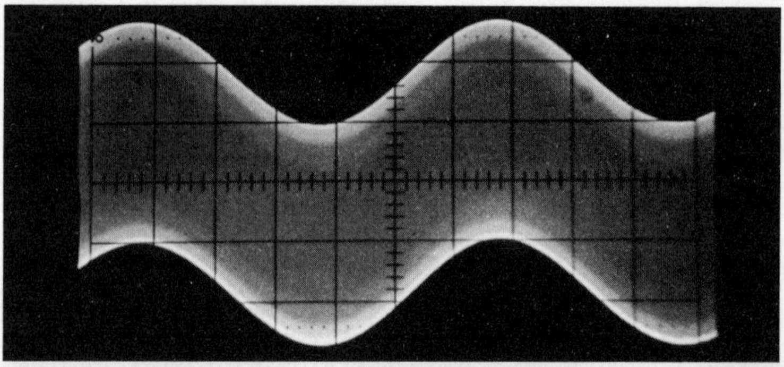

(a) Waveform at low sweep speed showing low frequency superimposed on high frequency. External triggering

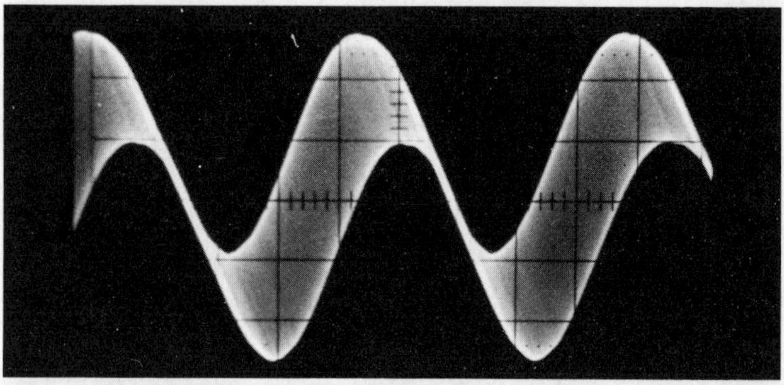

(b) At higher sweep speed showing high frequency waveform. A.C. internal triggering

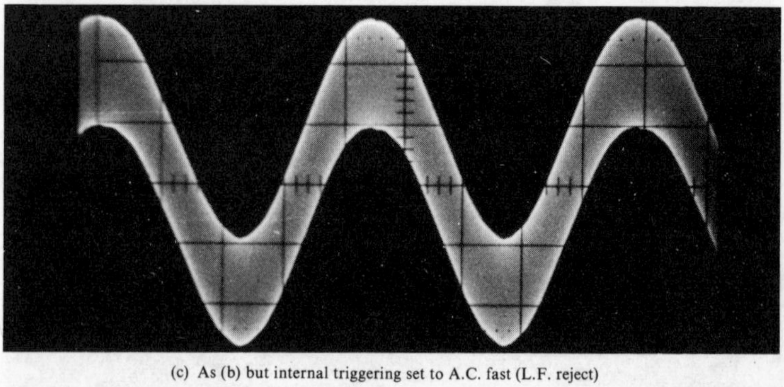

(c) As (b) but internal triggering set to A.C. fast (L.F. reject)

FIG. 13.5(a)–(c). SUM OF TWO SINE WAVEFORMS OF VERY DIFFERENT FREQUENCIES

ing is now internal using A.C. Because of the low frequency signal the instant of triggering varies, giving the result shown. What happens is shown in figure 13.6 at (a), p. 146, when it is seen that the point on the cycle where triggering occurs varies, and hence a multiple trace is obtained. At (c) the triggering was set to A.C. fast (or L.F. reject) which removes most of the low frequency signal from the triggering circuit. Since the waveform fed to the triggering circuit is now ONLY the higher frequency [as shown in figure 13.6(b)] the time-

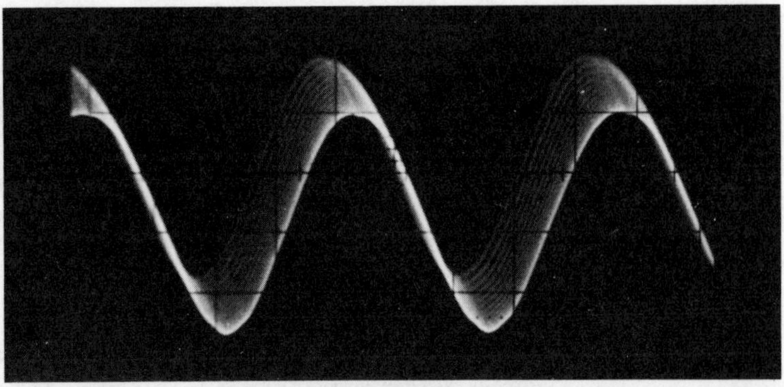

(d) As (b) but different setting of level control

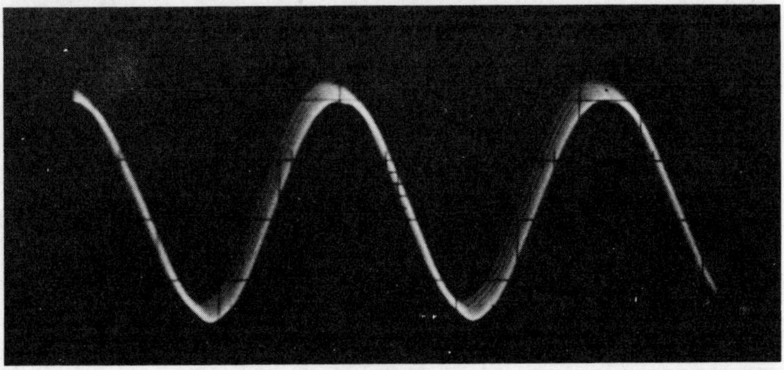

(e) As (d) but different setting of level control

FIG. 13.5(d)–(e). SUM OF TWO SINE WAVEFORMS OF VERY DIFFERENT FREQUENCIES

base is triggered at the same point on each cycle. However, the low frequency signal, which is still applied to the Y-deflection, moves the trace up and down. This is shown in figure 13.5 at (c). At (d) and (e) is shown the waveform obtained by two different settings of the level control, the triggering and other conditions being as (b). The reason for this result is shown in figure 13.6 at (c) and (d), when it is seen that by raising the trigger level the timebase only triggers on certain cycles and at (d) only a few cycles when the waveform has its maximum positive amplitude. Because it is triggered at fairly long intervals the brightness decreases, but does not show in the photographs as it has been corrected for by altering the brightness control.

(3) Television waveforms. These are examples to show the excellent results that can be obtained by using the facilities of a modern high-frequency oscilloscope. To look at line pulses in a composite waveform the sweep speed must be such that the total sweep time is one or two lines. The time of a line (on 625 lines) is 64 μs, hence, say, a sweep speed of 10 μs/div would display a line over 6·4 division. Such a display is shown in figure 13.7 (p. 147). The level control is set so that the timebase is triggered by the synchronizing pulses, and by using positive slope it is triggered by the leading edge as shown. Since this waveform corresponds to an actual television picture the vision portion is blurred, as it consists of a number of different lines superimposed.

To look at individual lines or field synchronizing pulses the sweep speed must be reduced so that most of one PICTURE period is covered by one sweep.

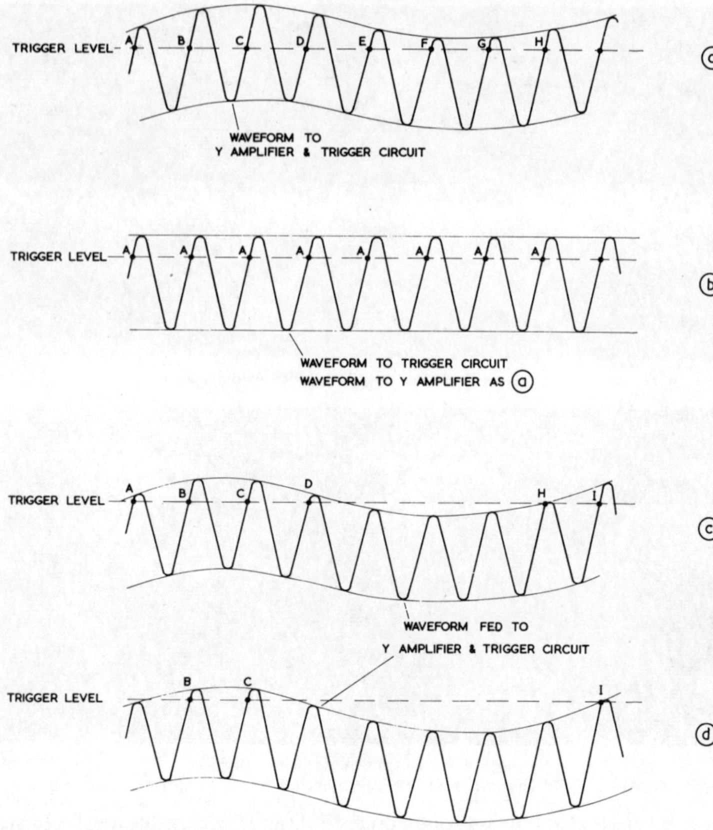

FIG. 13.6. EXPLANATION OF WAVEFORM OBTAINED IN FIGURE 13.5

(a) Instant of triggering varies as the high frequency waveform moves up and down

(b) Waveform fed to triggering circuit when set to A.C. fast (L.F. reject)

(c) Effect of varying level control so that triggering only takes place on a few cycles

(d) As (c) but higher trigger level

It could be set to display more than this, but when the expanded trace is used the brightness is reduced, hence the faster the sweep of timebase A the better (the faster it repeats the better, but it cannot repeat at a frequency more than 25 times/second). The picture time is $\frac{1}{25}$th second or 40 ms, therefore a sweep speed of, say, 5 ms/div would be suitable. The variable sweep control may be used to obtain the required sweep speed as the calibration is not required. Such a display is shown in figure 13.8(a). Unless the oscilloscope has a TV field (or TV frame) synchronizing position, the timebase is best triggered externally. If a television receiver is being used there is no difficulty as such pulses can easily be obtained from the field timebase or field output stage. This display shows little of the synchronizing pulses, and individual lines cannot be resolved. To get a better display the delayed sweep feature should be used. (If this is not available some improvement is possible by using the maximum X expansion). Having made sure that the trigger is steady, the oscilloscope is then switched to 'A intensified by B', and the sweep speed of B

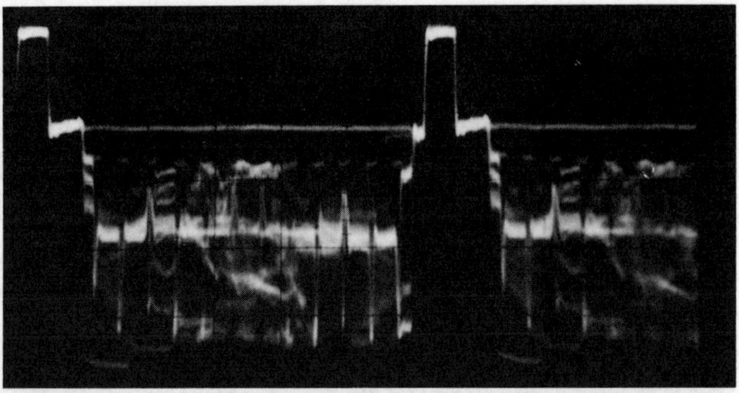

FIG. 13.7. TELEVISION WAVEFORM SHOWING SUPERIMPOSED LINES

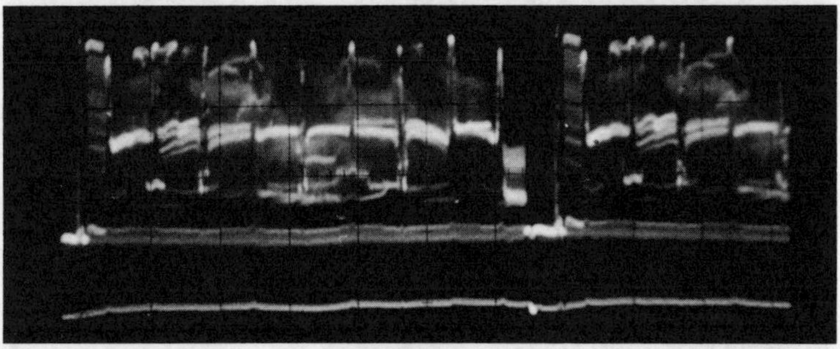

(a) Timebase A only

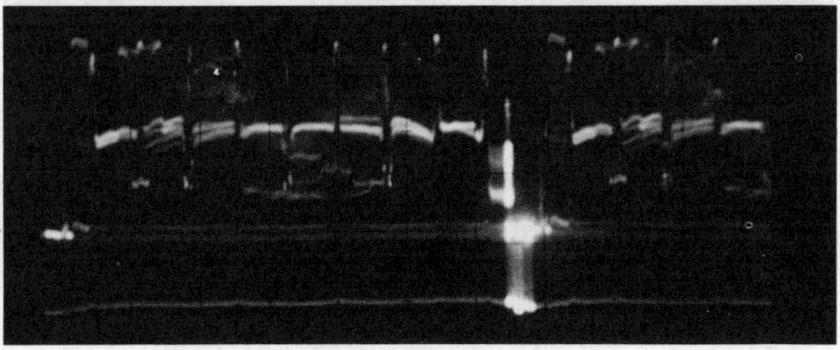

(b) Timebase A intensified by timebase B

FIG. 13.8(a)–(b). TELEVISION WAVEFORM FROM 625-LINE RECEIVER

adjusted so that it covers a few lines. The position of the intensified portion is moved by the DELAY TIME MULTIPLIER to the position required, as shown at (b). The oscilloscope is now switched to B delayed by A and the result is (c), where a clear display of the equalizing pulses and field pulses is shown. As the sweep speed is high but the repetition rate low (25/s) the brilliance may have to be increased. One field is shown at (c). It is pure chance which one is displayed as it depends on which field the oscilloscope happens to trigger. The other field may be obtained by disconnecting the trigger pulse for a brief time, and

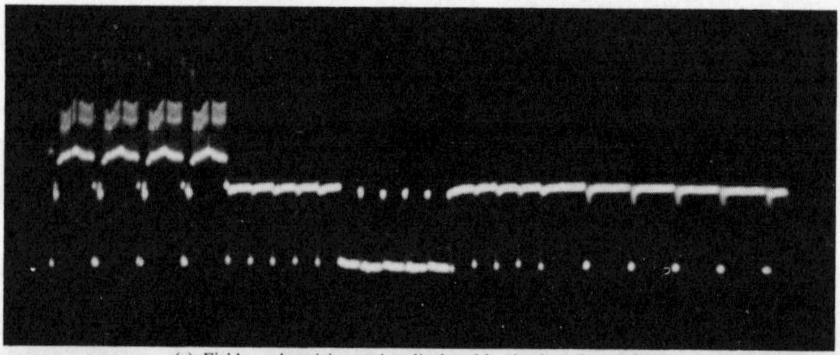

(c) Field synchronizing region displayed by timebase B, one field

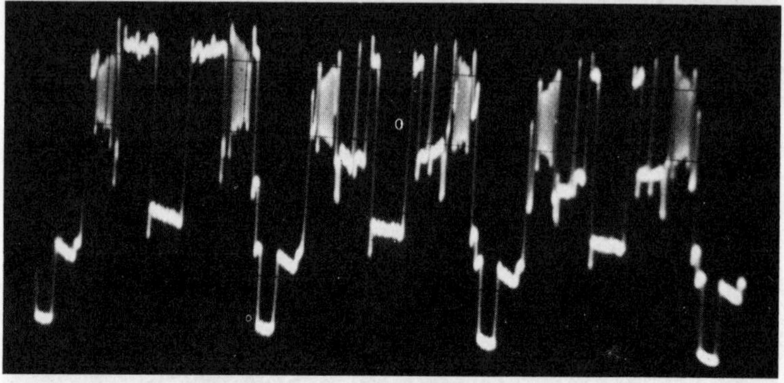

(d) A few lines displayed by timebase B

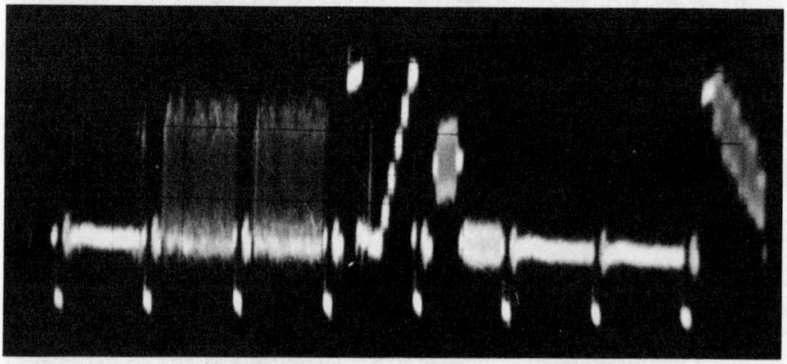

(e) Ceefax and test waveform displayed by timebase B

FIG. 13.8(c)–(e). TELEVISION WAVEFORM FROM 625-LINE RECEIVER

repeating until the required set of pulses is obtained. A few lines are shown at (d), obtained by using the DELAY TIME MULTIPLIER control and increasing the sweep speed of timebase B. At (e) the CEEFAX waveform is shown on the left-hand side (2 lines) and the colour test signal on the right-hand side (2 lines). These were obtained from a receiver on a BBC transmission. The CEEFAX waveform is blurred because it was continually changing and it changed during the exposure time.

On an oscilloscope with a bandwidth of at least 50 MHz it is possible to display the actual modulated signal at intermediate frequency. The same procedure is used, but the Y-input is fed from the last stage of the i.f. amplifier. Waveforms are shown in figure 13.9. At (a) is the display by timebase A, the television signal being 625 lines, and therefore the modulation is negative. At (b) is the field synchronizing pulse region where the equalizing and field pulses can easily be seen. A few lines of an actual picture are shown at (c).

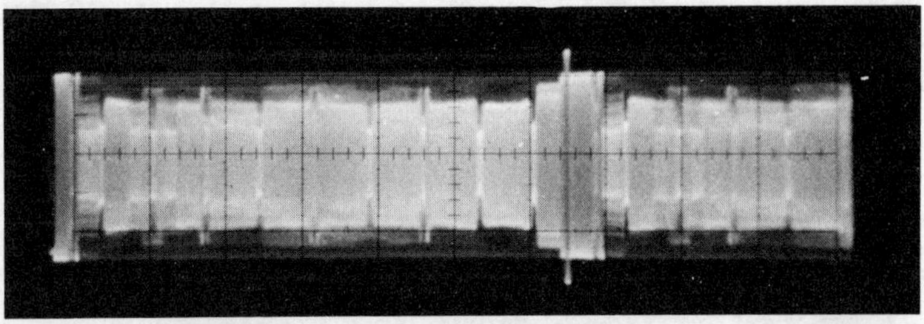

(a) Timebase A only showing about $\frac{3}{4}$ of a picture period 625-line

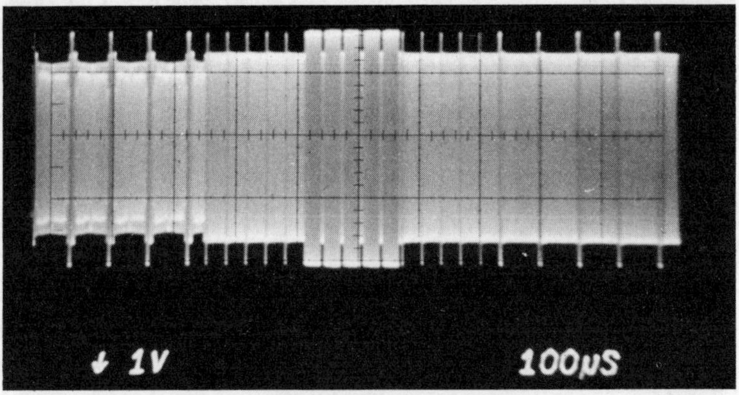

(b) B delayed by A showing field synchronizing pulses, expanded trace, 625-line

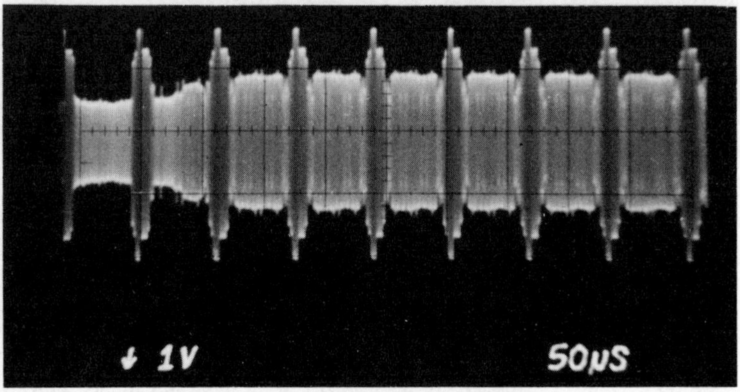

(c) B delayed by A showing a few lines of a picture, expanded trace, 625-line

FIG. 13.9(a)–(c). TELEVISION I.F. WAVEFORM

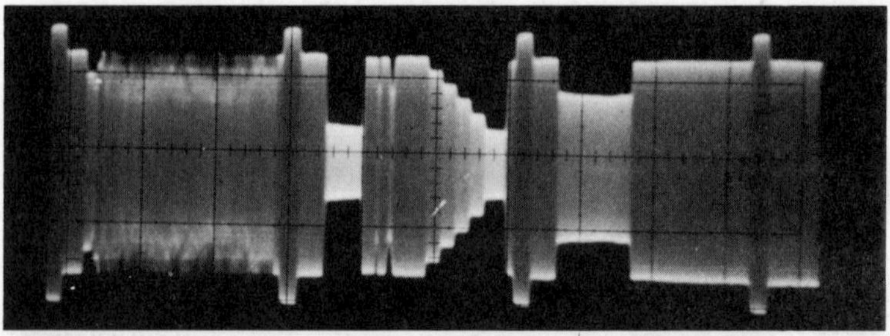

(d) B delayed by A showing Ceefax and test signals, expanded trace, 625-line

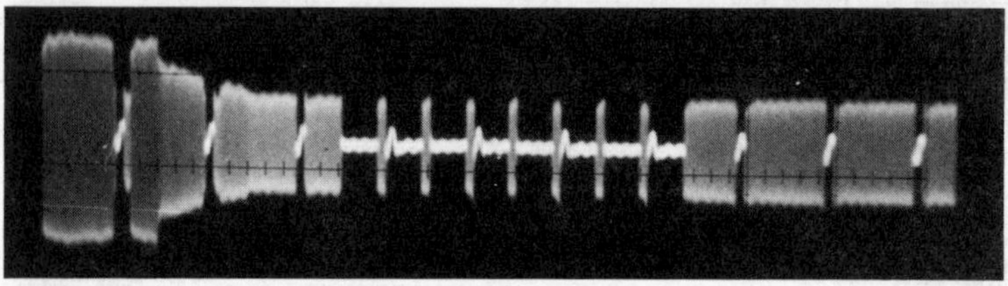

(e) 405-line i.f. waveform: B delayed by A showing field synchronizing pulses (positive modulation, 405-line system)

FIG. 13.9(d)–(e). TELEVISION I.F. WAVEFORM

The Ceefax transmission and test signal are shown at (d). If it is a 405-line transmission the modulation is positive and the synchronizing pulse region of such a transmission is shown at (e).

The effect of delayed timebase operation is shown in the case of integrated synchronizing pulses in figure 11.4.

(4) Audio frequency amplifier waveforms. Figure 13.10 shows the output of an audio frequency amplifier with overloading just occurring. Since this clipping is on one half-cycle only it means that the balance control (it is a quasi-complementary amplifier) was not set correctly, *i.e.* the steady output voltage was not half way between the two supply lines.

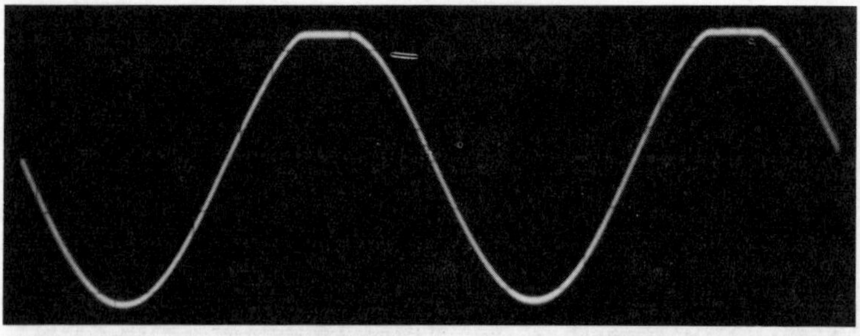

FIG. 13.10. OUTPUT OF AUDIO FREQUENCY AMPLIFIER SHOWING OVERLOADING

In figure 13.11 is shown the result of square wave testing of an amplifier. The frequency was 1 kHz, and (a) shows the output with the tone control set to the central position. A good square wave output is obtained because the response is uniform over a large frequency range. If the response is not uniform there is distortion of the square waveform. In the following waveforms the response is made non-uniform by altering the setting of the tone controls. The output waveform with the top response increased is shown at (b), and at

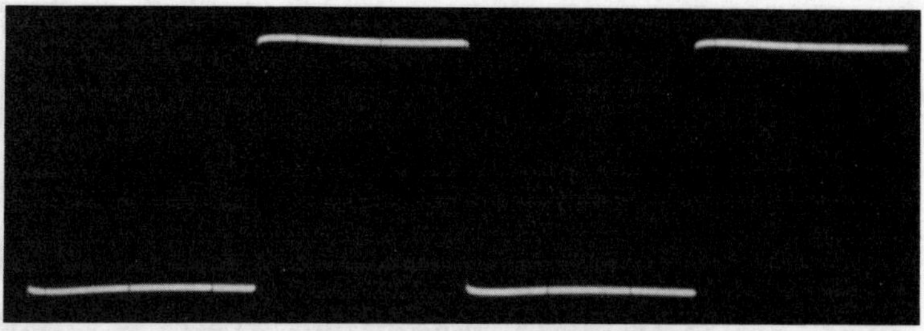

(a) Output with tone controls in central position, 1 kHz input

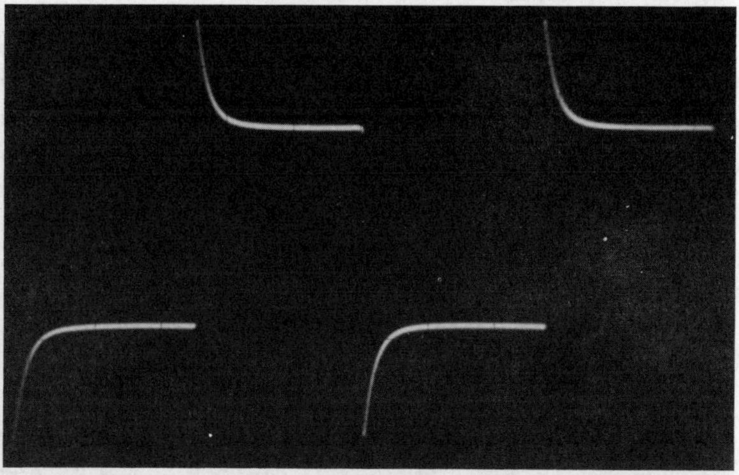

(b) Output with increased top response, 1 kHz input

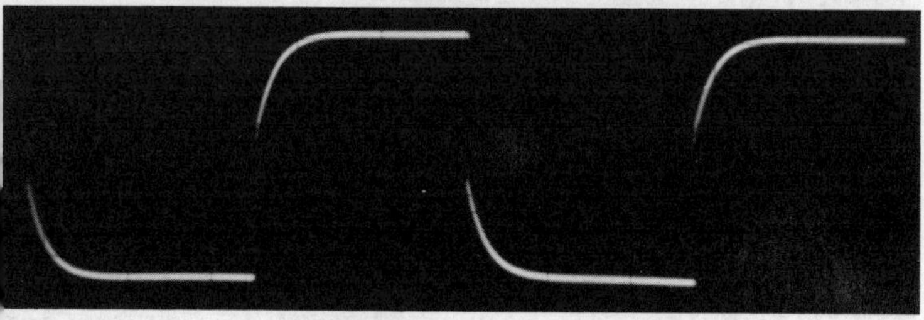

(c) Output with decreased top response, 1 kHz input

FIG. 13.11(a)–(c). SQUARE WAVE INPUT TO AUDIO AMPLIFIER

(c) it is reduced. At (d) is shown the effect of increasing the bass response, and at (e) the effect of reducing the bass response. Distortion does occur at the frequency limits of the amplifier, even with the response set to level. That at (f) is at 10 kHz when it is seen that there is some rounding of the square wave due to the limited high frequency response; and (g) is the same result at 100 Hz, when a slope occurs on the waveform due to the limited response at very low frequencies.

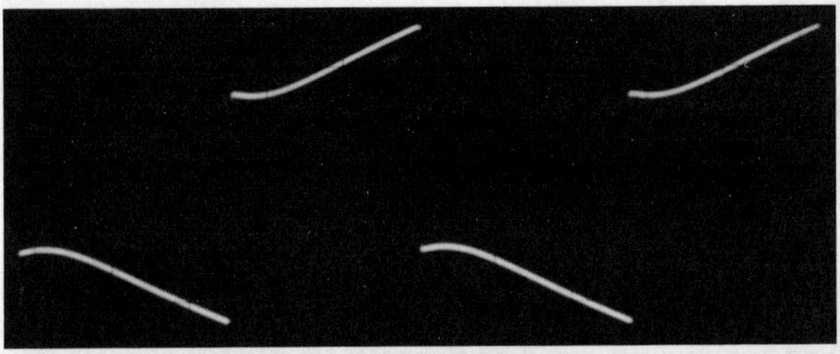

(d) Output with increased bass response, 1 kHz input

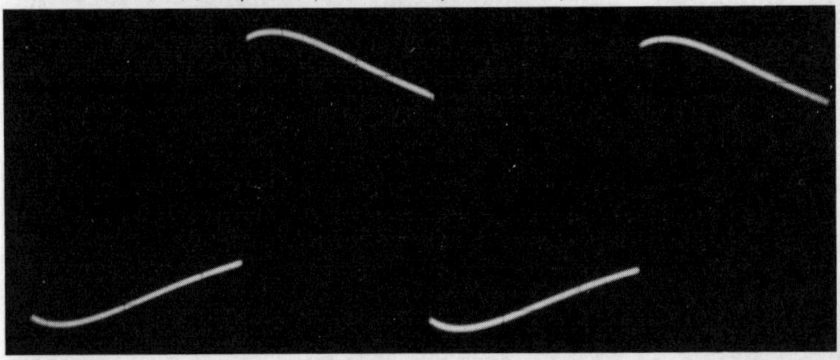

(e) Output with decreased bass response, 1 kHz input

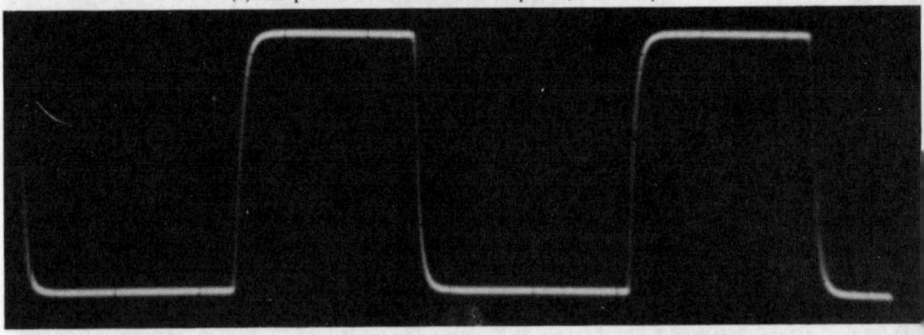

(f) Output with input of 10 kHz, tone controls in central position

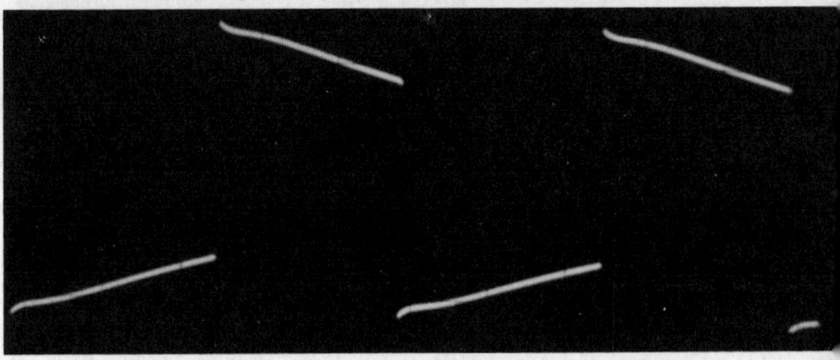

(g) Output with input of 100 Hz, tone controls in central position

FIG. 13.11(d)–(g). SQUARE WAVE INPUT TO AUDIO AMPLIFIER

FREQUENCY MEASUREMENT or TIME MEASUREMENT

The time of 1 cycle can be read off from the graticule by finding the number of divisions between two corresponding points on the waveform (*i.e.* one cycle) and multiplying by the appropriate time/div of the timebase setting. If there is a variable control it MUST be set to CALIBRATE. Greater accuracy is obtained by making the cycle cover as many horizontal divisions as possible, but NOT using the variable sweep speed control. To obtain a more accurate result the oscilloscope is switched to 'A intensified by B' and the B timebase sweep speed increased until only a small part of the trace is intensified. The TIME DELAY MULTIPLIER is now adjusted until the bright portion corresponds to a definite point on the cycle on the left-hand side of the screen. For example, if a sine waveform the zero crossing point, and if a pulse type waveform a point having a sharp rise or fall. The reading of the TIME DELAY MULTIPLIER is noted, say 2·63. The TIME DELAY MULTIPLIER is now moved until the brightened part is at a *corresponding* part of the waveform one cycle later. The reading is now noted, say 8·32. The sweep speed of the A timebase is noted, say 100 μs/div. The time of the cycle is therefore $(8{\cdot}32-2{\cdot}63)\times 100\ \mu s = 569\ \mu s$. The frequency is therefore

$$\frac{1}{569}\times 10^6 = 1757 \text{ Hz.}$$

FREQUENCY COMPARISON

Frequency comparison may be done on a double-trace oscilloscope by feeding one signal to one input and the other frequency to the other input.

If only an approximate comparison is required, the timebase should be set to be triggered by ALT signals (when this is possible).

The number of cycles on one trace can be compared with those on the other, which is explained in Chapter 10. This can be used for any two frequencies and one need not be a multiple of the other.

The same method may be used to adjust one frequency until it is some exact multiple of the other, and adjusted until the ratio is correct so far as can be seen on this display. The timebase should now be switched so that it is triggered by signal A. Display A will remain locked, but display B will move (unless one is an EXACT multiple of the other). The frequency of B can now be varied until display B is stationary and having the correct ratio to display A. For example, if a ratio of 3 to 1 is required there should be 3 cycles of B corresponding to 1 cycle of A. This is a very accurate method but difficult if the frequencies are high and/or the sources of the frequencies are not very stable.

A single-trace oscilloscope can be used in a similar way. The lower frequency is fed to the Y-amplifier and to the external trigger input (the timebase being set to manual trigger and correctly adjusted), and the timebase adjusted to display at least one complete cycle. The number of graticule division corresponding to one cycle is noted. This signal is now removed from the Y-input and the other signal (assumed of higher frequency) fed to the Y-input; but the first signal must still be fed to the trigger input so that the timebase is still triggered by the first signal.

The second frequency is now adjusted until the waveform is stationary, or nearly so, and such that the required number of cycles occupy the same number of divisions on the graticule as one cycle of the other frequency.

An alternative way is to feed one signal to the Y-plates and the other to the X-plates when Lissajous figures are produced. Some are shown in figure 13.12. As one frequency changes relative to the other (assuming the frequencies are not locked to each other) the figures change. Exact multiples up to a ratio of about 10 are fairly easy to see, depending on how stable the frequencies are. The ratio is given by the number of loops contacting one side relative to the number of loops contacting a side at right angles to the first. It is important not to miss any of the loops. In certain phase relationships the loops are superimposed. Ratios other than exact multiples can be detected in the same

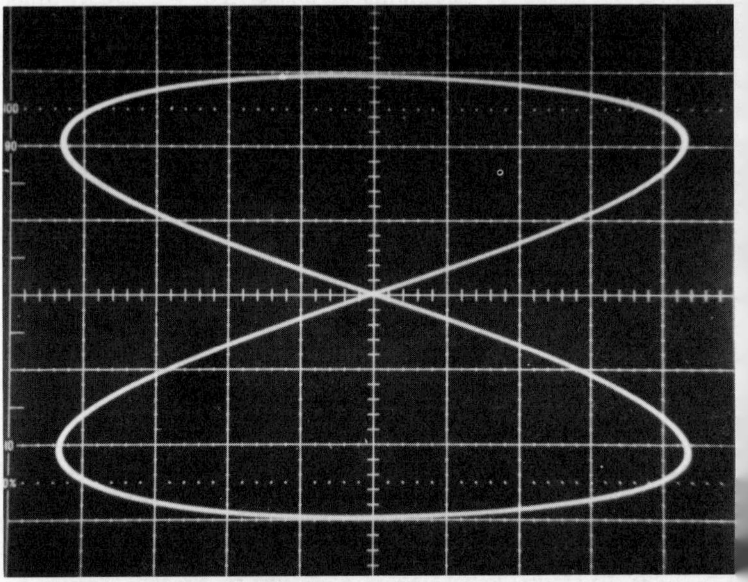

(a) Ratio 2/1

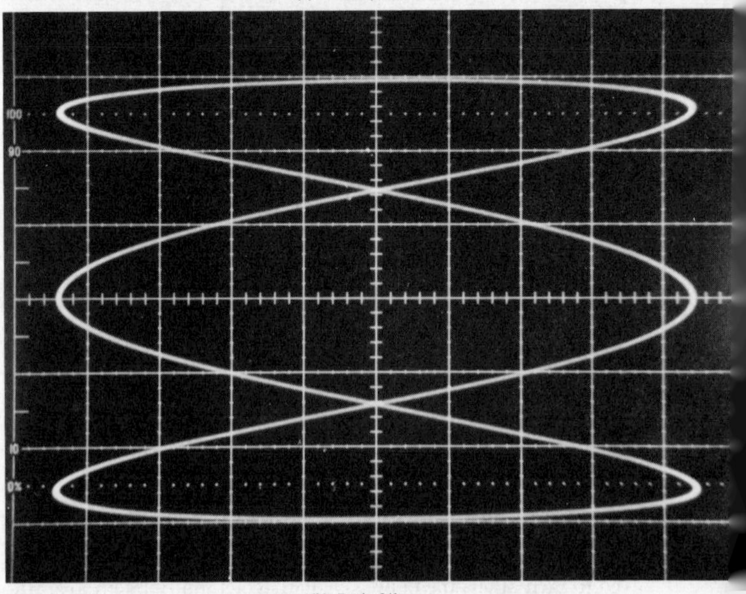

(b) Ratio 3/1

FIG. 13.12(a)–(b). LISSAJOUS FIGURES

way, but stable frequencies are necessary for the more complex figures. Very accurate comparison is possible using this method.

In figure 13.12 at (a) the ratio is 2/1, while at (b) it is 3/1. At (c) the ratio is

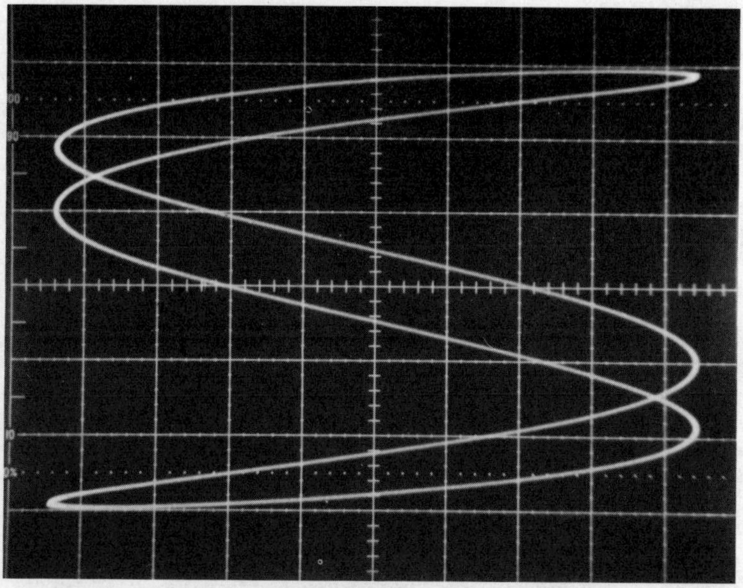

(c) Ratio 3/1

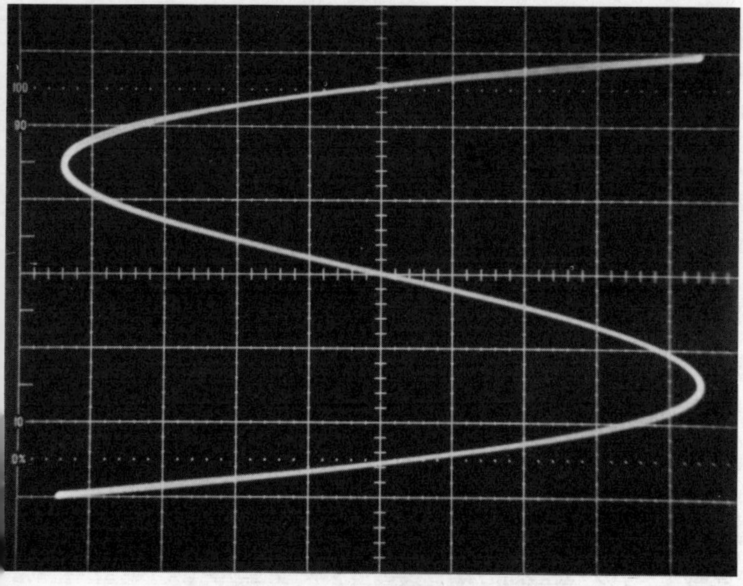

(d) Ratio 3/1

FIG. 13.12(c)–(d). LISSAJOUS FIGURES

still 3/1, but the phase relationship is different. At (d) is shown a misleading result, this actually being 3/1 because the two traces as in (c) are superimposed. At (e) is a ratio of 3/2, while at (f) is shown a ratio of 10/1, which is only possible

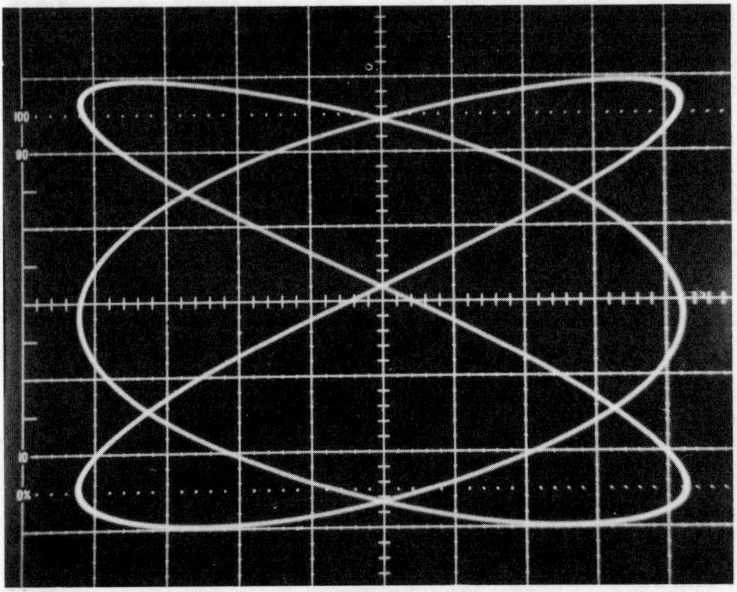

(e) Ratio 3/2

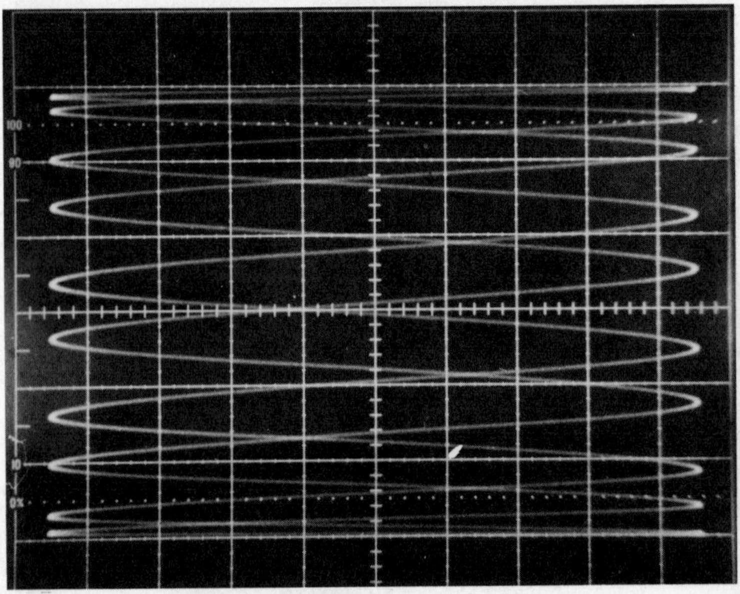

(f) Ratio 10/1

FIG. 13.12(e)–(f). LISSAJOUS FIGURES

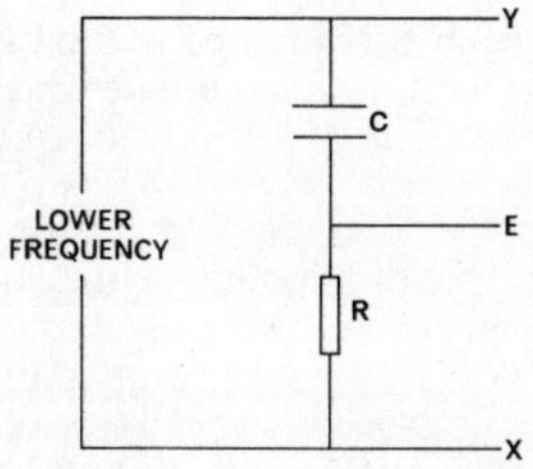

Reactance of C = R

FIG. 13.13. CIRCUIT TO PRODUCE CIRCULAR TRACE

to use in practice if the two frequency sources are stable.

An alternative is to generate a circular trace, as in figure 13.13, from the lower frequency. The higher frequency is fed to the Z-modulation socket of the oscilloscope so that the circle is modulated in brightness. If the ratio of the frequencies is x then x bars will appear. Higher multiples are possible this way if the frequencies are stable, but care is needed to avoid being confused by bars caused by ratios other than exact multiples which give a similar trace. A display is shown in figure 13.14, the ratio being 17/1.

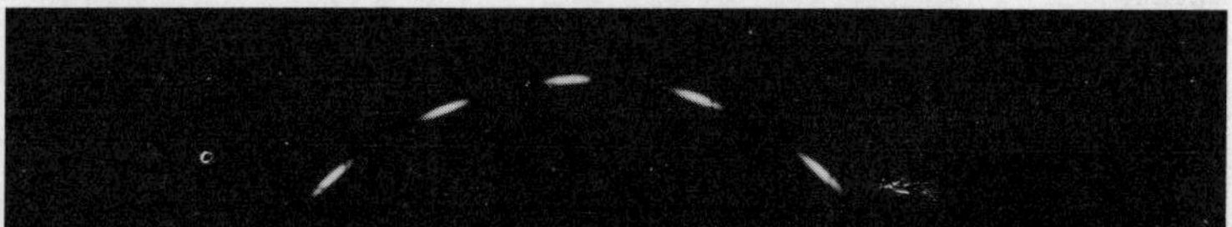

FIG. 13.14. Z-MODULATION OF CIRCULAR TRACE BY HIGHER FREQUENCY

PHASE DIFFERENCE MEASUREMENT

This is most easily done using a double trace (or multitrace) oscilloscope. It is preferable to use the d.c. input so that phase shifts are not produced by the C-R input circuit as they are when the oscilloscope is switched to a.c. It is particularly important not to select D.C. on one input and A.C. on the other. It is equally important not to use a probe on one circuit and not on the other. Each channel must be connected by similar means, say two identical probes. When a switched beam oscilloscope is used either a chopped trace (at low frequencies) must be used, or an alternate trace *provided the timebase is triggered by one signal only*. The ALTERNATE trace triggering *must not* be used because the phase relationship between the two traces is then lost (see figure 10.5). The two traces are superimposed one on top of the other. The time difference of two corresponding points of the waveforms is noted and compared with the time of one cycle. The most common point to use is the zero crossing point, but, of course, the waveforms must be positioned so that they are exactly symmetrically placed vertically in relation to the graticule line used as the zero reference. An example is shown in figure 13.15(a). The total number of divisions for one cycle is 6·2 divs, and the number representing

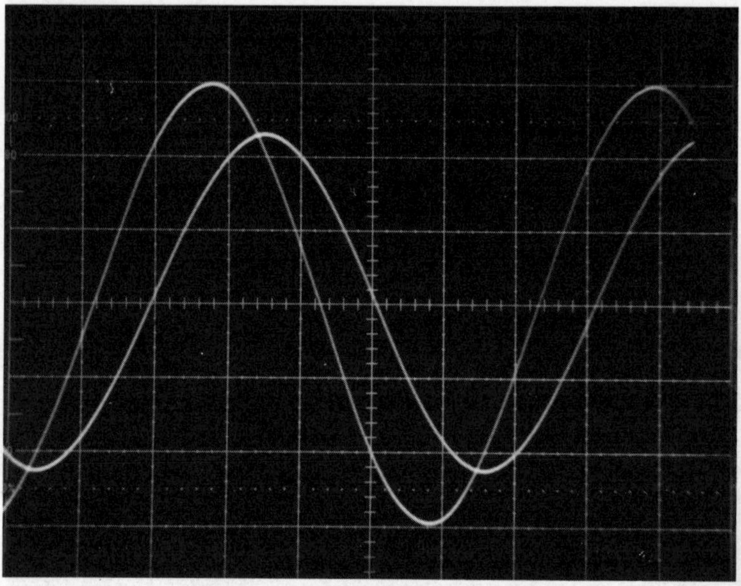

(a) Normal display

FIG. 13.15. MEASUREMENT OF PHASE ANGLE

the phase difference is 0·8 divs. The total number for one cycle represents 360°, and hence the phase angle is

$$\frac{0{\cdot}8}{6{\cdot}2} \times 360 = 46{\cdot}5°.$$

Greater accuracy can be obtained by expanding the trace by, say, 10 to measure the time difference between the two traces if the difference is small. This is shown in figure 13.15(b). A result within 2 or 3 degrees is possible using this method.

If a dual-time base is available, intensified markers and the TIME DELAY MULTIPLIER may be used as explained earlier in the chapter for the measurement of frequency. Two measurements need to be made: the time of a complete cycle; and the time difference between corresponding points on the two waveforms.

The phase angle of sinusoidal voltages can be measured without using a linear timebase or a double-trace oscilloscope. In this case one input is fed to the Y-amplifier and the other to the X-amplifier. For accurate results the X and

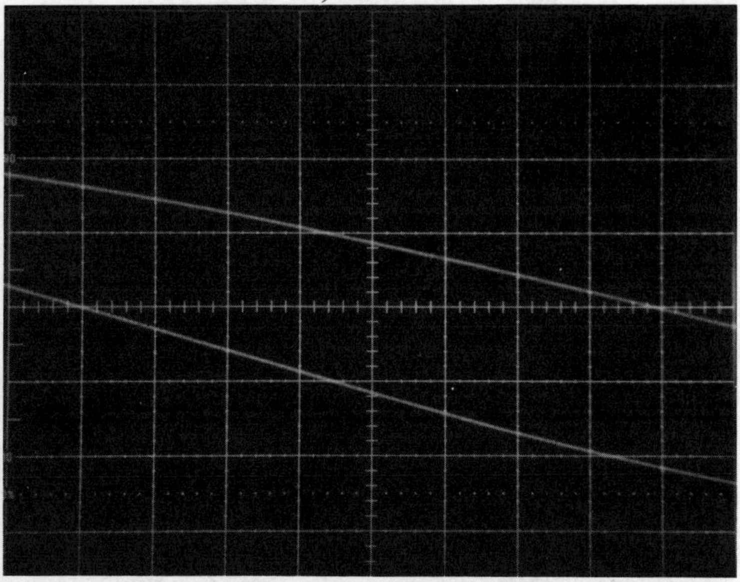

(b) Use of × 10 horizontal expansion

FIG. 13.15. MEASUREMENT OF PHASE ANGLE

Y amplifiers should be similar, or have similar phase-shift characteristics. Some double-trace oscilloscopes are made so that one of the Y-amplifiers can be used as an X-amplifier. If there is no phase difference between the inputs a tilted line will result, as in figure 13.16(a). The slope will depend on the

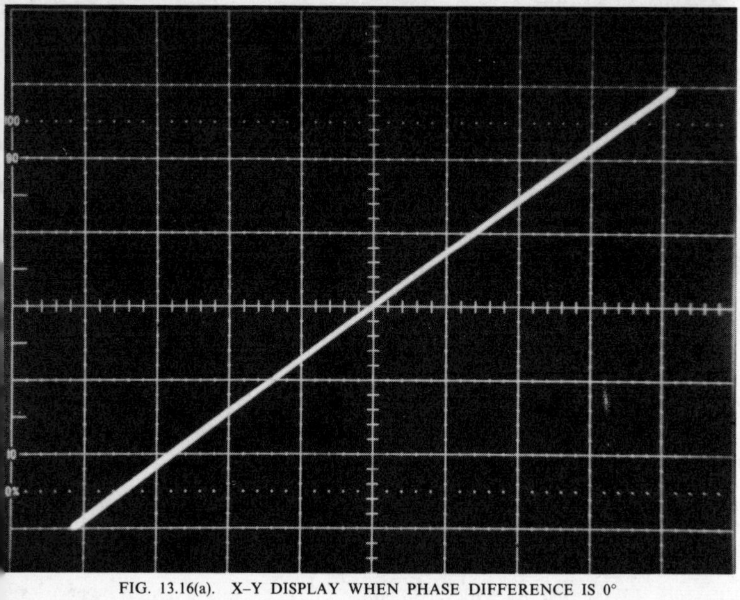

FIG. 13.16(a). X–Y DISPLAY WHEN PHASE DIFFERENCE IS 0°

relative amplitudes of the two inputs, and the X and Y sensitivity settings of the oscilloscope. If there is some phase difference (about 24°) then the trace will be an ellipse, as in figure 13.16(b). The ellipse will get thicker with increase in phase difference, as shown at (c), for an angle of about 57°, and at 90°

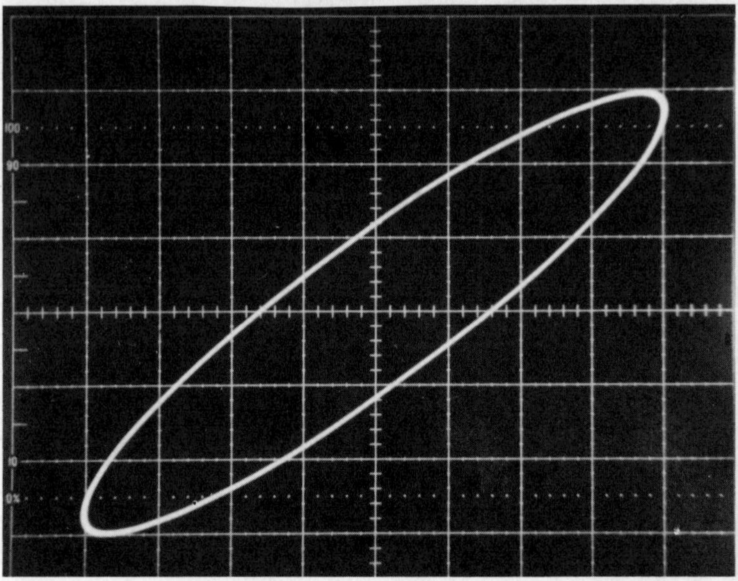

FIG. 13.16(b). X–Y DISPLAY WHEN THERE IS A PHASE DIFFERENCE (APPROX. 24°)

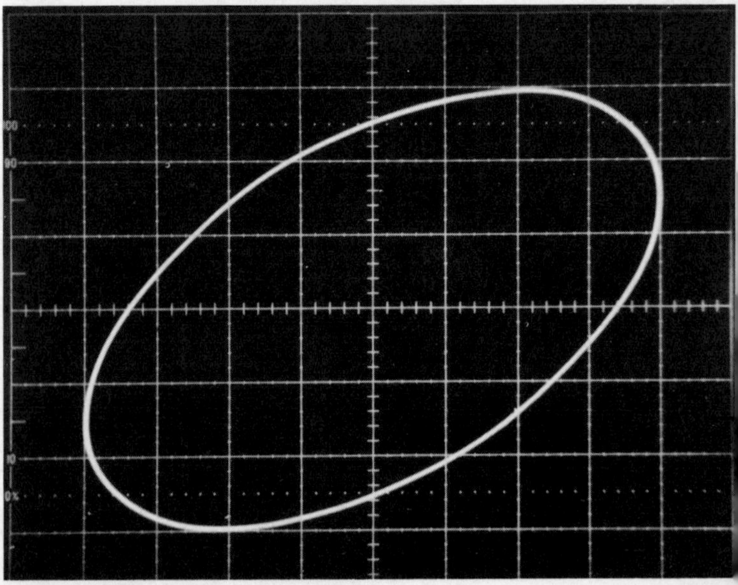

FIG. 13.16. (c) X–Y DISPLAY WHEN THE PHASE DIFFERENCE IS GREATER (APPROX. 57°

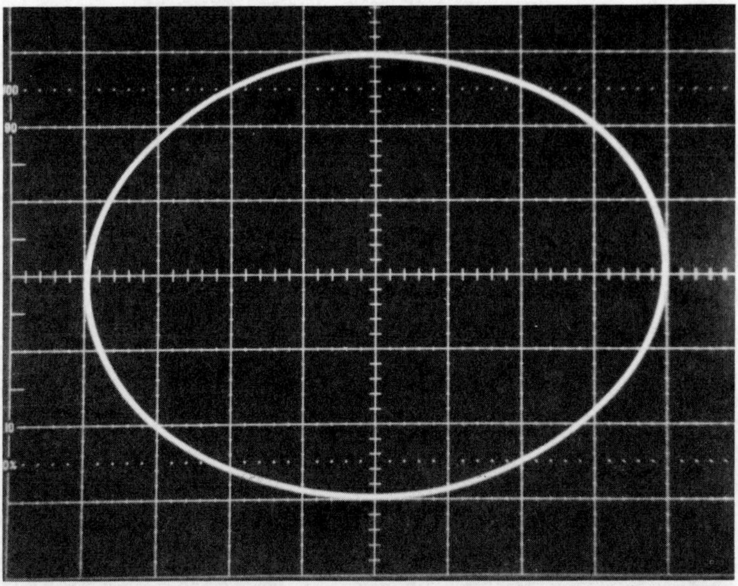

FIG. 13.16(d). X–Y DISPLAY WHEN THE PHASE DIFFERENCE IS 90°

phase difference an ellipse is produced as at (d). A circle is produced if the X and Y deflections are of equal magnitude. The phase angle can be obtained in two ways:

(a) With reference to figure 13.17: $\sin \phi = \dfrac{C}{A}$

(*i.e.* the amplitude of X, when Y is zero, compared with the maximum amplitude of X).

(b) With reference to figure 13.17: $\cos \phi = \dfrac{B}{A}$

(*i.e.* the amplitude of X, when Y is a maximum, compared with the maximum value of X). This is most accurate at large phase angles where B changes more rapidly with change of angle.

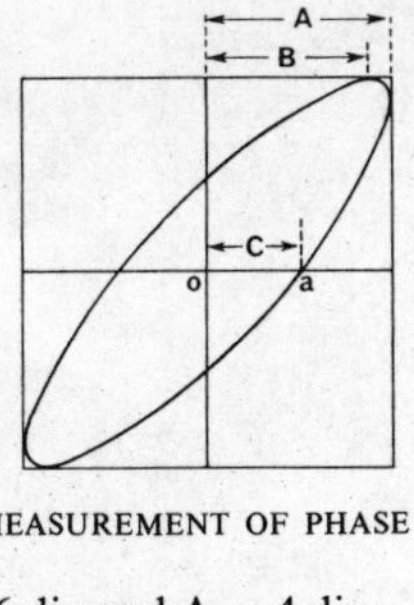

FIG. 13.17. MEASUREMENT OF PHASE DIFFERENCE

In figure 13.16(b) C = 1·6 div and A = 4 div.

Thus $\sin \phi = \dfrac{1{\cdot}6}{4} = 0{\cdot}4$ and $\phi = 24°$.

In figure 13.16(c) C = 3·4 div and A = 4 divisions

hence $\sin \phi = \frac{3{\cdot}4}{4} = 0{\cdot}85$ and $\phi = 58°$. Using the alternative method B = 2·2 divs and A = 4 divs

hence $\cos \phi = \frac{2{\cdot}2}{4} = 0{\cdot}55$ and $\phi = 57°$ (both to the nearest degree).

The shape of the ellipse is the same for both leading and lagging phases.

MODULATION DISPLAY

The X-Y type of display can be used for indicating the percentage modulation of a signal generator or transmitter, and for showing distortion. The modulated r.f. output is fed to the Y-plates and the modulating voltage to the X-plates. Such displays are shown in figure 13.18. At (a) the percentage modulation is about 50%. If V_2 is the maximum value of the display (on the left-hand side) and V_1 the minimum value (on the right-hand side) then the percentage modulation is given by

$$\frac{V_2 - V_1}{V_2 + V_1} \times 100 \quad \%$$

At (b) is shown the result of almost 100% modulation. This also shows some distortion because the sides are not straight. The loop effect is due to

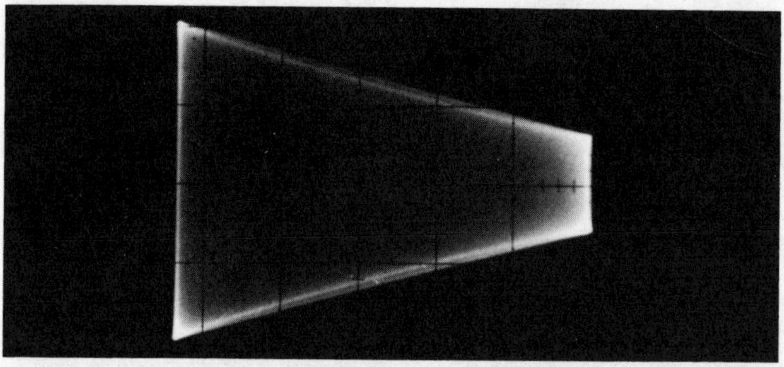

FIG. 13.18(a). MODULATION APPROX. 50%

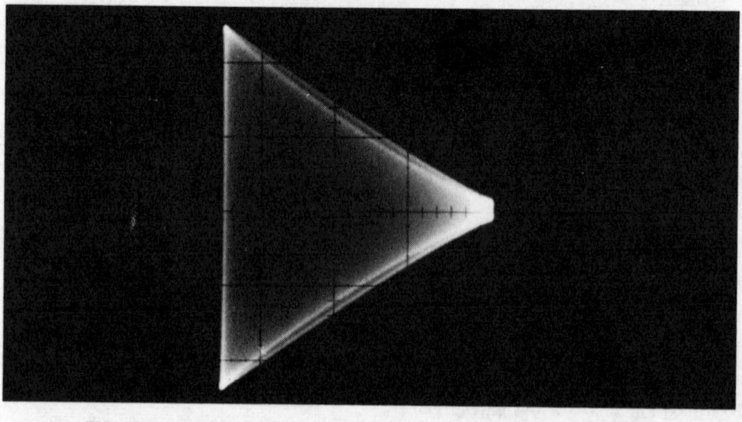

FIG. 13.18(b). MODULATION APPROX. 100% WITH SOME DISTORTION

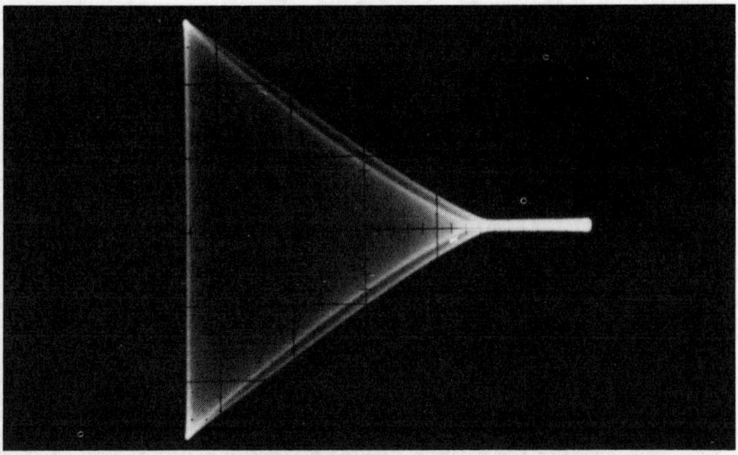

FIG. 13.18(c). OVERMODULATION SHOWING SERIOUS DISTORTION

phase shift and not usually of importance. At (c) is the effect of over-modulation with corresponding distortion.

DISPLAY OF CHARACTERISTICS OF DEVICES

It is possible to display the characteristics of many devices on the screen of an oscilloscope. For example, take the circuit of figure 13.19. The voltage across the device is fed to the X-amplifier and produces a horizontal voltage

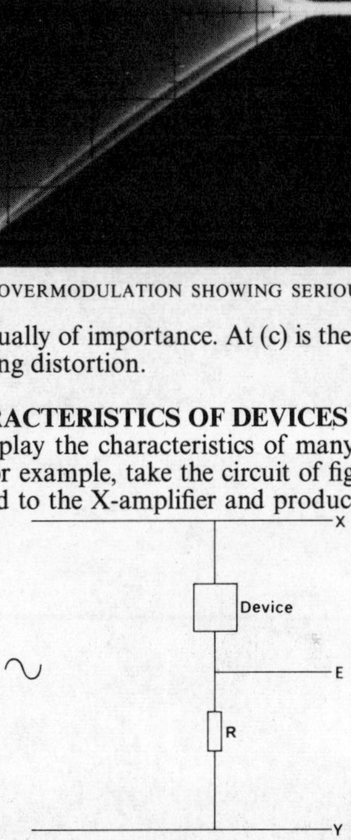

FIG. 13.19. CIRCUIT FOR DISPLAYING CHARACTERISTICS OF DEVICES

axis. The voltage fed to the Y-plates is that across R and proportional to current. Thus the vertical axis becomes a current axis. When an alternating voltage (earth-free) is connected as shown, a continuous display of the characteristic of the device is given. The frequency should be low (50 Hz is convenient) to avoid phase shifts due to capacitances. Obviously, the supply voltage must not be such as to damage the device (with either polarity) and R should be of such a value as to limit the current so that no damage is done to the device.

Figure 13.20 shows some characteristics obtained in this way. At (a) is the characteristic of a silicon diode. The centre-line and the second major graticule line are the zero axes. The bottom right-hand figures indicate the volts/division of the X axis, the arrow only indicating that the INVERT switch was operated. The top left-hand figures indicate the volts/division of the vertical axis, and since the resistor R was 1000 Ω the vertical scale is 2 mA/div. At (b) is a similar characteristic for a germanium diode where it will be seen that the

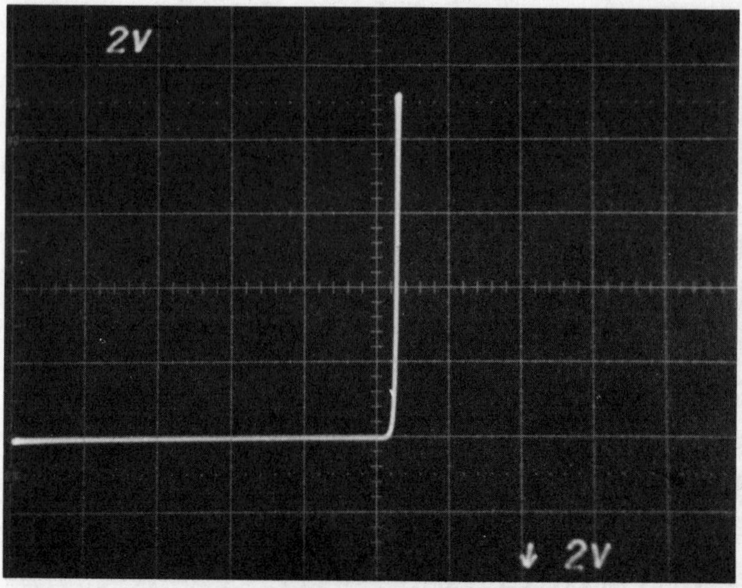

(a) Silicon rectifier

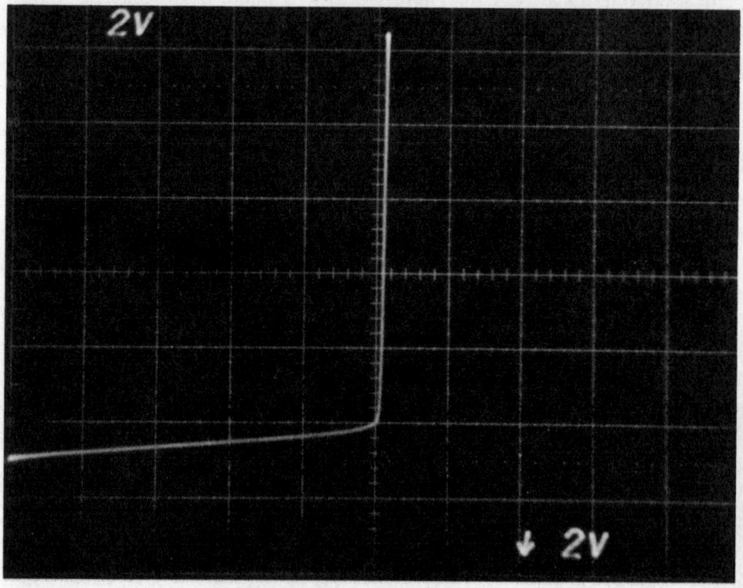

(b) Germanium rectifier

FIG. 13.20(a)–(b). DISPLAY OF DEVICE CHARACTERISTICS

voltage drop is less, but the reverse current is greater. Diagram (c) is the same as (b), but with increased voltage and current so that the diode is starting to avalanche in the reverse direction. At (d) is shown the characteristic of a voltage dependent resistor (VDR). This method is only possible for the

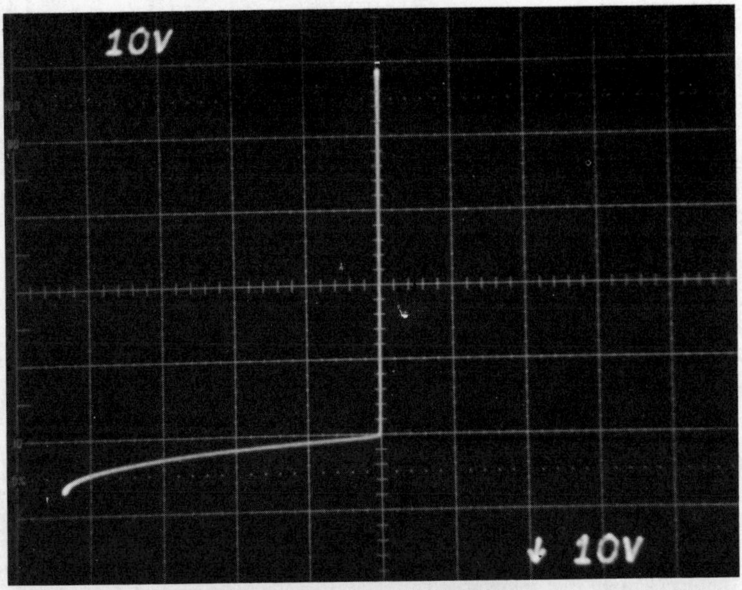

(c) Germanium rectifier with larger voltage

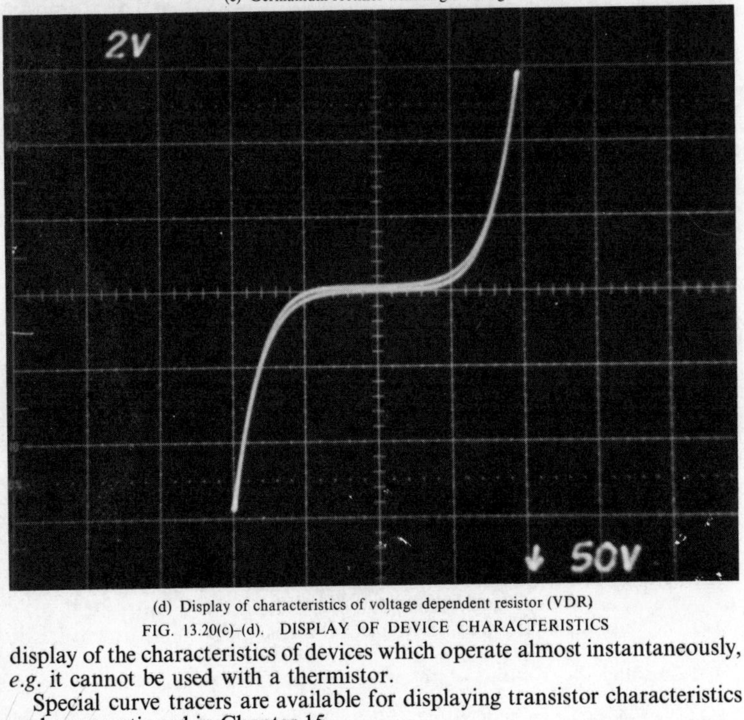

(d) Display of characteristics of voltage dependent resistor (VDR)

FIG. 13.20(c)–(d). DISPLAY OF DEVICE CHARACTERISTICS

display of the characteristics of devices which operate almost instantaneously, *e.g.* it cannot be used with a thermistor.

Special curve tracers are available for displaying transistor characteristics and are mentioned in Chapter 15.

USE OF OSCILLOSCOPE WITH SWEEP GENERATOR

By using a sweep generator (or wobbulator) the oscilloscope is changed to a display of voltage against frequency. A sweep generator is an f.m. generator which produces an output which can be frequency modulated over the range of frequencies to be displayed. By arranging that the X-deflection of the oscilloscope corresponds to the frequency variation, the horizontal scale becomes frequency. There are two methods of operation:

(a) The sweep generator may produce a voltage output proportional to the frequency variation, which is fed into the X-amplifier of the oscilloscope. Often a sinusoidal output at mains frequency is used to frequency modulate the output, hence a sine wave is fed to the X-amplifier. A disadvantage here is that the frequency is first swept in one direction and then in the other. This often produces two separate traces which can be confusing.

(b) The normal timebase of the oscilloscope is used and the sawtooth waveform is fed to the sweep generator, and it is this voltage which frequency modulates the output. Many oscilloscopes have a socket which gives this sawtooth output. An advantage is that the frequency modulation is linear and occurs only in one direction, the flyback being suppressed by the oscilloscope. The frequency used should be low, say 25–50 Hz. Errors occur if the frequency is too high, since high-Q circuits have not sufficient time to reach their maximum response and they do not die away quickly enough.

Although a sweep generator is normally calibrated as regards centre frequency and the frequency sweep may be approximately known, the frequency corresponding to various parts of the trace must often be accurately known. This can be done by using another signal generator of known frequency which will produce a marker pulse or pip where it beats with the output of the sweep generator.

A result is shown in figure 13.21, which is the response of a television i.f. amplifier. At (a) the marker has been placed at the vision carrier frequency of 39·5 MHz, and at (b) at 35 MHz. The dip on the left-hand side corresponds to the sound carrier.

RASTER TYPE DISPLAY

An unusual display is shown in figure 13.22 using linear sweep generators for both X and Y deflections. This enables one to have a long time-scale, although parts are missing due to the flyback and hold-off time. In this figure a musical signal has been added to the Y-deflection so producing the long time-scale. The vertical sweep generator was operating from bottom to top rather than the usual top-to-bottom direction. Z-modulation can be used in place of vertical modulation, the device then resembling a television display.

USE OF ISOLATING TRANSFORMERS

As already stated, the common or E terminal of most oscilloscopes is connected to the frame of the oscilloscope, which should be earthed. This can cause difficulties when displaying waveforms from the mains. It is possible to remove the earth from the frame of the oscilloscope and get satisfactory results, but this can be VERY DANGEROUS, not only to the user but also to others near the equipment. One does NOT expect the frame of an oscilloscope to be alive at, say, 240 volts. This difficulty can often be overcome either by:

(a) Feeding the equipment under test off an isolating transformer of, say, 1:1 ratio; or

(b) Feeding the Y-input of the oscilloscope through an isolating transformer.

The extent to which this can be done depends on the waveforms to be

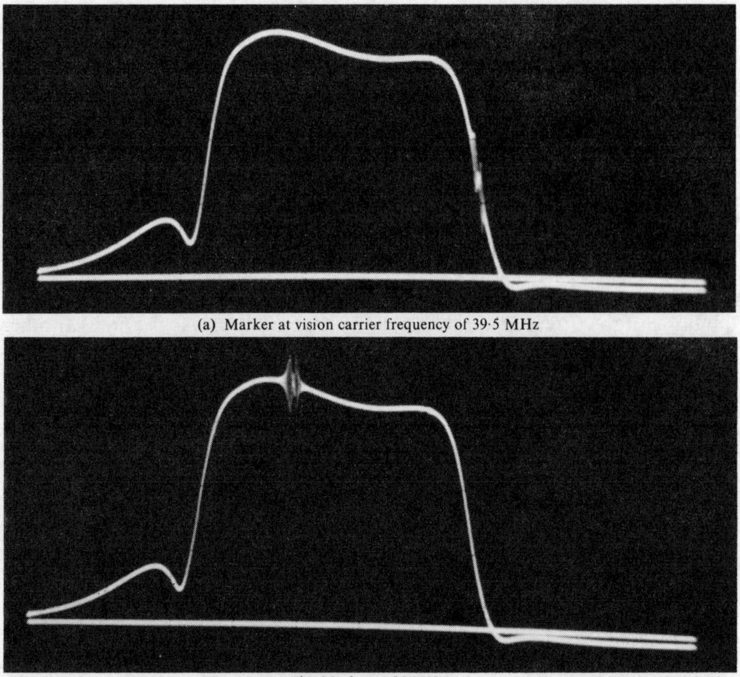

(a) Marker at vision carrier frequency of 39·5 MHz

(b) Marker at 35 MHz

FIG. 13.21. RESPONSE CHARACTERISTICS OF TELEVISION I.F. AMPLIFIER

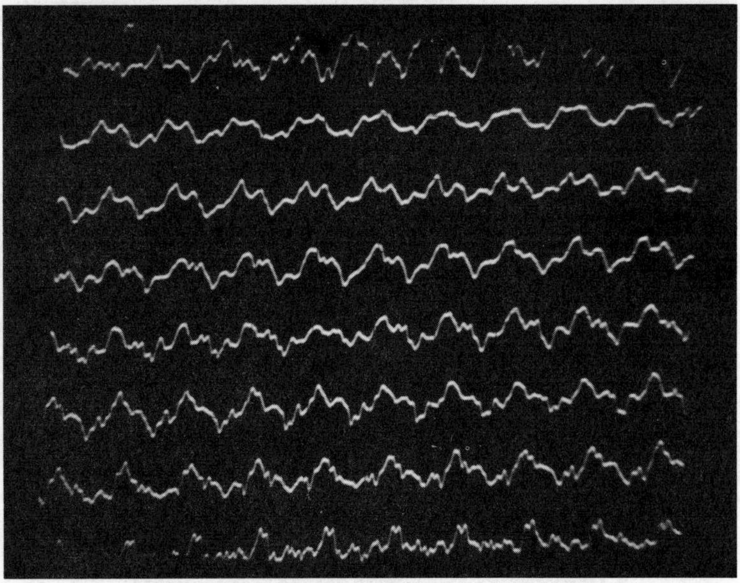

FIG. 13.22. RASTER DISPLAY OF MUSIC WAVEFORM

examined, as one must, of course, be careful that distortion is not produced by the transformer.

When using a multitrace oscilloscope a similar difficulty arises in that all the inputs have a common terminal (assuming differential amplifiers are not being used). In some cases the problem may be solved by rearrangement of the components in the circuit or, again, an isolating transformer in one of the Y-inputs can sometimes be used.

NON-ELECTRICAL USES

This can be mentioned only briefly as there are so many uses in a wide range of subjects. The oscilloscope has many applications in mechanical engineering where pressures, velocities, accelerations and vibrations are important. These are measured by suitable transducers that convert the mechanical quantities into corresponding electrical voltages. Ultrasonic testing is another application. Sound measurements can also use an oscilloscope, the sound wave being converted into an electrical signal by a microphone. The oscilloscope is now used for many purposes by the medical profession such as electrocardiograph displays and measuring blood pressure. It is also used for displays from computers.

14

Photography

THERE are various difficulties connected with photographing an oscilloscope trace. At low frequencies and with repetitive traces there are few problems; but at high frequencies, especially with high-speed transients where maximum writing speed is required, photography becomes more difficult. In principle, photographing an oscilloscope trace is similar to photographing anything else, but there are differences. A normal camera can be used, but it must be capable of focusing down to a short distance so that the screen of the oscilloscope covers as much of the film as possible. This may be done by moving the lens farther away from the film, as is done in a double extension camera (using bellows), or by using extension tubes with, say, a 35 mm camera. Alternatively, a suitable supplementary lens can be used. The advantage of a supplementary lens is that its effective f number remains the same, whereas with extension tubes the effective f number is increased by a factor

$$\frac{S}{S-1}$$

where S is the ratio of object distance to focal length of the lens.

A 35 mm reflex camera is convenient as the trace can be focused and the camera positioned correctly. If many photographs are to be taken some form of hood to keep out ambient lighting is necessary. The hood may act as a support for the camera. Without a hood it is difficult to avoid reflections from the face of the tube, unless working in almost total darkness. Some means are essential of seeing the trace with the hood in place so that the trace can be adjusted correctly before a photograph is taken. The question of exposure is a difficult one because the trace brightness varies according to the sweep speed and is also adjustable; it is largely a matter of trial and error, at all times trying to adjust the trace to give the same brightness. A small change in visual brightness can correspond to 1 or 2 stops. The definition is improved if brightness of the trace is kept to a minimum. One might expect that a photograph should produce exactly what is seen on the screen. This is not possible because the range of brightness that can be seen is of the order of $10^6/1$, whereas the range that can be recorded on film is only about 100/1. It is sometimes difficult to gauge the correct exposure because portions of the trace may vary greatly in brightness. For example, a square waveform will have bright top and bottom portions, but if the rise time is short the vertical traces will be much less bright. Generally, over-exposure is better than under-exposure because the dim portions of the trace will then be recorded. One may have to select the exposure according to that portion of the trace which is the most important. Portions badly over-exposed will have a bright background due to reflected light from the trace, and the definition will be reduced because a larger proportion of the spot will be recorded. This is because the brightness of the spot increases gradually from one side up to a maximum at the centre, and decreases down to zero as in figure 14.1. Photographs showing the effect of various exposures are shown in figure 14.2. At low frequencies exposure time is important. The time of the opening of the shutter *must be longer* than the time taken for the spot to move across the screen, otherwise only a portion of the trace will be recorded. It is generally desirable to have several traces across the screen during the exposure time or the exposure will be uneven, *e.g.* there may be $1\frac{1}{2}$ traces and hence half of the screen is brighter than the other. The effect of varying exposure times is shown in figure 14.3 (pp 172 & 173).

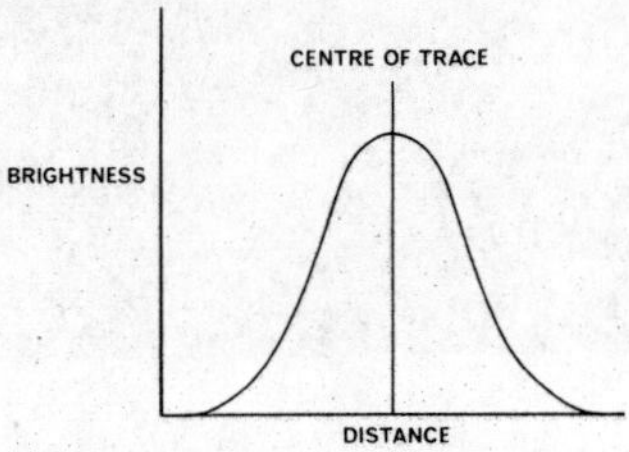

FIG. 14.1. VARIATIONS OF BRIGHTNESS ACROSS THE SPOT

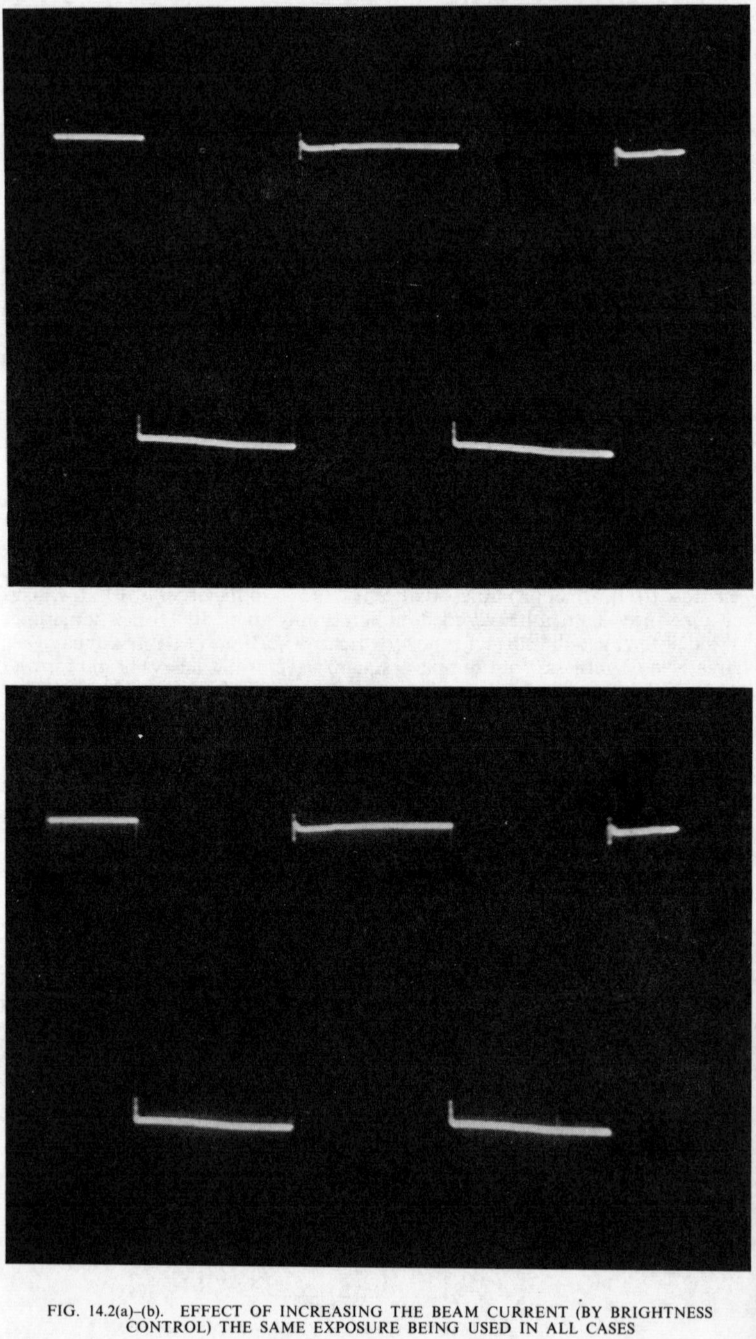

FIG. 14.2(a)–(b). EFFECT OF INCREASING THE BEAM CURRENT (BY BRIGHTNESS CONTROL) THE SAME EXPOSURE BEING USED IN ALL CASES

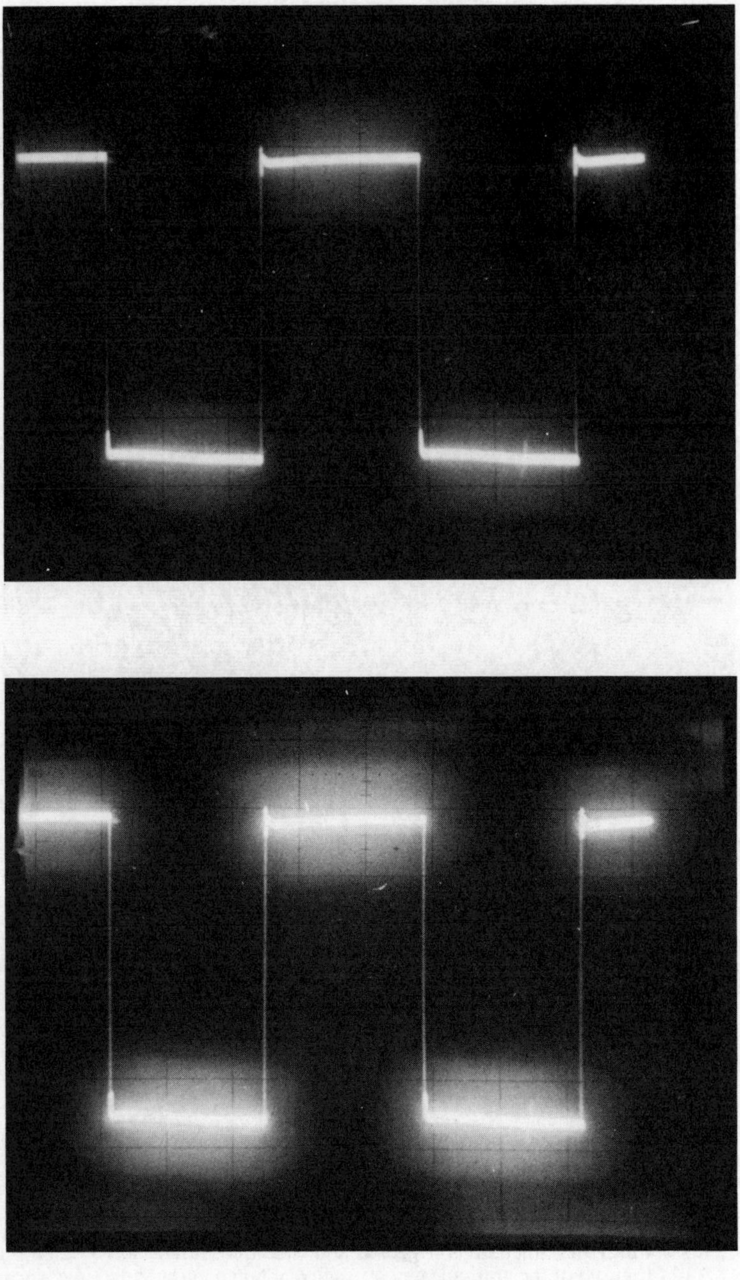

FIG. 14.2(c)–(d). EFFECT OF INCREASING THE BEAM CURRENT (BY BRIGHTNESS CONTROL) THE SAME EXPOSURE BEING USED IN ALL CASES

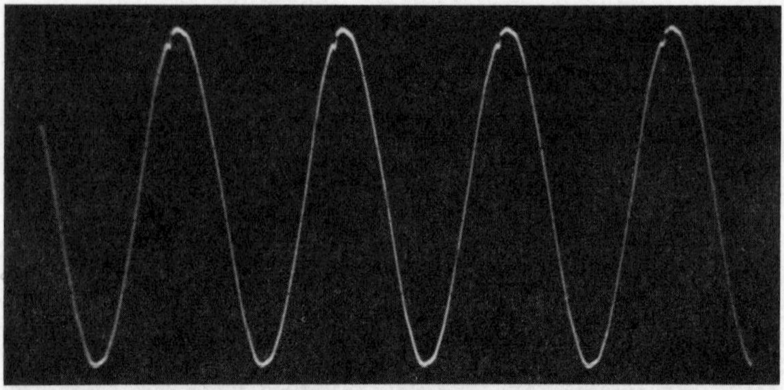

(a) Shutter time $\frac{1}{2}$ second

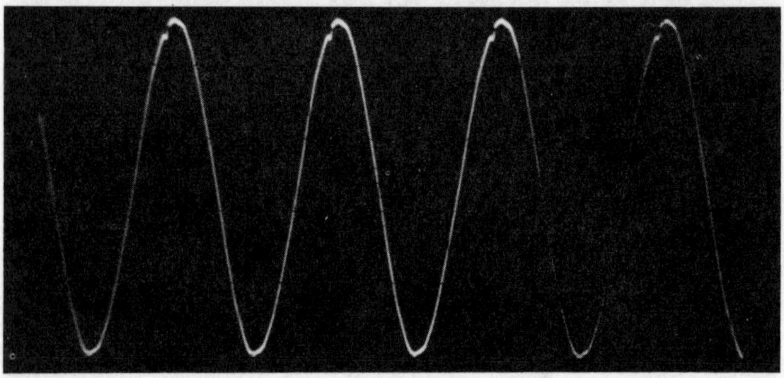

(b) Shutter time $\frac{1}{5}$ second showing uneven trace

FIG. 14.3(a)–(b). EFFECT OF SHUTTER OPENING TIME. FREQUENCY OF WAVEFORM 30 Hz

Provided the trace has a steady value and there is no jitter, better results are obtainable by using a long exposure and a less bright trace. However, if there is jitter the exposure time may have to be reduced.

If the graticule is to be included it must be illuminated; again, the setting of the illumination must be by trial and error. The separate edge-lit graticule may give better results than the internal graticule, which is difficult to light without also illuminating the screen. The normal screen is green and will photograph satisfactorily with the use of orthochromatic or panchromatic materials. If a coloured filter is normally used in front of the tube it should be removed, as it will only reduce the amount of light falling on the film. Removing a filter may reduce the exposure required by several stops. Special films are made for photography of oscilloscope traces, but they are not essential for simple traces.

When a transient is to be photographed the single-sweep timebase must be used. In this case the oscilloscope is set up and the trace will not be visible except during the single scan. Hence the camera shutter is opened (using the 'bulb' setting), the timebase is triggered and then the shutter closed again.

A photograph of the graticule will normally be taken at the same time when dealing with repetitive traces. In transient recordings it may be better to record the graticule separately *with the camera in exactly the same position.*

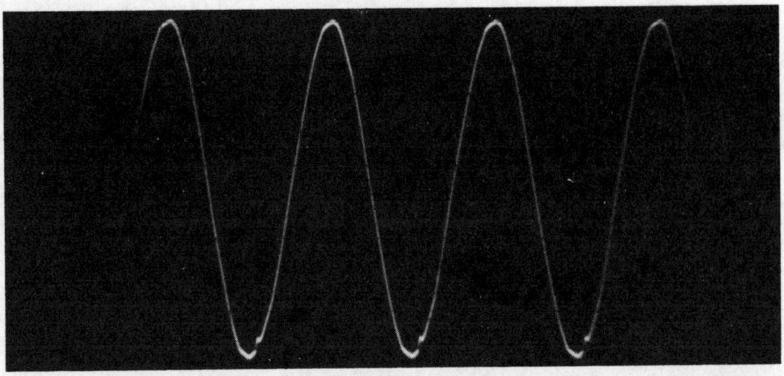

(c) Shutter time $\frac{1}{10}$ second showing only part of trace

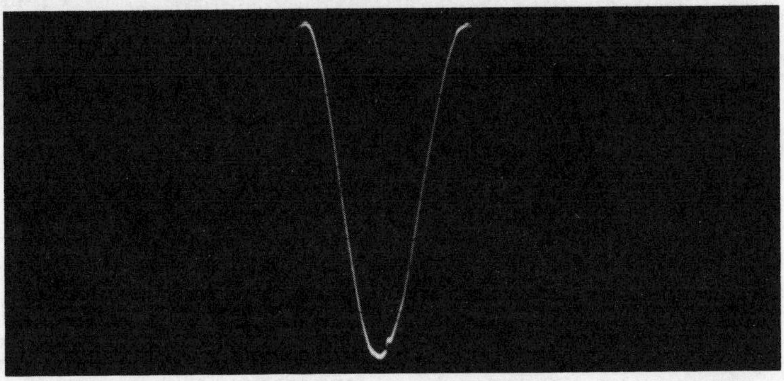

(d) Shutter time $\frac{1}{25}$ second showing only part of trace

FIG. 14.3(c)–(d). EFFECT OF SHUTTER OPENING TIME. FREQUENCY OF WAVEFORM 30 Hz

This enables the correct exposure to be found for both the trace and the graticule.

Special oscilloscope cameras are invariably used for high speed recording, particularly high-speed transient recording, and may also be used for repetitive traces. These cameras are too numerous to be described in detail and only their general principles will be considered. They may use 35 mm film or larger roll film, sheet film or film packs. A negative image is produced on the film, so any number of positive copies may be obtained by contact printing or enlargement. An alternative is Polaroid film, which is processed in less than one minute to produce a positive on paper. Multiple copies are not made so easily but, of course, this system has the advantage of being able to see the results very quickly. Normal film takes a considerable time to process. Polaroid film tends to be expensive and on some cameras it is possible to take a number of photographs on a single sheet, so reducing the cost. A Polaroid film is now available which produces a film negative. Many cameras are made with different backs to accommodate the various types of film.

Some cameras have easy arrangements for focusing. Tektronix, for example, use two vertical bars of light projected on to the cathode-ray tube face. The focus control is adjusted until the bars coincide, the camera then being in focus. Hewlett Packard use a split image rangefinder. Means for determining the exposure may be provided: Tektronix use a projected spot which is adjusted to the same brightness as the trace.

Suppose that we wish to increase the recorded writing speed, *i.e.* the maximum trace speed that can be recorded, commonly quoted in cm/μs or cm/ns. The writing speed is simply the speed at which the spot traverses the screen, and will normally vary between different parts of the trace. In a sine waveform the maximum vertical writing speed is

$$S = 2\pi fA$$

where A is the peak amplitude of the sine waveform. This expression neglects the horizontal component of velocity, but unless this is comparable with the vertical velocity it will have little effect on the total velocity. It is seen that the writing speed is increased if the amplitude is large, therefore when working near the recording limit some improvement is possible by reducing the amplitude of the trace.

The writing speed is really only of importance in the case of transients; with a repetitive trace a long enough exposure can be given to obtain the required trace. The only problem that may arise is jitter of the waveform which will blur the photographic image.

The writing speed depends on:

(a) The brightness of the trace; and
(b) The camera and film.

(a) THIS DEPENDS ON:

(i) The screen phosphor. The blue P11 phosphor has a spectrum that fits the sensitivity spectrum of the film better than the green P31 screen. The writing speed can be approximately doubled by changing the screen from P31 to P11.

(ii) The accelerating voltage. This is determined by the oscilloscope and when high writing speeds are required an oscilloscope with a high accelerating voltage (normally PDA voltage) is required, say up to 24 kV.

(iii) The beam current. The higher the current the brighter the trace, but the trace becomes thicker, hence there is a limit to the usable beam current.

(iv) Coloured filter. Remove any colour filter in front of the screen as this only reduces the light output and offers no advantages to photography. An increase of writing speed of several times is possible.

(b) THIS DEPENDS ON:

(i) The f number of the lens. The relative speeds are given in Table 14.1. Obviously, the smaller the f number the better, but the greater the cost. Also, the camera must be more carefully focused as the depth of focus is reduced.

TABLE 14.1

f number	Speed relative to f2·8
1·0	× 8
1·4	× 4
2·0	× 2
2·8	× 1
4·0	÷ 2
5·6	÷ 4
8	÷ 8
11	÷ 16
16	÷ 32

(ii) Image size. If the size of the image on the film is made smaller the speed is increased, because the same amount of light is concentrated over a smaller area.

(iii) Film speed. The faster the film speed (usually now quoted as an ASA number) the greater the writing speed. With normal photography, doubling the ASA speed means reducing the exposure by a factor of 2. Owing to the limited spectrum of light from the cathode-ray tube this relationship may not exactly hold. In some cases the speed of films for oscilloscope recording is not quoted. Increasing the ASA speed of Polaroid film from 3000 to 10000 will increase the writing speed some 2 to $2\frac{1}{2}$ times.

(iv) Controlled fogging. Some increase in writing speed can also be obtained by controlled fogging. This may be done before making the exposure, during or after the exposure, but done during the exposure gives the greatest increase in speed. An increase of 3 times for Polaroid films of speed 3000 ASA is possible and 2 times for film of speed 10000 ASA. Some cameras (Tektronix) can have automatic fogging devices fixed to them for this purpose. Hewlett Packard use a flood gun to illuminate the screen in some of their oscilloscopes. Figure 14.4 shows the effect of prefogging on a Polaroid film.

(a) No prefogging

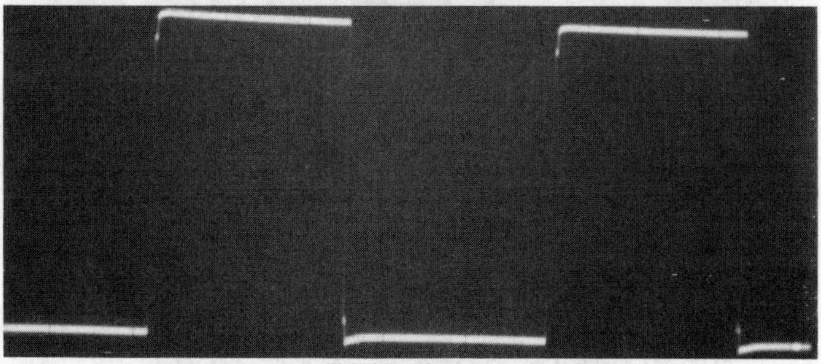

(b) With prefogging

FIG. 14.4. EFFECT OF PREFOGGING ON POLAROID FILM

The writing speed for an oscilloscope is often quoted, but it must be noted that this is not just a function of the oscilloscope but embraces the camera and type of film used. Hence, any figure for writing speed must include data on the camera and film. Speeds up to the order of 10 cm/ns are possible.

Where a large number of transient phenomena have to be photographed an electrically operated shutter is a help. This may be arranged so that the

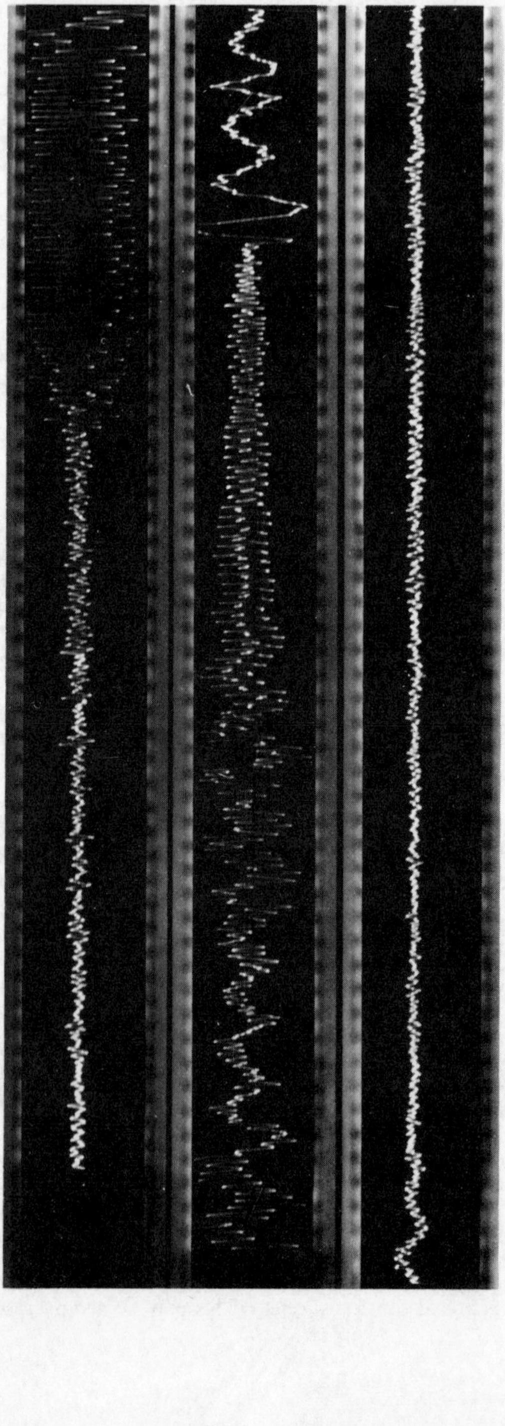

FIG. 14.5. RECORDING MADE ON MOVING FILM (BIRD CALLS)

shutter is triggered and opened, which also arms the oscilloscope sweep. Shortly after the sweep has taken place the shutter is closed.

Photographs can be taken from storage oscilloscopes without any particular difficulty. By using colour positive film effective slides can be made from oscilloscope traces. These are particularly effective if a coloured graticule is used that is a different colour to the trace. Obviously, such film is too expensive for normal use.

When the transient is long, using the normal sweep of the timebase may not give sufficient detail. This can be overcome by using a moving film and deflecting the beam in the Y-direction only. Motor-driven cameras are available for this purpose. Usually a number of film speeds are provided by changing the gear ratio. With this method the film is started, the shutter is opened and the transient started in rapid sequence. At the end of the transient the shutter is closed and the film stopped. Although this is the only method to use in some cases it can be very expensive on film, particularly if the frequency of the waveform is high. The afterglow of the screen must be short or a blurred trace will result. An example of a record on moving film is given in figure 14.5.

15

Complex and Special Oscilloscopes

LOGIC CONTROL

As oscilloscopes become more complex with additional controls, difficulties are encountered in fitting all the controls on the front panel and yet keeping the equipment within a reasonable size. Miniature components and often concentric controls are used. At high frequencies in particular another difficulty arises concerning the leads to the various controls. In many cases any wires carrying high frequency voltage cannot be long as they cause interference or may pick up interference, and they increase the circuit capacitances. One solution is to have a suitable mechanical extension from the knob to the actual switch or potentiometer, which is placed as near as possible to the correct position. However, this is a method that has obvious limits.

Tektronix now use in their 7000 series a complex logic system which provides appropriate command signals from the various plug-ins to the main frame, etc. Thus, many of the switches are operating only the logic circuit so that appropriate switching action can take place in the circuit itself, which is often an integrated circuit. The logic circuits also feed clock pulses for beam switching, etc. This is a versatile system and helps in the circuit design when a large number of plug-ins are available to fit various frames.

DIGITAL READOUT

Also in their 7000 series, Tektronix have introduced an optional readout facility. This is of alphanumerical information produced by the normal beam, the read out being time shared with the normal display. Consider a 7603 main frame used with two dual amplifiers (giving four traces) and, say, a dual time-base unit. The readout gives the sensitivity (*i.e.* volts/div or mV/div) for each of the four channels together with the sweep setting, *i.e.* μs/div. There are six places where the read out can be displayed as shown in figure 15.1. Positions 1 to 4 display the channel sensitivity settings, and position 5 displays the sweep

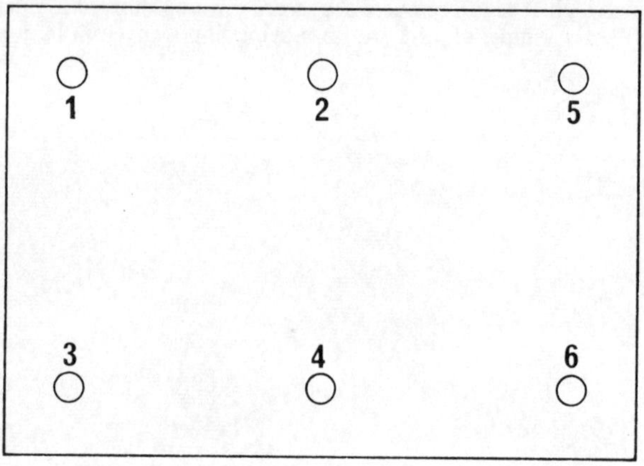

FIG. 15.1. POSITIONS OF READOUT ON SCREEN

speed of the main timebase A. If the delayed timebase B is used, the sweep speed is indicated in position 6. When both timebases are used for display, as in mixed sweep, timebase A is shown in position 5 and timebase B in position 6. If the variable gain control is used on any channel then the symbol > ('greater than') is displayed before the sensitivity figure drawing attention to the fact that the channel calibration cannot be used. If the INVERT switch is used on any channel then an arrow, ↓, is placed before the corresponding sensitivity figure. Similarly, if the variable sweep control is used then > is placed before the sweep speed figure, drawing attention to the fact that the time scale is no longer correct. If the × 10 horizontal expansion button is operated, the sweep speed figure display is automatically altered. If the correct probe is used then again the sensitivity figure is altered so that it reads the sensitivity at the probe tip. When four traces are being displayed it is not always easy to know which is which. This is overcome by the use of a button on the amplifier plug-ins, one associated with each channel. If correct probes are used, a button on the probe head serves the same purpose. When the button is pressed the trace is moved upwards by a few millimetres and the sensitivity figure of this particular channel changed to read IDENTIFY.

A typical display is shown in figure 15.2, using four traces and one timebase at (a). At (b) one of the trace identify buttons has been pressed and mixed sweep has been used. The brightness of the readout display can be varied relative

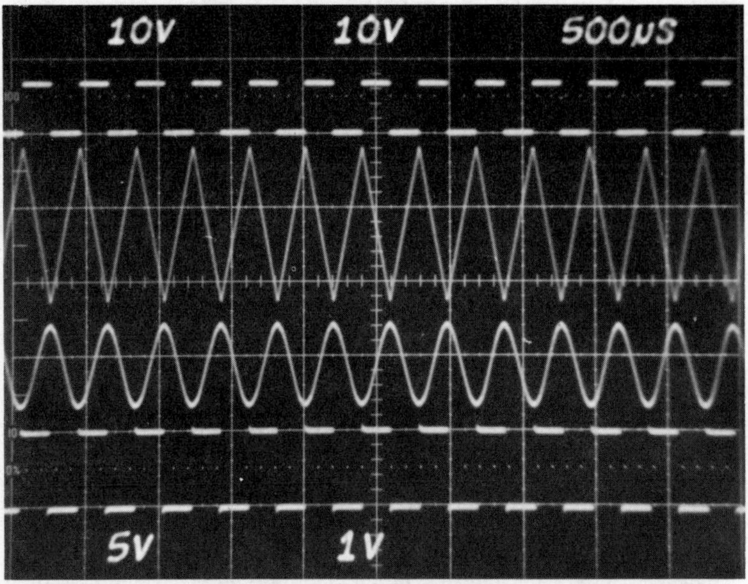

(a) Single timebase with sweep speed of 500 μs/div. Sensitivity of traces from top to bottom: 10 V/div 5 V/div 10 V/div and 1 V/div

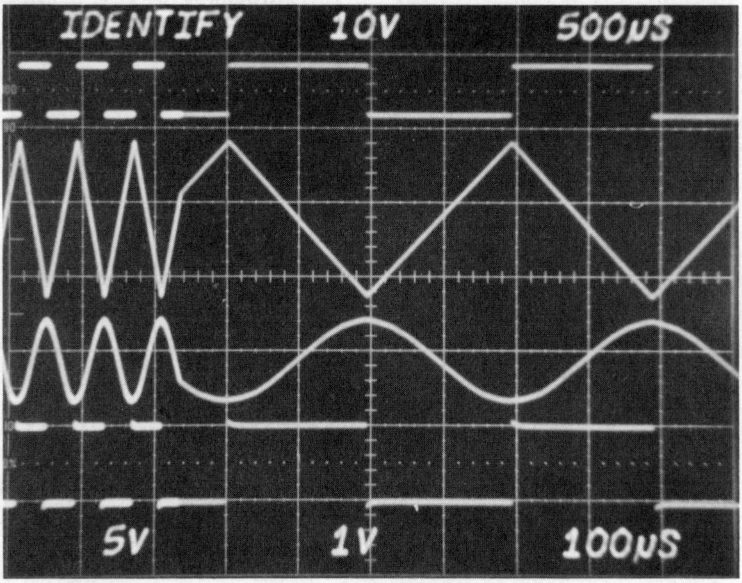

(b) Same waveforms and sensitivities as (a) but mixed sweep with speeds of 500 μs/div (A) and 100 μs/div (B). The identify button has been pressed for the top trace as indicated by READOUT and upper trace is moved upwards 0·2 div

FIG. 15.2. FOUR TRACES WITH READOUT

to that of the trace (by a knob concentric with the normal brightness control) and also switched off.

One might think this an unnecessary and costly complication, even rather a gimmick, but it does help in the prevention of errors. Errors can be extremely expensive, *e.g.* if a piece of research has to be repeated because of incorrect data being recorded. It is very valuable when photographs are taken as the settings of all the controls are there on the photograph and no errors can result. Most of the readout circuitry is on a single board, which need not be purchased with the main frame as it can be added later.

This facility has other uses. A 'readout unit' plug-in is available which enables additional data to be written on the screen. This has a series of buttons and enables the display of any capital letter, together with figures 0 to 9 and a number of other symbols, *e.g.* μ, + and Ω. This is useful for displaying test data, such as test number, or date of test. As it takes up one plug-in space it can only be used on the 7603 (with only three plug-in spaces) by removing an amplifier. It can be used with some of the other 7000 series *e.g.* 7704 with both amplifiers as there is space for four plug-ins. The readout facility is used still further with digital counters, plug-in, etc. to be described later. It is also used with the digital processing oscilloscope.

A number of other plug-ins are available to fit the 7000 series, which, at first, appear to have no connection with an oscilloscope. These also use the read out facilities. Only two will be described, briefly.

(a) Digital Multimeter

This measures d.c. voltage, current and resistance using the readout facilities of the main frame. The display is $3\frac{1}{2}$ digits (*i.e.* four figures, but the first figure limited to 0 or 1). There are four voltage ranges, four current ranges, and five resistance ranges. There is also a temperature range using a probe. Polarity is automatically indicated together with units like kΩ, mA and C.

(b) Counter/Timer

This is a sophisticated counter/timer making use of the readout facilities. Also, the measurement being taken is displayed on the screen by using the B sweep as a gating pulse. The ranges in use and the actual time are displayed on the screen using the readout.

Tektronix also make a digital multimeter (with temperature measurement) to fit on the top of some of their models, including a storage oscilloscope, the display now being on LEDs (light emitting diodes). This enables measurements to be made of voltage, resistance, time and temperature. Time can be measured on a waveform by the use of a bright marker which is moved between the two points of interest.

A 275 MHz oscilloscope is made by Hewlett-Packard which gives a read out of voltage and time. The readout is now on a $3\frac{1}{2}$ digit LED display. For time measurement the waveform is displayed in the normal way and two intensified markers are moved to the two points of the waveform where the time difference is required. The difference is calculated on a microprocessor and displayed by the LEDs. Instead of time for a complete cycle the frequency can be displayed. Greater accuracy can be obtained by the use of a dual delayed sweep facility. The oscilloscope will also measure the mean d.c. voltage and it can also be used to measure the voltage between any two points on the waveform so that it is unnecessary to count the graticule marks. Compensation can be made when a 10/1 probe is used.

DIGITAL PROCESSING OSCILLOSCOPES (TEKTRONIX)

Again, only brief details can be given, but basically this is a combination oscilloscope and computer. The input signal can be fed to the oscilloscope and then processed, *e.g.* differentiated, and the result then displayed on the screen of the oscilloscope with the necessary sensitivity, sweep speed and other information produced by the readout facilities. The facilities that can be made available by such complex equipment are almost limitless and depend largely on what one is prepared to pay. It is outside the scope of this book.

An interesting recording oscilloscope using cathode-ray tubes is manufactured by Medelec Ltd. When recording from a normal cathode-ray tube a camera is used as described in Chapter 14. The optical efficiency of this arrangement is poor, even with an aluminized screen. The light is emitted in the forward direction from the phosphor (and by reflection from the aluminium coating), but over a large angle as shown in figure 15.3. Only a small proportion (*e.g.* 2%) of the light is collected by the camera lens. Further, except

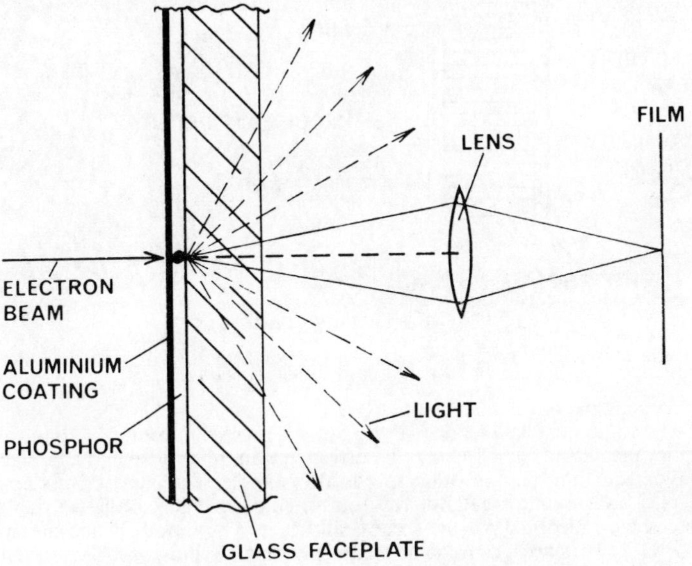

FIG. 15.3. PHOTOGRAPHY USING NORMAL OSCILLOSCOPE TUBE SHOWING SPREAD OF LIGHT AND ONLY SMALL FRACTION REACHING FILM

on the axis of the camera lens, some distortion may occur because the light has to travel through the thickness of the glass faceplate where refraction takes place. The idea of this equipment is to use fibre optics to concentrate the light directly on to the photographic light-sensitive paper without a lens. If a normal screen were used the definition would be poor owing to the spread of light, as in figure 15.3. The faceplate is now made in the form of a large number of glass fibres each consisting of a centre portion of high refractive index glass sheathed by a low refractive index glass. Light is now transmitted directly along the fibre by total internal reflection, as shown in figure 15.4. To prevent cross-coupling between fibres a black light-absorbent layer surrounds each fibre. The fibres are 75 microns in diameter with almost 200 fibres per mm^2. Since the screen is 100 mm × 80 mm there are more than $1\frac{1}{2}$ million fibres on the faceplate, giving good definition. This results in a light improvement up to 100 times compared with a normal tube and camera. It is a PDA tube with a total accelerating voltage of 6 kV and has a blue (P11) screen. The brightness is such that the trace can be recorded on slow-speed recording paper (as used in ultraviolet recorders) and developed by ambient lighting. Full chemical

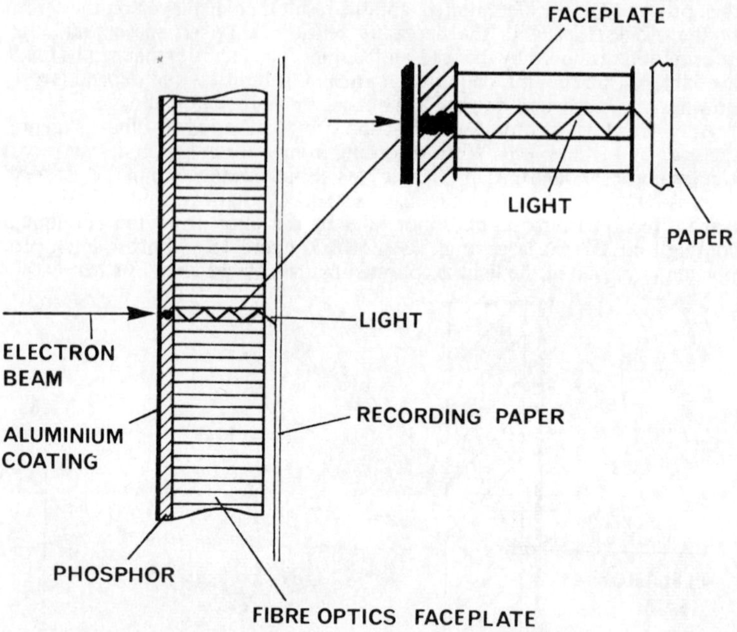

FIG. 15.4. FIBRE OPTICS FACEPLATE AND CONCENTRATION OF MOST OF LIGHT ON TO RECORDING PAPER (Medelec Ltd)

processing paper can also be used.

The equipment is so made that the paper is pressed in contact with the fibre optics faceplate. The paper can be driven at a number of speeds from 0·5 to 100 cm/s as normal, but other speeds are possible. The paper covers up the display so a monitor cathode-ray tube is connected in parallel so that the trace being recorded can be seen at all times. The monitor tube measures $14 \times 11{\cdot}4$ cm, hence it can also be used as a normal oscilloscope. There are four channels with beam switching using the chopping method. Sensitivity figures from 10 mV/div to 50 V/div and normal oscilloscope facilities are provided. The timebase runs from 50 μs/div to 0·5 s/div with an $\times$ 10 expansion control. The bandwidth of the Y-amplifier is d.c. to 100 kHz, the upper limit being due to the limited writing speed of the paper. Z-modulation facilities are available.

The equipment can be used in a number of ways:

(a) Paper stationary. Single sweep transients.
(b) Paper stationary. Multiple trace records.
(c) Paper stationary. X-Y plotting.
(d) Moving paper with signal giving horizontal deflection and paper moving vertically. This is similar to normal paper recorders.
(e) Moving paper but normal horizontal timebase and vertical deflection which results in a raster display similar to that shown in figure 13.22 except that there is no limit to the length of the raster. Some information is lost during flyback but it is small. This display method can be most useful.
(f) Moving paper with normal horizontal timebase but Z-modulation resulting in raster display. This can produce a facsimile type display.

COLOURED TRACES

An oscilloscope using a 15″ cathode-ray tube is produced by Telonic Industries which has three channels and the displays are in red, green and blue, which makes identification much easier. Coloured horizontal and vertical reference lines can be added.

CURVE TRACERS

In Chapter 13 the use of an oscilloscope to display the characteristics of a device was described. It is more convenient to have an oscilloscope together with all the supplies, and such a unit is called a 'Curve Tracer'. By using, for example, a step-function generator to change the base current of the transistor under test a whole family of characteristics can be produced. A typical characteristic is shown in figure 15.5. Readout facilities are provided on some models, and a storage tube may be used so that the characteristics can be stored to determine the effect of temperature, etc. Detailed descriptions of these units are beyond the scope of this book.

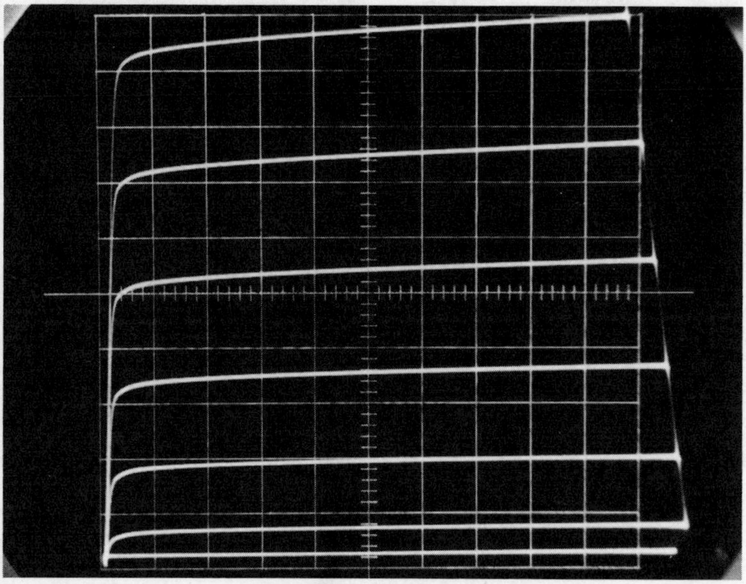

FIG. 15.5. DISPLAY FROM CURVE TRACER. TRANSISTOR CHARACTERISTIC WITH COLLECTOR CURRENT VERTICALLY AND COLLECTOR VOLTAGE HORIZONTALLY FOR DIFFERENT BASE CURRENTS

VECTORSCOPE

This is a special type of oscilloscope used in connection with colour television. All normal oscilloscope displays are in rectangular co-ordinates, *i.e.* X and Y deflections at right angles. It is possible however to make an oscilloscope to display a signal in terms of polar co-ordinates, *i.e.* as a line of varying amplitude and varying phase angle. The latter corresponds to the normal vector or phasor; it is difficult to display and is rarely used in other applications.

In a colour television signal two colour or chrominance signals are used which are modulated on to two carriers of the same frequency, but 90° out of phase with each other. The two signals are then known as V and U signals. The saturation of the colour is determined by the amplitude of the resultant, and the hue by the phase of the resultant of vector addition of V and U. It is therefore useful to display various colours in this way, for which a vectorscope is used. A typical display is shown in figure 15.6.

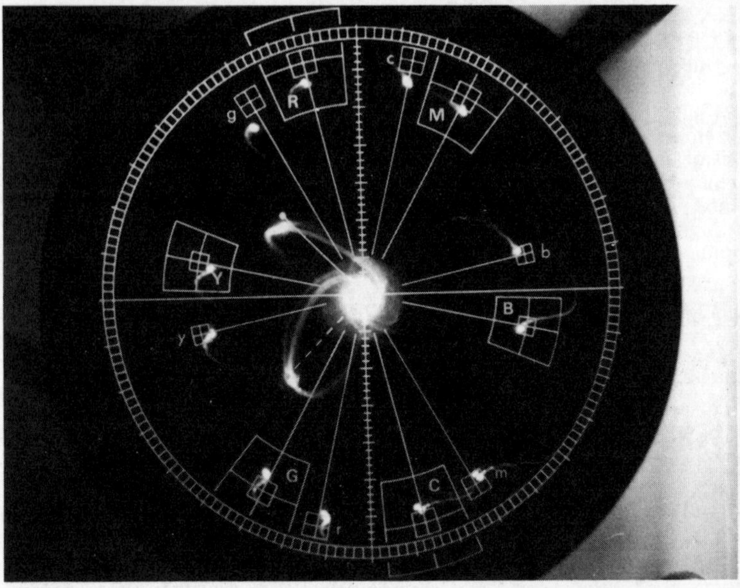

FIG. 15.6. VECTORSCOPE DISPLAY FOR PAL SIGNAL ON COLOUR BARS, SHOWING SOME ERRORS IN ADJUSTMENT

LARGE DISPLAYS

There are many special types of oscilloscopes that cannot be covered, such as display units used with computers. Special units are also made for the medical field, *e.g.* patient monitors.

16

Storage Oscilloscopes and Transient Recorders

STORAGE OSCILLOSCOPES

WHEN the phenomena being examined is repetitive at a rate, say, above 20 times per second a continuous trace is obtained on a normal oscilloscope. If a record is required a sketch may be made, which is laborious and of indifferent accuracy. The usual method is to photograph the trace, as explained in Chapter 14. Where transient phenomena are concerned then, depending on the speed of the transient, it is difficult to see the trace. If the transient is high speed and requires a high sweep speed the trace is almost impossible to see. With a normal oscilloscope the only way of seeing the trace is to photograph it. (In some cases a long afterglow tube may help to see it). This requires a camera and, unless of a Polaroid type, the results are not available immediately and one cannot be certain for some time that a satisfactory result has been obtained. For example, if it is required to find out what happens when changes are made to a circuit many photographs may have to be taken – an expensive and time-consuming operation. It is particularly time consuming if one wishes to find out the effect of one change before making another. If the transient occurs at random then a large number of photographs may be required before the required information is obtained.

With a very slow transient, or slow repetitive trace, it is difficult to see the shape of the waveform or display, although this can be remedied to some extent by using a tube having a long afterglow screen.

The storage oscilloscope removes these difficulties. The transient is written on the screen and the information stored so that a display is seen for a considerable time after the transient has passed. This stored information is easily removed when required and another trace stored. Thus it is possible to see and examine the transient, and the effect of changes in a circuit can be seen immediately. If a record is required the stored trace can be photographed, but, of course, this will only be done when the stored trace is the one that is actually required. In slow-speed transients and repetitive phenomena, again a stored trace is obtained and can easily be examined. As will be seen later, most storage oscilloscopes can also be run as variable persistance oscilloscopes, and this mode of operation may be used for slow phenomena.

If the transient phenomenon is one that occurs at random it can be stored by the oscilloscope and, as explained later, it can be displayed several hours after its occurrence.

There are two basic types of storage tubes: the transmission tube; and the direct bi-stable tube. The way in which these tubes operate is complex, and the following explanations are somewhat simplified. There are also variations in the exact methods of operation used by different manufacturers.

Before describing these tubes some information about secondary emission must be given. When a material is bombarded by high-speed electrons (called 'primary electrons') their energy is transferred to some of the electrons (called 'secondary electrons') in the material and are emitted. Depending on the direction of the electric field near the material, these secondary electrons either travel away from the material or return to it. The number of secondary electrons emitted depends mainly on the material and the velocity of the primary electrons. A typical relationship between secondary emission ratio (*i.e.* ratio of number of secondary electrons to number of primary electrons) and the velocity of the primary electrons (expressed in terms of accelerating voltage) is given in figure 16.1. There are two important points on this character-

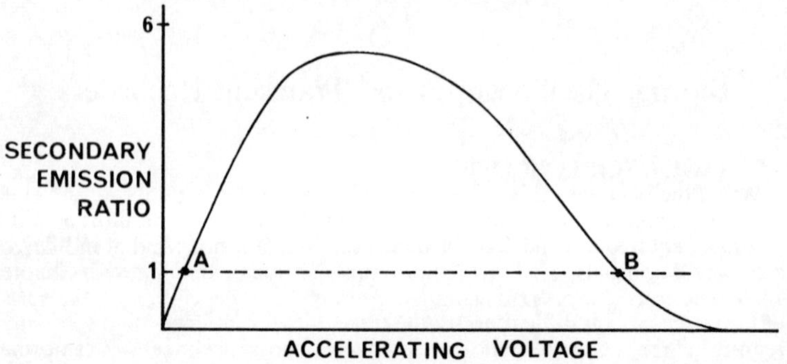

FIG. 16.1. RELATIONSHIP BETWEEN SECONDARY EMISSION RATIO AND ACCELERATING VOLTAGE

istic, A and B, where the secondary emission ratio is unity; they are called the first and second crossover points, respectively. For accelerating voltages below A and above B, the secondary emission ratio is less than unity; for voltages between A and B the ratio is greater than unity. The importance is that if the ratio is less than unity, there are more primary electrons arriving at the surface than leaving it, therefore an insulated target charges in a negative direction (since it is collecting electrons). If the ratio is greater than unity (and all the secondary electrons are attracted away) there are more electrons leaving the target than arriving, hence it loses electrons and so charges in a positive direction.

Suppose we consider a target T bombarded by primary electrons from an electron gun with final anode voltage V_G, as shown in figure 16.2. The target is largely surrounded by a collector C kept at a voltage V_C (in relation to the gun's cathode). The target is fed with a variable voltage V_T. The first important point to stress is that the velocity of electrons arriving at the target is settled ONLY by the voltage V_T and not by the voltages V_G on the electron gun. If the final anode voltage V_G of the gun is higher than V_T, the electrons will be accelerated in the gun but decelerated again as they move towards the target T. The second important point to stress is that the secondary electrons emitted from the target will go to the collector C if the field direction is from target to collector, but if it is in the other direction they will return to the target T.

[When referring to the direction of the electric field in this book this is always taken as the direction in which an electron (negative charge) will travel, whereas the true definition is the direction in which a positive charge would move].

We will now see how the characteristics given in figure 16.1 control the target current in the circuit arrangement of figure 16.2. This is shown in figure 16.3 in which all voltages are measured with respect to the cathode of the electron gun. Point A of figure 16.3 corresponds to point A of figure 16.1, the first crossover point. The term 'effective secondary emission ratio' will be used, which is the ratio of secondary electrons leaving the target to the number of primary electrons. As explained this will depend on the secondary emission ratio and on the direction of the field around the target.

With switch S closed and starting at a voltage V_T slightly higher than that corresponding to point A of figure 16.3, the actual secondary emission ratio (SER) is greater than unity (from figure 16.1). Since the collector voltage V_C is greater than target voltage V_T (see figure 16.3) the field direction is from the target T to the collector C. Thus the secondary electrons will travel to the collector. Because the actual SER is greater than unity more electrons leave the target than arrive at it, and an ELECTRON CURRENT flows from the supply V_T

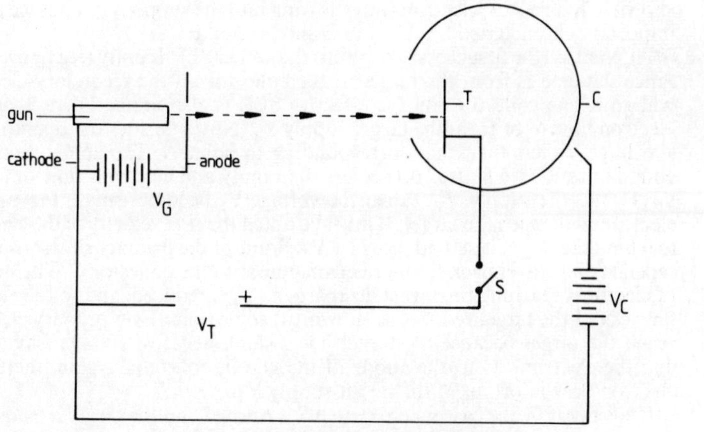

FIG. 16.2. TARGET BOMBARDED WITH PRIMARY ELECTRONS

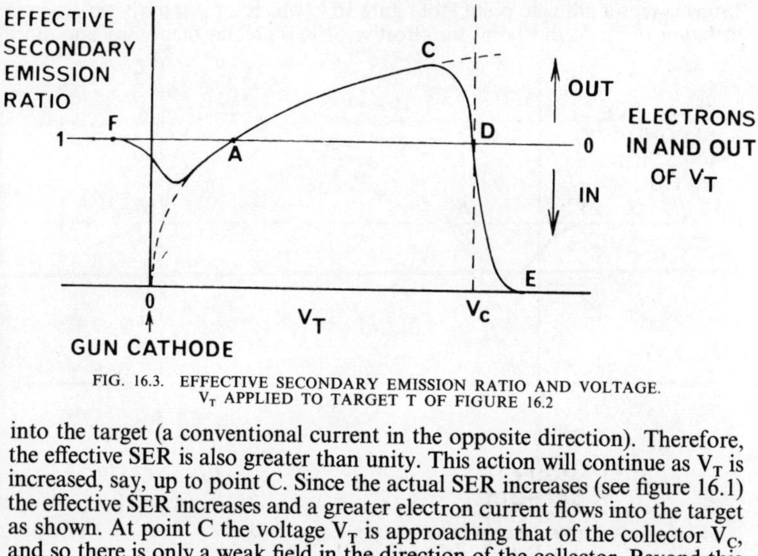

FIG. 16.3. EFFECTIVE SECONDARY EMISSION RATIO AND VOLTAGE. V_T APPLIED TO TARGET T OF FIGURE 16.2

into the target (a conventional current in the opposite direction). Therefore, the effective SER is also greater than unity. This action will continue as V_T is increased, say, up to point C. Since the actual SER increases (see figure 16.1) the effective SER increases and a greater electron current flows into the target as shown. At point C the voltage V_T is approaching that of the collector V_C, and so there is only a weak field in the direction of the collector. Beyond this point, although the actual SER is increasing (see figure 16.1), the effective SER decreases as some of the electrons now return to the target. At point D the target voltage $V_T = V_C$ and there is no electric field between the target and collector. Under this voltage condition it will be assumed for simplicity that as many secondary electrons go to the collector as there are primary electrons arriving at the target, *i.e.* the EFFECTIVE SER is unity. Thus, there is now no electron flow to or from the target supply V_T.

If the target voltage V_T is now raised above V_C the direction of the field will be towards the target, and eventually all electrons will be returned to the target. Thus the EFFECTIVE SER ratio is less than unity (although the actual SER is greater than unity assuming V_C is less than that corresponding to point B of figure 16.1). The direction of electron flow from the target is now reversed as shown in figure 16.3, electrons now flowing into the supply V_T. Thus we get the important characteristic A, C, D to E and beyond E.

At point A (the first crossover point) the actual SER is unity (see figure 16.1). Since the field is from the target to the collector all the secondary electrons will go to the collector and the effective SER is also unity. There is now no electron flow to or from the target supply V_T. Now consider the operation for a voltage V_T less than that corresponding to point A. The actual secondary emission ratio (see figure 16.1) is less than unity and electrons flow out of the target into the supply V_T. When the voltage V_T becomes small, few primary electrons will reach the target. It must be noted that the velocity of the electrons reaching the target is settled ONLY by V_T. Some of the primary electrons will be repelled and travel back to the electron gun or to the collector C. The number of electrons reaching the target decreases as V_T is reduced, and so the electron flow out of the target reduces as shown. At some point F no primary electrons reach the target because the target has a small negative voltage (say -5 V) and the electrons go to the anode of the gun or collector. Again, there is no electron flow in or out of the target supply V_T.

Consider now the case where switch S is opened and the target is isolated. If the target were made of insulating material it would behave in exactly the same way as this isolated target. Suppose that we start with an initial potential of the target corresponding to point P of figure 16.4 (which is the same basic diagram as figure 16.3). At this point the effective SER is greater than unity and more

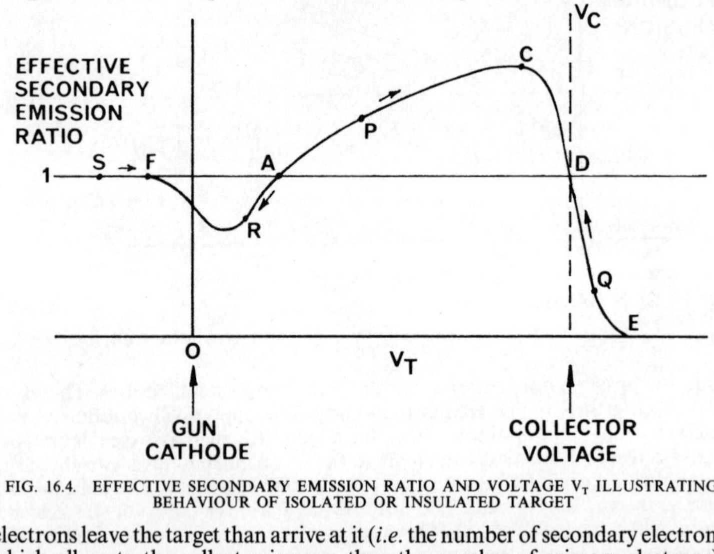

FIG. 16.4. EFFECTIVE SECONDARY EMISSION RATIO AND VOLTAGE V_T ILLUSTRATING BEHAVIOUR OF ISOLATED OR INSULATED TARGET

electrons leave the target than arrive at it (*i.e.* the number of secondary electrons which all go to the collector is more than the number of primary electrons). Thus the target is losing electrons and charging in a positive direction, as indicated by the arrow. This action will continue until point D is reached. At D the effective secondary emission ratio is unity, and the number of electrons arriving equals the number leaving, and the target voltage remains fixed.

Suppose that the initial target voltage V_T was that corresponding to point Q. The effective SER is now less than unity, more electrons will arrive than leave and the target therefore collects electrons and charges in a negative direction, as indicated by the arrow. This action will continue to point D, where the effective SER is unity and the target therefore remains at a fixed voltage. Thus, provided the initial target voltage is above that corresponding to point A, the target will charge in either one direction or the other to a stable voltage corresponding to point D, which, near enough in practice, is equal to the voltage V_C of the collector. This stable point is of great importance.

Now consider the target with an initial voltage corresponding to point R. The effective SER is less than unity, the target collects electrons and the voltage moves in a negative direction, as indicated by the arrow. This action will continue until the target reaches a voltage corresponding to point F, where the effective SER is unity. At this point there are no electrons reaching the target, so it remains at this voltage. Suppose that the initial potential of the target was more negative than that at point F, say point S. Ideally it would remain at this voltage since no electrons would reach the target. In practice, the insulation of the target will not be perfect and it may move in a positive direction due, say to leakage from the positive collector. More important, owing to the imperfect vacuum, the primary electrons from the electron gun will cause ionization of the gas and the production of positive ions. These ions will be attracted to the target, as it is now the most negative electrode, and will be collected by it. Thus the target charges in a positive direction, as shown by the arrow, until it reaches the point F. At point F positive ions may also be collected, but it will settle to such a voltage that the number of positive ions equals the number of electrons collected, therefore point F is another stable point of great importance.

Although the effective (and actual) SER equals unity at point A, this is NOT a stable point because the slightest change of voltage from this point causes the target to go to one or other of the stable points F or D. Thus we have a bistable device with only two stable voltages for the target, corresponding to F and D.

TRANSMISSION STORAGE TUBES

The basic principles of this type of tube will first be considered; figure 16.5 is a simplified diagram. The tube has a gun G_1 and deflecting system D, which are the same as those in a normal oscilloscope tube. This gun is called the

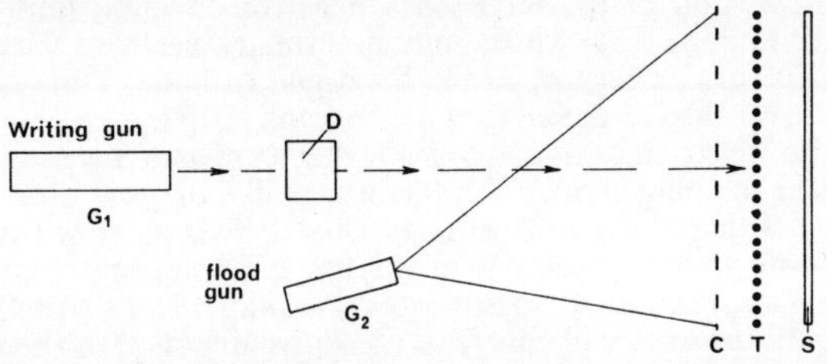

FIG. 16.5. SIMPLIFIED DIAGRAM OF TRANSMISSION TYPE STORAGE TUBE

'writing gun'. There is a fairly coarse metal mesh C called 'the collector', and a fine mesh T called the 'target'. This target consists of a metal mesh, but has a coating of insulating material on the side facing the gun G_1. The aluminized phosphor screen S is maintained at a high positive potential in a similar way to a PDA tube. There is a second gun (or guns) G_2 called the 'flood gun', which is operated at a lower voltage than G_1, its purpose being to flood the target T (and screen S) with relatively low-velocity electrons. Collimating electrodes (not shown in the figure) are fitted on the side of the tube to produce a uniform flooding of the target.

The insulated coating on the target mesh corresponds to the target of figure 16.2 with switch S open, and the collector C (figure 16.5) to the collector C of the figure 16.2. Descriptions of these tubes cause some confusion as regards electrode voltages, and the point of reference is not always made clear. In this chapter ALL VOLTAGES will now be with reference to the flood-gun cathode unless otherwise stated. The flood-gun cathode is at approximately the final anode potential of the writing gun and the mean potential of the deflecting plates. It is therefore around earthly potential, but the voltage depends on the Y-amplifier circuit. Under normal conditions the collector will be positive but only at a low voltage (say 150 V) relative to the flood-gun cathode. Consider a small portion of the target mesh as in figure 16.6. Suppose that the target insulating material in parts P and Q is initially below that of the first crossover

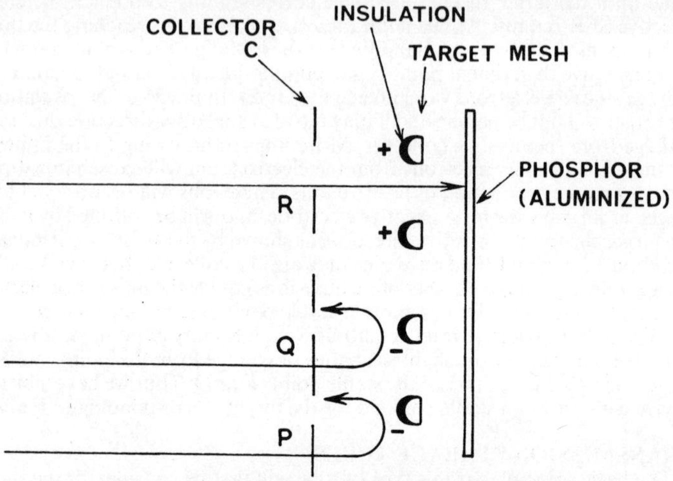

FIG. 16.6. EFFECT OF TARGET VOLTAGE ON ELECTRONS

point A (figure 16.4). As previously explained, the flood-gun electrons will charge this portion of the target in a negative direction until the voltage corresponds to point F, at which point no primary electrons fall on the target as it is negative with respect to the flood-gun cathode. Thus the flood-gun electrons are repelled as shown in figure 16.6 and go to the collector C, which is positive. The mesh of the target is so made that it prevents any electrons passing through it to the phosphor screen, *i.e.* it acts like the grid of a valve with a voltage higher than the cut-off voltage. Thus there is no glow from the phosphor. Suppose that a portion R of the target has an initial charge so that its voltage is above the first crossover point A (figure 16.3 or 16.4). As already explained, this portion will charge in a positive direction to the upper stable point D, equal to the collector voltage. The effective SER is now unity, hence as many electrons leave the target as fall on it. Some of these electrons that are leaving and some of the flood-gun electrons will now pass through the target mesh (the target mesh, *i.e.* the metal conducting part of it, is maintained positive at, say, +100V) and are accelerated to the phosphor screen and so illuminate the screen. The portions P and Q are negative; they repel electrons and prevent flood-gun electrons getting through the target mesh so the corresponding portion of the phosphor will be dark. This charge condition will remain for a long period, and is the condition when the trace is being READ, *i.e.* seen on the screen after it has been written by the electron beam on the screen.

The writing process will now be considered. Suppose, by some means to be described, that the whole target is given a voltage below the first crossover point A (figure 16.2) and that there is no beam from the writing gun. The flood gun electrons will now charge the target to a voltage corresponding to point F, and no electrons (or only a few) will fall on the phosphor so it will be dark. Suppose also that the writing beam is now switched on and a single trace is made across the screen. The cathode of the writing gun is at a high negative voltage, say −1500 V, therefore the writing beam consists of high velocity electrons of much higher electron density than from the flood gun. Although the target (insulating material) may be a few volts negative with respect to the flood-gun cathode it is highly positive with respect to the writing-gun cathode, and the effective velocity of these writing electrons is well above that of the first crossover point A. Hence the writing beam charges the portions of the target on which it falls to the upper stable point D. This is the WRITING condition. Assuming that the writing beam is now cut off, we have the READ or DISPLAY conditions, as already explained. The written portions of the target are positive (point D), and flood-gun electrons illuminate the phosphor screen while the other portions are negative and prevent electrons reaching the screen. Thus a permanent display appears on the screen of that which has been traced out by the writing gun.

The display can be ERASED by giving the target mesh (sometimes called the 'backing electrode', *i.e.* the conducting metal mesh) a positive voltage for a short period. By capacitance coupling this raises the potential of the insulating material above the first crossover point A, and the target insulation material charges to the stable point D. The whole screen will now appear bright. If the target mesh is now reduced in voltage below the first crossover point, the target will charge in a negative direction to the stable point F and the screen will appear dark in readiness for the writing of another trace by the writing beam.

This method of operation is known as 'bistable operation', since there can be only two states of the target – either at point F or point D. There is therefore either a bright phosphor or a dark one; there is nothing in between, what is commonly called 'half-tone.' Generally this is undesirable: if the writing speed of the trace varies, some parts will be written and some not, *i.e.* the writing speed has to be slow enough for the writing beam to charge the target above point A if a trace is to be seen.

However, the tube is not normally operated in this way, but in a modified way so that a half-tone display can be stored. Although the basic construction is the same, detailed changes are necessary and the cut-off characteristic of the target mesh is now not so great. In figure 16.7(a) is shown the basic characteristic of a storage tube as in figures 16.3 and 16.4. The number of flood-gun electrons getting to the screen with a target potential corresponding to F will

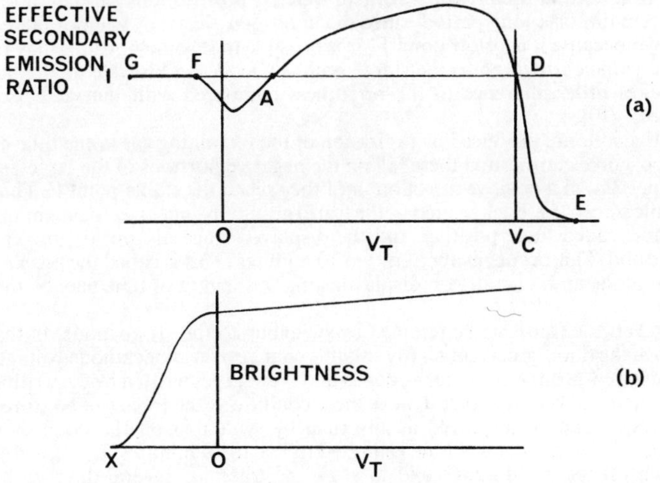

FIG. 16.7. HALF-TONE METHOD OF OPERATING TRANSMISSION STORAGE TUBE

depend on the construction. In the half-tone tube this is relatively large, there being little difference in brightness between voltages corresponding to points F and D, as shown at (b). However, by making the target more negative than point F the number of electrons and the brightness drop fairly rapidly to zero at X at a voltage corresponding to point G. Thus, in order to cut off the flood-gun electrons from the screen the target must be made more negative than the lower point F, *i.e.* point G, which is negative with respect to the flood-gun cathode.

The first operation is setting the target to point G. This is done by the application of a suitable pulse to the target mesh (backing electrode). Assume that all the target is at the stable point F. The target mesh is now given a positive pulse (relative to the flood-gun cathode) of a low value (say +8 V). It must not be so large as to raise the potential to that of point A (the crossover point). By capacitance coupling the insulation of the target will also be pulsed positive and cause electrons to fall on it; but, because it is operating below point A, the insulated target is charged back to point F, the stable point. (During this period the screen will be bright). If the +8 V on the target mesh is suddenly removed (*i.e.* the end of the positive pulse) the insulated section (which has capacitance to the conducting mesh) will move in a negative direction by an amount equal to the reduction in mesh voltage, 8 V as in the example given. Thus the target is at point G and if this voltage is of suitable value, the flood-gun electrons are cut off from the phosphor which now appears dark. Neglecting any effects of ionization the target would remain in this state since no flood-gun electrons can reach the negative target.

Suppose now that a single trace is made by the high-velocity writing beam. Since the writing beam causes the target to operate above the first crossover point A, more secondary electrons than primary electrons are emitted and the target charges positive. The amount that the target charges positive will depend on how many primary electrons fall on a particular section, *i.e.* on the beam current and writing speed. Thus, when the writing beam has passed the various portions of the target, the target will be at potentials between, say, G and F. Thus the various portions of the phosphor screen will have varying brightnesses depending on the potential of the target section, as seen in figure 16.7(b). Thus a half-tone (*i.e.* various tonal values) display is produced similar to that of a normal oscilloscope. If there were no positive ions this condition would remain for a long period, since the flood-gun electrons will not fall on the target because it is below point F. It is possible that some portions may be charged sufficiently to go to the stable point D, as in the bistable mode, but this makes little difference to the brightness compared with that at F [see figure 16.7 (b)].

Positive ions are produced by ionization of the remaining gas in the tube by the flood-gun electrons and these fall on the negative portions of the target, so charging them in a positive direction until they reach the stable point F. Thus the whole screen (*i.e.* background of the trace) gradually increases to maximum brightness called 'fade positive' and the displayed trace disappears into the background. This occurs in about five to 10 minutes. This is called the DISPLAY or READ mode and, of course, a display lasting this length of time may be sufficient.

However, the trace can be retained by switching to the STORE mode. In this condition the flood-gun is cut off (by suitable change of grid or cathode voltage) hence very few positive ions are produced and the charges written by the writing gun are retained on the target. Under these conditions the trace can be stored for hours; it can be displayed at any time by switching on the flood gun. However, the TOTAL display time is limited to five to 10 minutes. It is possible to switch off the oscilloscope and to store the trace for several days, to be displayed again by switching to the READ mode. Certain precautions are necessary so that the trace is not erased by transient conditions that occur when switching off or on.

If the voltage corresponding to point G is made more negative than the brightness cut-off point X, it will take longer for the target to charge to point F, hence the rate of rise of the background is reduced. However, the more negative point G the greater must be the charge given to the target by the writing beam, therefore the writing speed is reduced (for a given beam current).

There can be an intermediate stage. If the flood-gun beam is reduced there will be fewer positive ions produced, and it will take longer for the ions to upset the charge on the mesh. Naturally, the brilliance will be reduced since there are fewer electrons to fall on the phosphor screen. Thus one can trade brilliance for storage time. This usually is not done by reducing the steady flood-gun current but by pulsing it on and off, and varying the ratio of the on-to-off periods. As the off period is increased the number of flood-gun electrons are reduced. This reduces the number of positive ions and the rate of charge of target, but also reduces the brightness of the display. By this means the display time can be increased by, say, a factor of 10, but with $\frac{1}{10}$ of the brightness.

A rather complex pulse is applied to erase the trace, as in figure 16.8. First, the target mesh is made positive (V_1) above the first crossover point A (say +200 V) so that all parts of the mesh charge up to the stable upper point D.

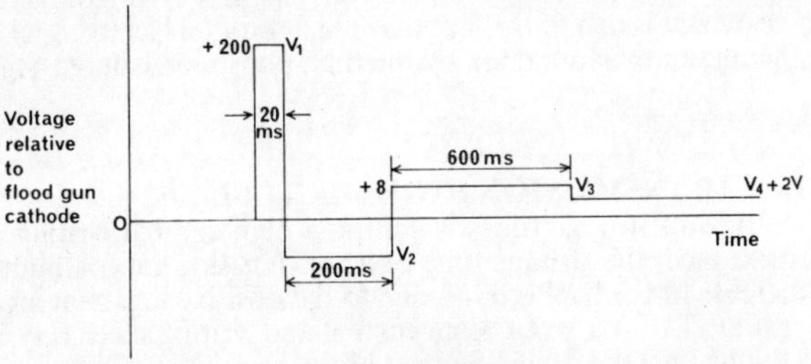

FIG. 16.8. ERASE PULSE FOR HALF-TONE TRANSMISSION TUBE (NOT DRAWN TO SCALE)

(This is essential because some parts of the screen may have been charged by the writing beam above A to the upper stable point D). The pulse may be of 20 ms duration. The voltage is then reduced to voltage V_2, which is below that of the flood-gun cathode and removes any positive ions in the area of the mesh (for, say, 200 ms). The mesh is now given a small positive voltage V_3 (say +8 V), as already explained. This causes the target to charge to the lower stable point F. This may have a duration of 600 ms. The voltage is then reduced to a small positive voltage V_4 (say +2 V) which takes the target to point G, and the screen is cut off ready for writing another trace. During this erase process the e.h.t. to the phosphor may be removed so that more flood electrons are available to charge the target.

If the tube is required to record an uncontrolled transient it can be placed into another condition, sometimes called AUTO. Until the transient occurs the flood gun is off and the background remains dark, the target remaining at point G. When the transient occurs the trace is written by the writing beam and the flood gun then automatically switches on to give a display. Alternatively, the flood gun can be maintained off, and when required it can be switched on manually.

The tube can also be operated as a variable persistence tube (similar to a long afterglow tube) with a persistence time from, say, 1 second to 1 minute. This can be most useful when viewing low-frequency phenomena. The tube is written in the normal manner, but is subject to short duration erase pulses. Short duration (approximately 1 μs) pulses of small magnitude, say 8 V, are applied to the target mesh. During the period of the pulse the target is taken to a potential corresponding to F, or above, and hence collects electrons, which tend to cancel the positive ions collected during the remaining period. Thus the background is charged back towards G, and the background does not come bright as is normal. The written portion being more positive will collect more electrons; it is also charged towards point G, *i.e.* it tends to be erased. For long persistence time the pulse rate is low, say 1 per 15 millisecond, but by increasing the pulse rate to, say, 1 per millisecond the persistence is reduced to, say, 1 second.

This may be used to examine repetitive waveforms, with superimposed random noise. The noise will decay rapidly, but the repetitive signal will remain relatively bright and free from noise.

The tube can also be used as a normal oscilloscope, the flood gun being cut off and the e.h.t. may also be reduced. The storage mesh is given a negative voltage, say -30 V. In practice one would not normally use the tube regularly in this manner owing to its cost.

The storage tube is very expensive, is somewhat fragile, and necessitates care so that the target is not burnt by a bright stationary spot. Its life tends to be shorter than that of a normal tube. The writing speed in the storage mode is not so high as a normal tube, say 1 cm/μs, depending on the tube and the flood-gun current. The big advantage of the transmission-type storage tube is its bright trace, which is due to the fact that the flood-gun electrons which do get through the mesh are accelerated by the high phosphor voltage, *e.g.* 6–7 kV.

TRANSFER TRANSMISSION STORAGE TUBE

A difficulty with storage tubes is getting a high enough writing speed. As already mentioned, the storage time of a transmission tube is limited by the gradual increase of the background due to the positive ions that are collected by the target mesh. For a given beam current and writing speed (*i.e.* number of electrons falling on a particular element of the mesh or the charge given to the mesh) the voltage change will be greater if the capacitance of each element is reduced. Thus, by modifying the mesh structure to give a low capacitance between the insulated portions and the conducting mesh the writing speed will be increased. However, in the same way, the positive ions will also cause a greater change of voltage and reduce the storage time. Thus high writing speed can be traded for storage time, but obviously at some stage the storage time will be too short to be of any use. To overcome this Tektronix manufacture a tube with two storage meshes; a fast one (*i.e.* nearer to the writing gun) with low capacitance, and a slow one next to the phosphor screen. The trace is written on the high-speed mesh (about 100 times faster than the slow mesh) and within a period of 0·1 second after the writing on the fast screen the image is transferred to the slow screen. This gives a long viewing time but, of course, at the high writing speed settled by the first mesh. Improved speeds are possible by increasing the accelerating voltage (which increases the secondary emission ratio) and by reducing the size of display.

To make the transfer a very high voltage pulse is applied to the second target (slow speed) with the effect that the flood-gun electrons now pass through those portions of the fast target (similar to the operation of a normal transmission tube). These fast electrons 'write' the display on to the slow target as with a normal beam. The second target is operated in the bistable mode, as already explained. After the pulse the second target is restored to its normal working voltage; the first target is made positive so that flood-gun electrons can now pass through, and the second target then operates as a normal bistable target. Since the target operates in a bistable mode the display can be viewed for a long period without deterioration, the high voltage on the aluminized screen giving a bright trace.

By erasing the trace on the fast target it is possible to produce multiple displays, which are written on the slow bistable target. The second target can be used normally as either a bistable or half-tone storage, the fast target now being connected to the collector mesh and so has a negligible effect. This special type of tube will give storage writing times of 1000 cm/μs.

DIRECT BISTABLE STORAGE OR BISTABLE PHOSPHOR STORAGE OSCILLOSCOPE

This is manufactured by Tektronix and does not have a transmission type storage mesh. The basic construction is shown in figure 16.9. The tube has a writing gun G_1 and flood guns G_2, as in the transmission tube. The screen consists now of a phosphor coating S which acts as the target, and a transparent

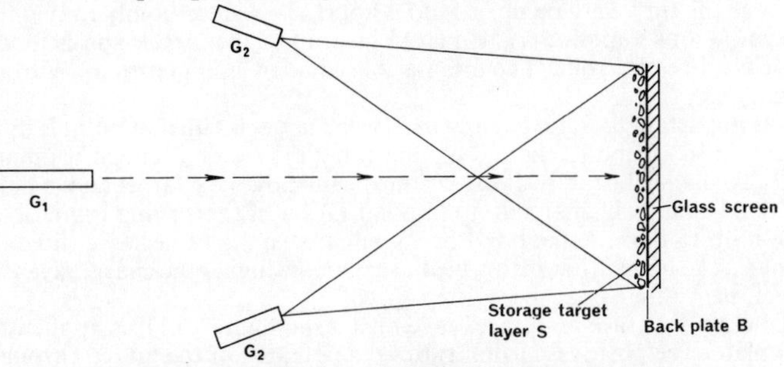

FIG. 16.9. PRINCIPLE OF DIRECT BISTABLE STORAGE TUBE

backplate B which acts as a collector. The storage target layer consists of a large number of insulated elements of phosphor. Although the phosphor material is basically the same as on a normal tube, the phosphor layer in this type of tube is critical. It must be porous so that particles are insulated from each other, and so that the backplate can collect the secondary electrons emitted. As previously, the cathode of the flood guns will be taken as the reference voltage. Suppose that the storage target layer is at a voltage below the first crossover point A of figure 16.3 and that only the flood gun is operating. The electrons from the flood gun will charge the target elements in a negative direction to point F slightly below the potential of the flood-gun cathode. Thus, assuming perfect conditions, no further electrons will reach the target and there will be no glow from the screen. If the writing gun is now operated the electrons from it will strike the target at a high velocity because the writing-gun cathode is at several kilovolts negative with respect to the flood-gun cathode. (The electron density of the writing gun is very high compared with that of the flood gun). These writing electrons will cause the screen to glow in the normal manner and produce a trace. However, if they are in sufficient numbers they will also charge the parts of the target they hit in a positive direction above the first crossover point A towards point D. Any part charged above point A will now collect flood electrons and it will be charged to the upper stable operating point D, which is approximately equal to the voltage on the backplate which acts as a collector. Thus, those parts written on by the writing beam are at potential D, say +300 V, and will attract flood-gun electrons and will glow. The unwritten parts will remain at roughly flood-gun cathode potential and repel electrons. This is shown in figure 16.10. Where the electrons bombard the written portion, so causing fluorescence, secondary electrons are produced (see figure 16.3) equal to the number of primary electrons (so that the target remains at a fixed potential). Thus an image of the trace, produced by the writing

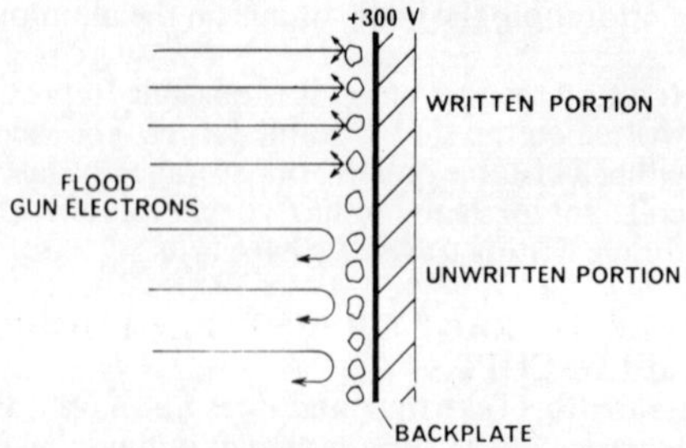

FIG. 16.10. EFFECT OF CHARGE ON SCREEN OF BISTABLE TUBE

gun, results and will continue theoretically so long as the flood gun is on. In practice the time may be up, say, to 4 hours. To reduce bombardment of the screen by ions a mesh may be placed in front of the target, and is made the most positive electrode. The mesh is placed a few millimetres away from the target.

It is important to note that any part of the target is either at point F or point D (figure 16.3), and so there is no half-tone; the screen is either lit up or not, hence the name 'bistable'. If the writing beam moves the target above point A, the flood beam will charge it up to point D; but if the writing beam does not take it up to point A the flood beam will charge it in a negative direction to point F. The higher the writing beam current the higher the charge given to the target, hence the faster the writing speed.

To erase the trace a voltage waveform, as in figure 16.11, is applied to the backplate. The positive portion raises the potential of the target, through the

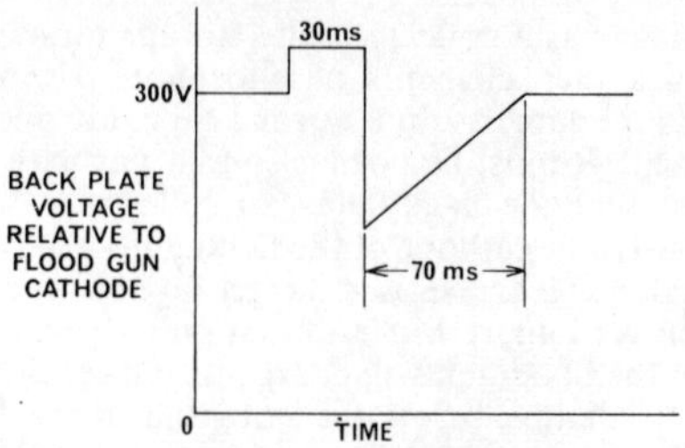

FIG. 16.11. ERASE WAVEFORM FOR BISTABLE TUBE

capacitance, above point A and the whole target is written by the flood gun, *i.e.* taken to point D. The negative portion takes the target below point A and the target is charged by the flood gun to point F. As the potential on the back-plate slowly rises, the target potential tends to rise with it; but provided the rise is not too rapid, the flood-gun electrons falling on the target maintain the potential below point A, and at the end of the pulse it is charged to point F ready for the writing of another trace.

Some improvement in writing speed can be obtained by using an ENHANCEMENT control. If the writing beam is not sufficient to charge the target above point A then, if nothing else happens, the flood gun charges this portion of the target back to point F and there is no trace. The idea of the enhancement is to apply a positive pulse of suitable amplitude to the backplate (or a negative pulse to the flood-gun cathode in some cases, which has the same effect), this pulse being applied after the trace. Suppose that the writing beam takes the target to point R (figure 16.4), *i.e.* below point A. If a positive pulse is now applied to the backplate BEFORE the flood gun has a chance to charge the written portion in a negative direction (and the pulse is large enough) it will take this portion of the target above point A. The flood gun will now charge this portion of the target in a positive direction to point D and writing has occurred. Provided the pulse is not too large the remainder of the target will remain below point A, and will be charged back to point F after the pulse. The mechanism is rather more involved and the magnitude and duration of the pulse are important. If the pulse is too large it will take the whole target above point A and the whole target is written. The magnitude of the pulse can usually be varied to give the best results.

Another method of effectively giving a faster writing speed, but applicable only to repetitive traces, is called the 'INTEGRATE mode'. In this case the flood gun is cut off during several scans of writing. Thus the first scan of the writing beam may charge the target to point R (figure 16.4) and as there are no flood-gun electrons it will stay at this potential. The next scan will be additive and may charge it above point A. If the flood gun is now switched on the electrons will charge this part of the target in a positive direction to point D. Hence, writing has been obtained by a writing beam which was not large enough to charge the target sufficiently in one scan.

Tektronix make a bistable storage oscilloscope with a split screen. The 6 × 10 cm screen is divided into two portions measuring 3 × 10 cm. Each half can be operated as a conventional or storage oscilloscope independent of the other half, which is particularly useful when two traces are produced by using a switched beam. One half, for example, can be used to store a reference trace while parameters of the circuit are changed and the trace shown on the other half in the normal way.

Erasure can be manual, or automatic erase can take place a predetermined time after a single trace has been written, the timebase being triggered only once until erasure takes place. The normal maximum writing time is of the order of 0·5 cm/μs and up to 5 cm/μs in the enhance mode. This type of tube is simpler, more robust and cheaper than the transmission tube. Split screen construction is not practicable in a transmission type tube. However, the brightness of the stored trace is much less than that of a transmission tube owing to the low voltage available to accelerate the flood electrons to the screen (hundreds of volts instead of, say, 6 kV as in the transmission tube). The display is a bistable one and has no half-tones, *i.e.* there is only brightness of fixed value or darkness. This makes photography rather easier as there are only these two fixed levels. In practice there is some slight glow from the background, depending on the voltage applied to the backplate. The higher the voltage on the backplate the brighter the trace, but the brighter the background. The actual operation of the target is complex due to some leakage through the insulation of the phosphor from the backplate.

The electrical performance of the storage oscilloscope such as bandwith, and timebase facilities, has little to do with the storage facilities, and hence these are similar to those of normal oscilloscopes and in the similar manner can vary considerably. In some cases the same plug-in units can be used as with the non-storage oscilloscopes.

TRANSIENT RECORDER

This equipment is included here as it performs a function similar to a storage oscilloscope. The signal fed to a storage oscilloscope is an analogue signal which is stored as such on the mesh. It would be convenient to have an analogue store where a transient could be fed into it and the information stored. If the information could then be fed out at a slower or faster rate it could be displayed on, say, a normal oscilloscope, particularly if a repetitive output could be obtained. A tape recorder is, of course, an analogue storage device, but of limited use for storage of transient waveforms. The idea of the transient recorder is to convert the analogue information into digital information using an analogue-to-digital converter (A/D converter) and then store the information in a digital store. This information can then be recovered and fed out at any speed using a D/A converter. The output can also be repetitive, and hence displayed as a continuous waveform on a normal oscilloscope.

Transient recorders of this type are made in this country by Data Laboratories Ltd and are intended to feed a normal oscilloscope. The output can also be fed to an X–Y recorder, tape punch or computer, but such applications will not be considered. Three models are at present produced: one going from d.c. to 100 kHz; one from d.c. to 3 MHz; and one to 20 MHz. The controls are similar to those of an oscilloscope and the input has a similar input impedance: 1 MΩ 48 pF. The sensitivity is controlled in 1, 2, 5 steps from 0·01 V to 50 V full scale (depending on the model), and the input may be d.c. or a.c. coupled. Triggering arrangements are generally similar to those of an oscilloscope and may be internal or external. The timebase arrangement is similar to that of a dual timebase oscilloscope, there being two sweep speeds set by two controls A and B. The sweep speeds are such as to produce a total sweep time of 200 μs to 10 s. There is also a digital delay control. The recorder may be used in three ways:

(a) Delayed Sweep. In this case the delay is used to give a delay between the trigger pulse and the start of the sweep. At the end of the sweep (settled by the number of samples that can be stored) the recorder changes to the display mode and feeds the oscilloscope. In the single-shot mode further triggering does not occur, and in the continuous mode further triggering can be prevented for a period determined by the B sweep.

(b) Switched Sweep. A trigger pulse now starts the recording at a rate settled by sweep speed A for a time settled by the delay control. The remaining information is stored at a rate settled by the sweep B. This corresponds to the normal mixed sweep trace. However, unlike a normal oscilloscope the speed of time sweep A can be faster than B. This is useful for showing the rise time of a transient that has a long tail.

(c) Pre-triggered Recording. This is a useful facility. It saves delay lines etc., and enables a signal to be obtained prior to and after triggering. The recorder is started and records information continuously at a rate settled by sweep control A, and the memory is filled up with information before the trigger pulse. The number of pre-trigger samples to be kept is determined by the delay control. When triggering occurs the remainder of the store is filled up with information taken after triggering.

The stored signal is fed to a normal simple oscilloscope and may be expanded with ×5 and ×10 controls provided, together with X and Y position controls. The display can also be inverted. The display time is 1 ms. The recorder uses a 1024 MOS digital memory (in the two lower frequency models), which is fed from a 5 MHz 8-bit analogue-to-digital converter. The number of samples taken and displayed is 1000.

Oscilloscopes using this principle are manufactured in America. It would appear that such devices will become more important, particularly if the cost can be reduced which is very likely as the cost of converters and stores is decreasing. This type of recorder is very versatile and can be used for recording high-speed transients (within the limits of its frequency response), and also slow-speed transient information. Both can be displayed as a continuous display on any simple oscilloscope. A digital storage oscilloscope is now produced by Gould Advance Ltd.

17

Sampling Oscilloscopes

SAMPLING oscilloscopes are also complex instruments and again only the basic principles can be considered.

The two basic types of sampling oscilloscopes are:

(a) Those using equivalent time sampling or non-real time sampling.

(b) Those using real time sampling.

Both work on the same principles, but it is convenient to consider them separately, and type (a) will be taken first.

(a) EQUIVALENT TIME SAMPLING

As the frequency of the signal to be examined gets higher there comes a limit which Y-amplifiers cannot be constructed to reach. This limit has increased over the years and at the present time (1975) is around 500 MHz. Perhaps even more important is the rise time. In Chapter 4 it was shown that there was a relationship between the maximum bandwidth (in this case, maximum frequency, -3 dB point) of operation of an oscilloscope and the rise time. This was

$$\text{rise time} \times \text{bandwidth} = 0{\cdot}35$$

hence a 500 MHz oscilloscope will have a rise time of about $0{\cdot}35/500\ \mu s =$ $0{\cdot}7$ ns. It was pointed out that to avoid large errors in the measurement of the rise time of a signal, the rise time of the oscilloscope should be small compared with that of the signal. Thus, without the use of a sampling oscilloscope one is limited (at present) to about these figures. By connecting directly to the plates of the tube it is possible to go up to about 1000 MHz (1 GHz) but, of course, a relatively large voltage is then required, so this method is limited.

By using a sampling oscilloscope it is possible to extend the bandwidth to, say, 20 GHz (20,000 MHz) with rise time of the order of 20 ps (1 pico second (ps) $= 10^{-3}$ ns), which is a great advance on the performance of a normal (real time) oscilloscope.

A sampling oscilloscope can be used only on a repetitive waveform, and the idea is to measure the magnitude of the waveform at a certain instant on the first cycle. On the second cycle it is measured at another instant, and so on. If sufficient instants or samples of the amplitude are taken they can be dis-

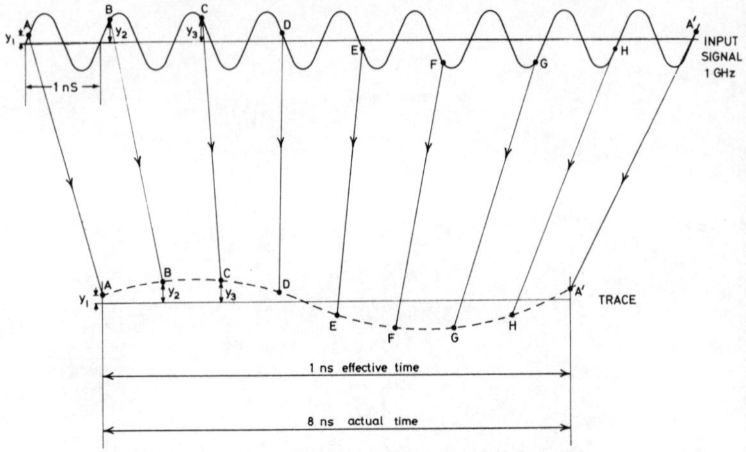

FIG. 17.1. PRINCIPLE OF SAMPLING OSCILLOSCOPE

played on an oscilloscope of very limited bandwidth. This is shown in figure 17.1. For simplicity, in both the explanation and the diagram it is assumed that a sample is taken from each cycle. This is not normally the case as the sampling frequency is quite low, say 100 kHz, hence sampling only occurs every x cycles where x may be large when the input frequency is high. Again for simplicity it will be assumed that a sample is taken during each cycle of the waveform. Consider an input frequency of 1 GHz (1,000 MHz) and hence the time of a cycle is 1 ns (1/1000 = 0·001 μs). At instant A a measurement of the magnitude y_1 of the waveform is taken over a very short period of time. The voltage measurement is displayed on an oscilloscope with, say, a total timebase sweep time of 8 ns. On the next cycle of the signal a sample is taken at B, which is slightly more than 1 ns in time from A. Let this amplitude be y_2. This is displayed on the oscilloscope as y_2 again approximately 1 ns from y_1 or approximately $\frac{1}{8}$th along the X axis, since the total X deflection time is 8 ns. This continues to point H and a complete cycle has been displayed on the oscilloscope as samples (in practice, as a series of dots similar to those shown). Only 8 samples have been used in order to simplify the figure. It is important to note that the actual timebase sweep time of the oscilloscope is 8 ns, but the trace indicates a waveform having a cycle time of only 1 ns. Thus the waveform is shown in non-real time or equivalent time. In fact, the timebase sweep setting would show a total sweep time of 1 ns (0·1 ns/div with 10 divisions). Thus the timebase time/div control shows the equivalent time and NOT the actual timebase speed. In practice this ratio is much greater than in this example.

There are two methods of sampling:

(i) That shown, known as sequential mode sampling, in which the samples follow one another, the second being slightly later in the cycle than the first.

(ii) Random mode sampling, *i.e.* the samples are taken at random.

The basic principles of the two methods are the same. Sequential sampling will be assumed for the present; the advantages of random sampling will be explained later.

(i) Sequential Mode Sampling

Perhaps the first query might be: Why can we take samples in this way for a time which must be a fraction of the cycle time of the signal and yet cannot amplify the signal directly? The answer is that it is possible to use very high speed diodes in a gate for sampling; these will operate at a frequency far beyond that possible by a transistor amplifier.

At least two methods may be used for taking and displaying the samples – namely open cycling sampling and error sampling feedback system.

The basic idea of the open cycle sampling system is shown in figure 17.2. R_1 represents the internal resistance of the signal source and D_1 is the fast sampling diode. This diode is normally reverse biased by the positive voltage

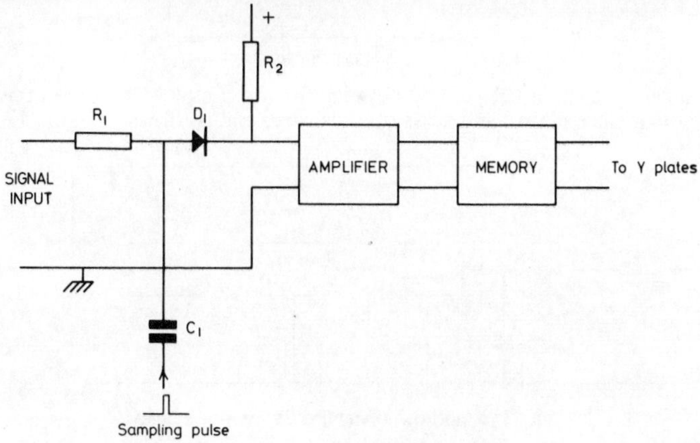

FIG. 17.2. OPEN CYCLE SAMPLING

fed through R_2 and a normal signal is not sufficient to turn it on. Very short duration positive sampling pulses are fed through C_1 which are large enough to turn on the diode. However, the voltage fed to the diode is the algebraic sum of the magnitude of the input signal, at this instant, and the magnitude of the sampling pulse (constant). Hence, the actual amplitude of pulse developed across R_2 will depend on the magnitude of the signal at the instant of sampling. Thus the pulses fed to the amplifier are amplitude modulated by the input signal. The amplifier serves two purposes; it increases the amplitude of the pulse; but it also increases the length of them to, say, 1 μs. This type of amplifier is commonly referred to as a 'stretch amplifier'. The pulses are then fed into what is called 'a memory', which produces an approximately square pulse proportional to the magnitude of the pulse into the amplifier (and so proportional to the signal amplitude at this instant), and of duration, say, 2 μs. This is fed to the Y-plates of the tube to give the Y-deflection. During this period the tube is unblanked (by a pulse derived from the sampling pulse) so as to produce a dot on the screen. As will be seen later, during this period the X movement is stopped and so a spot is produced and not a short line. At the end of this pulse the memory is reset ready for the next pulse. In this way a series of dots are produced representing the input waveform.

The basic idea of the error sampling feedback system is shown in figure 17.3. When a sampling pulse is applied to the gate, at A, the gate is closed, and hence the signal applied to the capacitor C_1. This capacitor, C_1, will charge through R_1 (which represents the internal resistance of the source and of the sampling circuit) and ideally it would charge to the input voltage at this instant. However, the time of gate closure is too short, and it may only charge to $\frac{1}{10}$th of the input voltage. This is known as the 'sampling efficiency' and in this case is only 10%. When the next sample occurs at B the input voltage is greater, and the capacitor is charged to a higher voltage, but again the change of

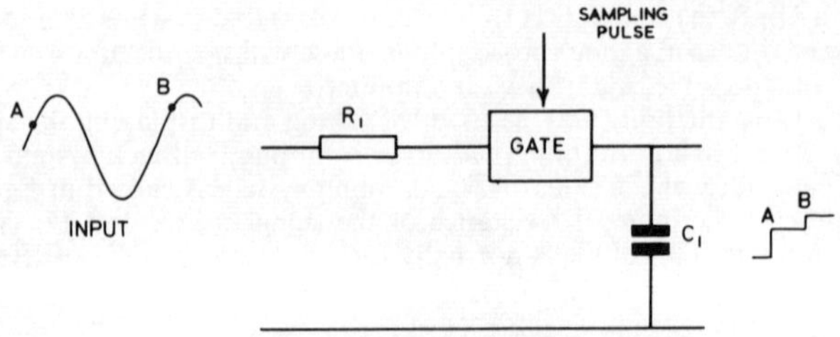

FIG. 17.3. SAMPLING GATE

voltage will be only 10% of that between the two samples. To correct this low sampling efficiency a feedback system is used, and is shown in figure 17.4.

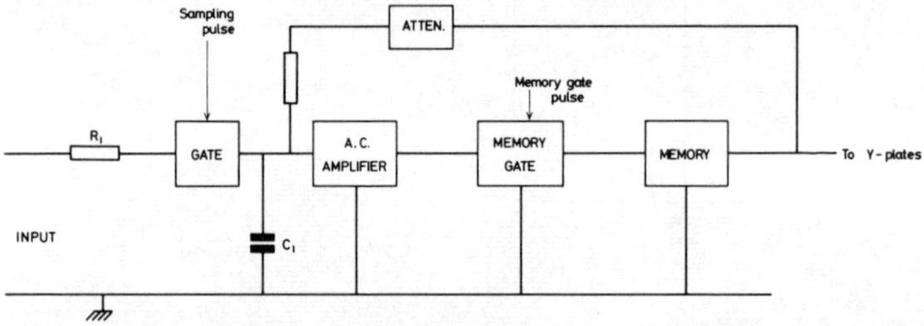

FIG. 17.4. ERROR SAMPLING FEEDBACK SYSTEM

The voltage across C_1 is amplified by the a.c. amplifier which also stretches the pulses in length. These are then fed to the memory circuit through a memory gate. The memory gate is opened for a longer period than the sampling gate, but fed with pulses derived from the pulse generator feeding the sampling gate. There is feedback from the memory to the gate output through an attenuator. The idea is to feed back a signal to C_1 from the memory so that it is charged to the value of the input signal. Suppose that the input signal is 0·1 V and that the sampling efficiency is 20%. In the absence of feedback, this will produce a voltage across C_1 of $0{\cdot}1 \times 0{\cdot}2 = 0{\cdot}002$ V. This is amplified, say, to 2 volts in the memory circuit. The attenuator is set so that the voltage fed back raises the voltage across C_1 to 0·1 V. Thus, when the second sample is taken the capacitor has a voltage equal to that of the signal when last sampled. The loop gain must, of course, be correct as otherwise C_1 is not charged sufficiently or is overcharged.

The gain of the a.c. amplifier is made variable (in steps) so as to give different sensitivity ranges, and so the attenuator must be coupled to this sensitivity control to maintain a constant loop gain.

When the next sample is taken the same action takes place. The voltage across the memory circuit is increased or decreased (it is not reset to zero between samples) in the same way as the input changes at the instant of sampling. The output from the memory feeds through suitable amplifiers to the Y-plates, giving the vertical deflection. As in the previous case the tube is unblanked to show a dot.

As in any other oscilloscope some means of triggering is required, which may be from the signal or from an external source. The actual arrangements for triggering are involved, and only basic principles are considered. Again, assuming that a sample is taken from each consecutive cycle, the sampling pulse generator must be triggered by the signal (or externally) in a similar way to a normal oscilloscope. The sampling pulse generator is a complex circuit required to produce the very short sampling pulses. It also has to produce other pulses, etc. Depending on which system is used it must provide reset pulses or memory gating pulses. It must also be arranged that the next sampling pulse is delayed slightly on the first and so on, until the whole waveform has been sampled. At the end of the sweep the process starts all over again. It must also operate in conjunction with the horizontal timebase or sweep generator. The sweep generator is not a sawtooth or linear ramp generator as in a normal oscilloscope but is a staircase generator producing a waveform as in figure 17.5. Thus the horizontal motion of the spot is in steps or jumps. During the period when the voltage is constant and the horizontal

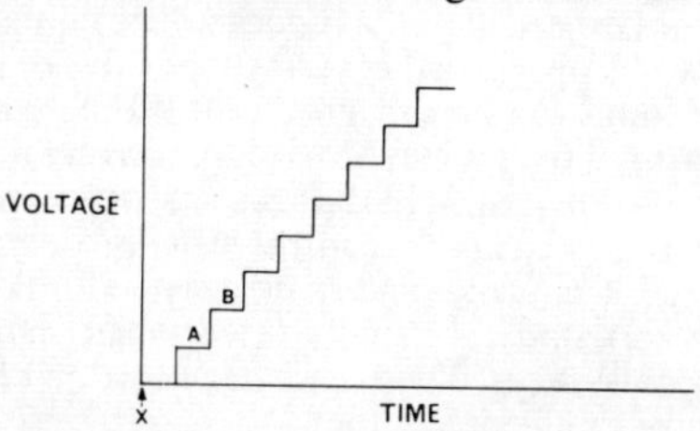

FIG. 17.5. STAIRCASE SWEEP WAVEFORM

motion is therefore zero, the sample is displayed by unblanking the tube. Hence the sample spot is displayed. Between samples the spot moves rapidly from one horizontal position to the next. It is this staircase waveform that normally controls the operation of the sampling pulse generator circuit. One method used is for the trigger circuit to start a fast ramp generator. The output of this is fed into a comparator together with the staircase waveform. As soon as the ramp voltage equals the staircase voltage a sampling pulse is produced.

A simplified block diagram is shown in figure 17.6. Consider the first step A (figure 17.5) of the staircase waveform. The fast ramp generator, starting at X (and assuming a positive ramp) will only run for a short time before it equals

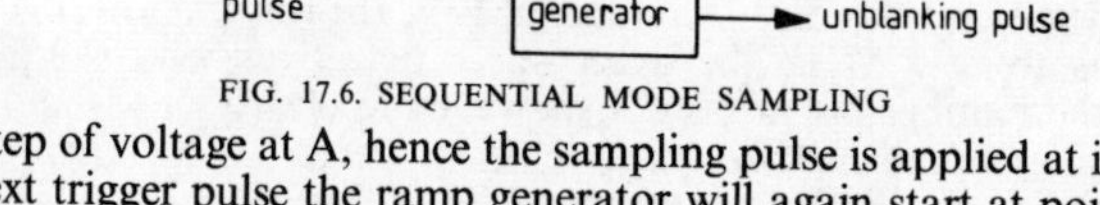

FIG. 17.6. SEQUENTIAL MODE SAMPLING

this low step of voltage at A, hence the sampling pulse is applied at instant A. On the next trigger pulse the ramp generator will again start at point X, but will now have to run for a longer period until it reaches the voltage at B of the staircase generator. Therefore, the sampling pulse (and other pulses) are delayed and the second sample taken at a time corresponding to B. This continues to the end of the staircase generator waveform (when the spot will be on the right-hand side of the screen) and the sweep generator is reset again. Other methods of producing this result may be used but will not be described.

Having considered the basic principles we will now consider the arrangements in more detail. Sampling units are often made as plug-ins to normal oscilloscopes, but the units must replace both the normal Y-amplifier and the normal timebase by the special staircase generator. The X and Y final amplifiers of the frame are used, of course.

The actual sampling head, commonly containing four diodes, is the vital part of the circuit which settles the rise time and bandwidth. This may be built in to the plug-in or a separate unit. The latter arrangement has the advantage that it can be placed next to the circuit under investigation and there is less chance of distortion of the waveform by the connecting leads to the sampling head. The sampling unit is usually designed with 50 Ω input impedance for matching to 50 Ω cables, but high impedance (1 MΩ) ones are available. Probes are also available. As the maximum reverse voltage of the diodes is relatively small the maximum signal that can normally be applied to the sampling head is usually about 2 V peak-to-peak. It is important that the maximum input is not exceeded or the diodes will be damaged. There is usually an OFFSET control which applies a d.c. voltage (positive or negative) so that any d.c. voltage on the input can be cancelled out. If the magnitude of this offset voltage can be measured then it can be used to measure the d.c. component of the input. There is also the normal VERTICAL POSITION control. As in a normal oscilloscope there is a sensitivity control usually operating in steps, also there may be a variable control. The range may be 2 mV/div to 200 mV/div. A control called SMOOTHING is usually fitted which can be used to reduce effects of noise but must be used with care or distortion of the trace can occur. It reduces the gain of the feedback loop.

It is quite usual to make dual sampling units similar to normal dual amplifier oscilloscopes, the two traces being time shared between samples. Such facilities as adding of the two inputs and subtraction by use of inverting switches are common.

Turning now to the timebase unit, this varies considerably between oscilloscopes, but often has many of the facilities (and others) provided by a normal timebase unit. A sweep TIME/DIV control is fitted and basically operates in the same way as a normal oscilloscope. It may be calibrated from say 100 μs/div to 1 ns/div. However, this is, of course, the EFFECTIVE timebase sweep speed and NOT the actual timebase speed which is quite low. A trigger level control and polarity control is also common and operates in the same basic way except that this is controlling the triggering of the pulse generator and the sweep generator only indirectly. A horizontal sweep magnifier may be fitted which expands the horizontal trace in exactly the same way as a normal oscilloscope, *i.e.* it increases the gain of the X-amplifier. Dual timebases may be used so that a trace by one timebase can be delayed by the other. A variable hold-off control may also be fitted.

Often the samples/div may be changed from, say, 5 to 1,000, the greater the number of samples the less the dot structure. This controls the number of steps in the staircase waveform. Figure 17.7(a), (b) and (c) show the effect of increasing the samples/div. Diagram (a) is 5; (b) is 10; and (c) is 50. These waveforms are at 1 MHz, the sweep speed (effective) being 0·2 μs/div. The effect of the number of samples is shown more clearly at (d) and (e). Both these are taken with the horizontal sweep magnifier at × 2, (d) having 5 samples/div and (e) having 1,000 samples/div.

There may be a TIME EXPANDER control. This expands a portion of the waveform but maintains the same number of samples/div. Incorporated with this may be a time delay control so that the actual portion of the waveform being viewed can be selected. The effect of the TIME EXPANDER is shown in figure 17.7(f), (g) and (h). These are all taken with a TIME/DIV setting of 50 ns/div and 20 samples/div. At (f) the TIME EXPANDER is at × 1, at (g) × 5 and at (h) × 20. In all cases the samples/div remain the same. At (i) is shown a 680 MHz sine waveform using a TIME/DIV of 1 ns/div, 100 samples per division and × 2 time expansion.

With a dual sampling unit facilities are sometimes made to produce an X-Y display by using one unit on the X-axis and the other on the Y-axis.

As there is a delay in the operation of the triggering circuit, if the trigger

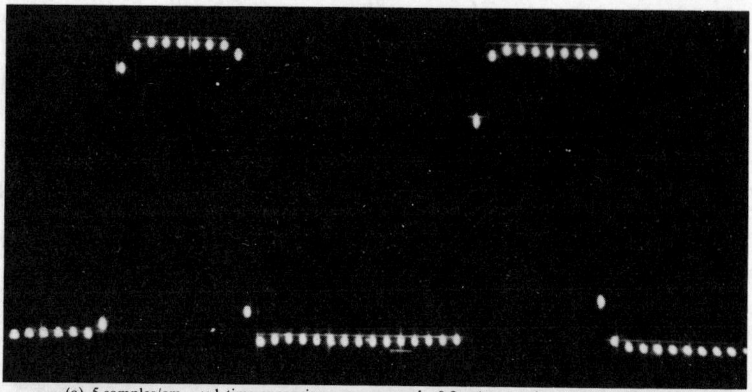

(a) 5 samples/cm × 1 time expansion sweep speed 0·2 μs/cm × 1 sweep magnification

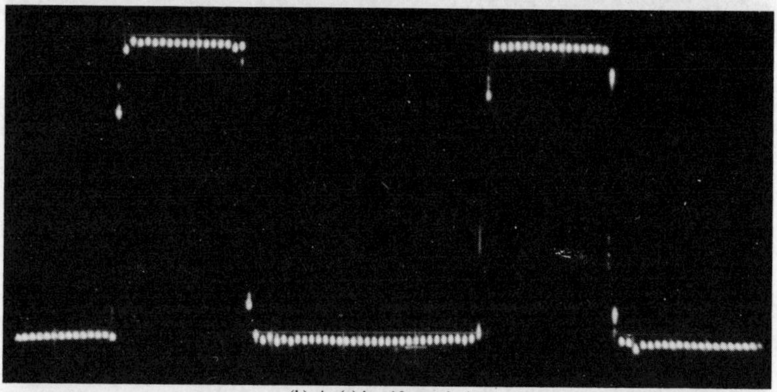

(b) As (a) but 10 samples/cm

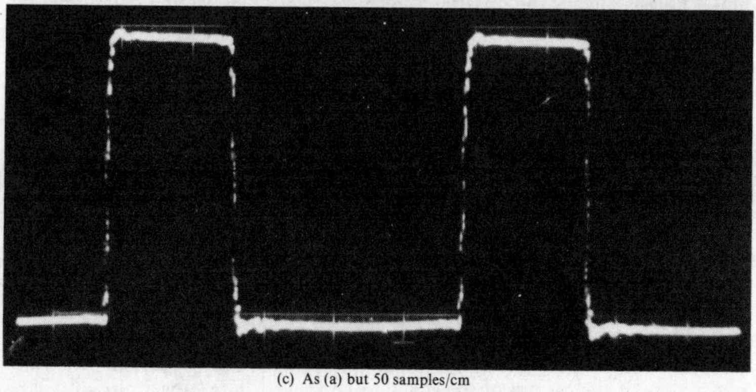

(c) As (a) but 50 samples/cm

FIG. 17.7(a)–(c). 1 MHZ SQUARE WAVEFORM

circuit is fed from the waveform under examination then the first part of the waveform will be missing (as in a normal oscilloscope with no Y-delay line). This can be overcome by triggering from a circuit that gives a trigger pulse just before the main waveform, if this is possible. Alternatively, the signal may be delayed before being applied to the sampling unit by a suitable delay

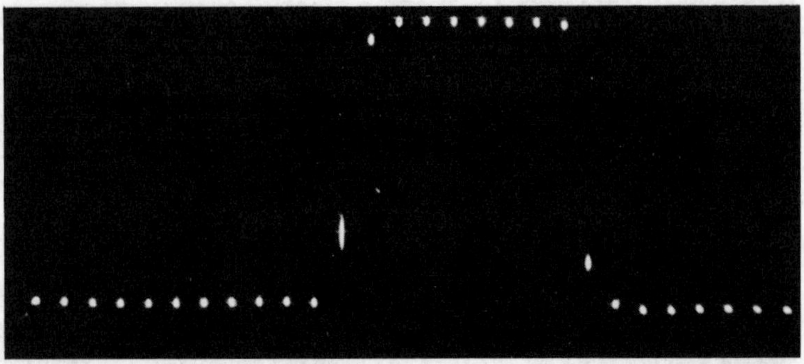

(d) 5 samples/cm × 1 time expansion sweep speed 0·2 μs/cm × 2 sweep magnification

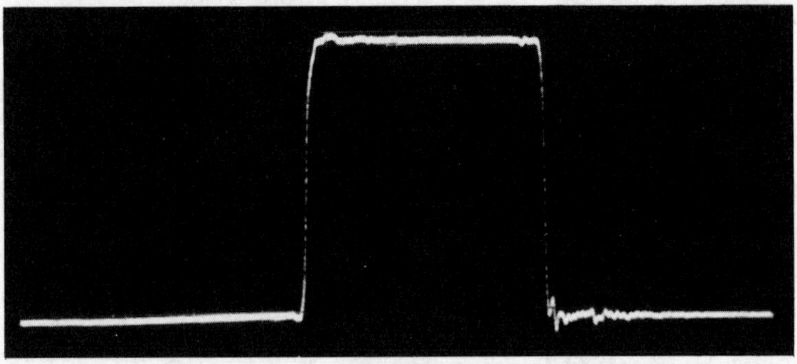

(e) As (d) but 1000 samples/cm

FIG. 17.7(d)–(e). 1 MHZ SQUARE WAVEFORM

line. There is always the possibility of distortion when using a delay line, the need for which can be removed by using random sampling as explained later.

When there is no triggering there is no display on the screen under normal conditions, and one is not sure whether the trace is off screen vertically or is not triggered. Some later units use an automatic timebase position which operates in a way similar to that of a normal oscilloscope with an AUTO trigger setting, *i.e.* it produces a horizontal line even when not triggered.

(ii) Random Mode Sampling

Only the basic ideas of random sampling will be discussed. In sequential sampling each sample is taken a little later in the waveform, and the samples appear in sequence across the screen. In random sampling the sampling pulses may appear in any order. Provided the correct X position is used it does not matter in what order the samples are taken.

Consider figure 17.8 which shows two cycles of the signal under different sampling conditions. The random sampling pulses are produced by a rate-meter around the instant of the signal to be viewed. The trigger circuit will be operated at A, but on the first cycle the sampling pulse is at time B [see (b)], which is after triggering. This sampling pulse is used to sample the waveform in the normal manner, giving then a Y-signal, y_1, which is stored in the memory. This trigger pulse, at B, is now delayed by a fixed time T, as shown at (c). When the trigger circuit is operated it starts what is called a 'timing ramp', as shown at (d). The value of this timing ramp, at the instant of the delayed trigger pulse,

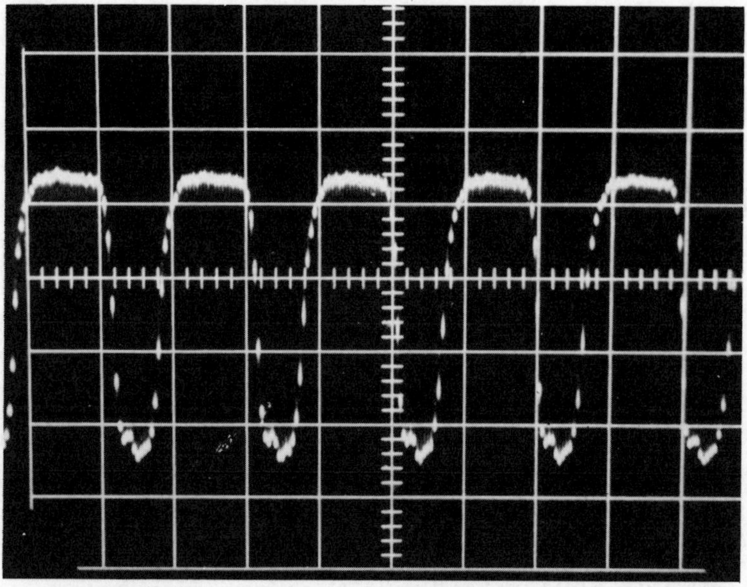

(f) 20 samples/cm × 1 time expansion sweep speed 50 ns/cm × 1 sweep magnification

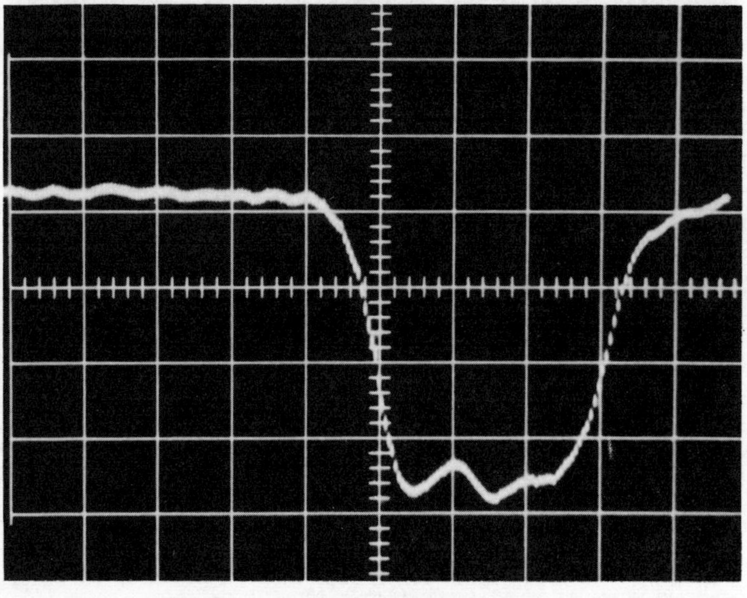

(g) As (f) but × 5 time expansion

FIG. 17.7(f)–(g). 10 MHZ SQUARE WAVEFORM

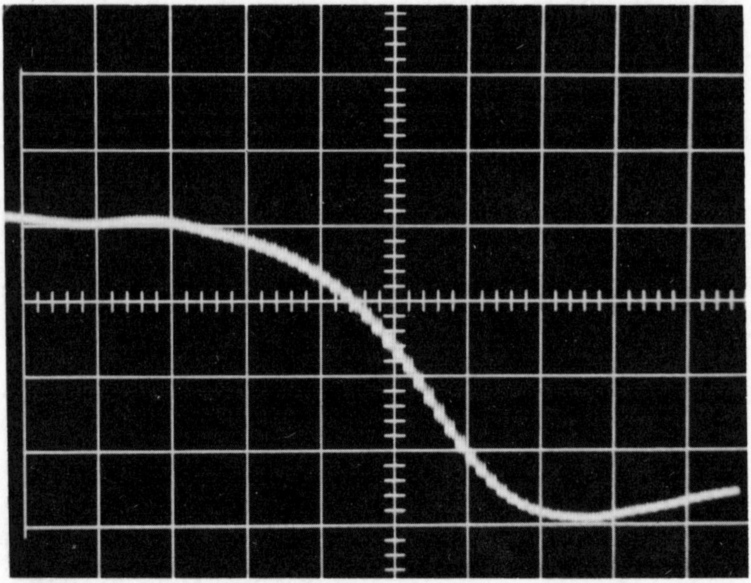

(h) As (f) but ×20 time expansion

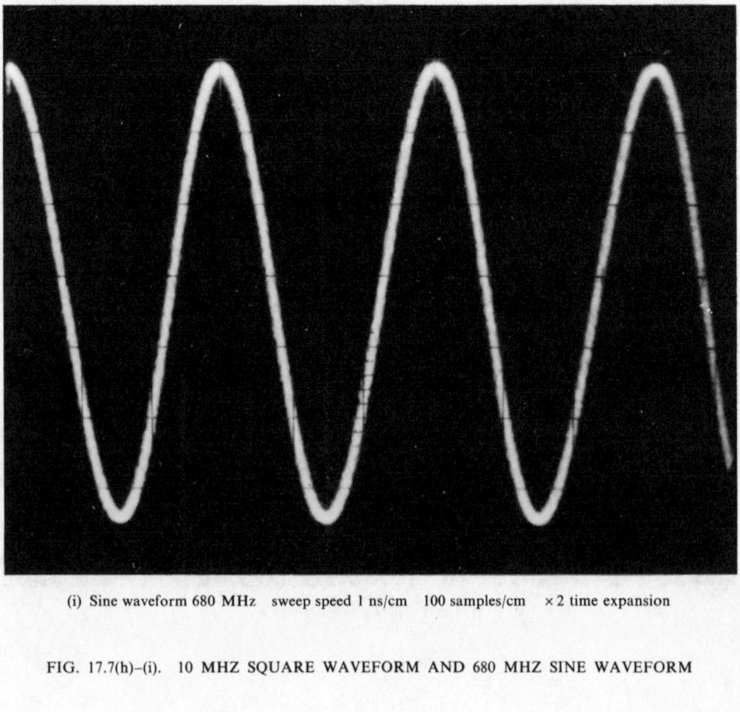

(i) Sine waveform 680 MHz sweep speed 1 ns/cm 100 samples/cm ×2 time expansion

FIG. 17.7(h)–(i). 10 MHZ SQUARE WAVEFORM AND 680 MHZ SINE WAVEFORM

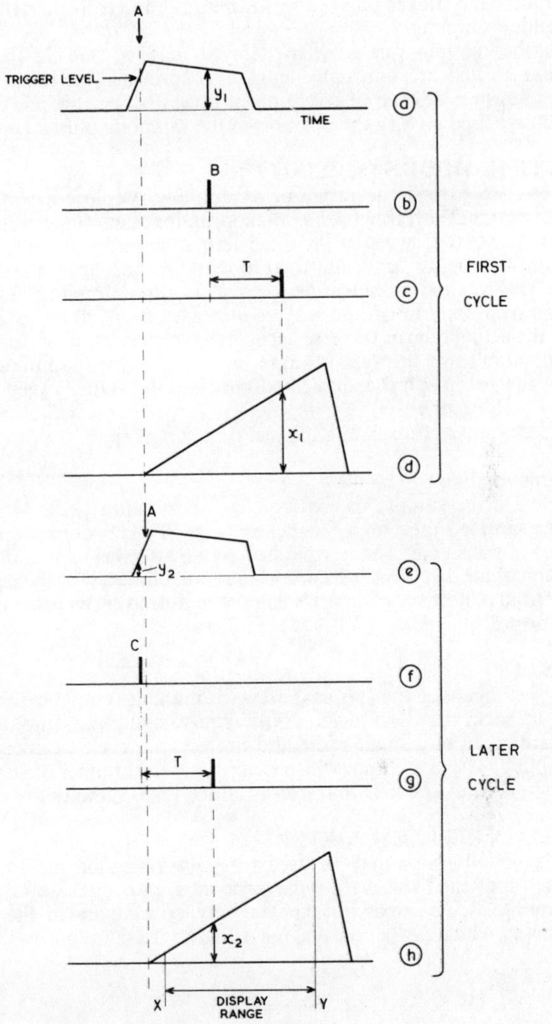

FIG. 17.8. PRINCIPLES OF RANDOM SAMPLING OSCILLOSCOPE

is obtained and stored. This value is x_1. The two signals y_1 and x_1 are then used to display a dot on the screen.

Now consider another cycle [figure 17.8(e)], where the random sampling pulse occurs at instant C [figure 17.8(f)], which is before the trigger circuit operates. This is used to sample the input waveform and obtain a value y_2. As before, this sampling pulse is delayed by the same time T as shown at (g). Again this is used to measure or sample the timing waveform shown at (h) and produce a value x_2. Again, y_2 and x_2 are used to place a dot on the screen. The action continues, so building up the display. It is important to note that a sampling pulse may occur before, at the same instant as, or after the instant of triggering. Thus the oscilloscope is able to display the whole of the waveform before and after the trigger point, and so there is no need for a delay line as with sequential sampling.

Some of the random pulses when delayed may be outside the display limits X and Y, and are unusuable pulses. To prevent this happening too often, and to ensure an even distribution of pulses between X and Y, a correcting circuit is used to vary the position of the sampling pulses.

(b) REAL TIME MODE SAMPLING

This is used to extend the range of a sampling oscilloscope in the low frequency direction. With either sequential or random sampling it is important to note that the ACTUAL sweep of the trace across the screen is much less than indicated on the time/div scale and used to read off times from the displayed waveform. This method is known as 'equivalent-time sampling'. Using this idea, if the sweep time/div on the scale is increased to much more than, say, 100 μs/div, the actual time or traverse across the screen is relatively long. There is now no point in using this type of sampling, and real time sampling is used. In this case the sweep speed as indicated becomes the actual sweep speed as in a normal oscilloscope.

There are two ways of operating the real time sampling.

(i) Free-run oscillator

In this case the sampler is fed from a free-running oscillator and the waveform sampled at a high frequency (*e.g.* 50 kHz) during each cycle (or cycles) of the signal. The normal horizontal sweep is used and the time per division of the display is the same as the actual time/div of the horizontal traverse. A large number of samples are taken during each cycle (or section of waveform displayed).

(ii) Clocked

Use is now made of the staircase waveform and a sample taken corresponding to each step. However, unlike equivalent time sampling many samples are taken of a single cycle and not one per cycle as in equivalent time sampling. The clock rate (which determines the number of steps in the staircase) may be at 100 kHz and may produce 1,000 samples per sweep.

TIME DOMAIN REFLECTOMETER

A sampling oscilloscope may be used for pulse reflection measurements. A pulse is fed into the cable by a pulse generator, part of the output going to the sampling oscilloscope. If there are any mismatches in the cable a reflection occurs, which can be seen and measured on the sampling oscilloscope.

18

Spectrum Analysers

SPECTRUM analysers are complex pieces of equipment and come in a wide range as complete instruments or plug-ins for normal oscilloscopes.

The idea of a spectrum analyser is to produce a display of voltage (Y) against frequency (X). To take an example, consider an amplitude modulated oscillator. The output will consist of three frequencies. The oscillator or carrier frequency f_c, together with two sidebands f_c+f_m and f_c-f_m, where f_m is the modulating frequency. If the output from the oscillator is fed into a suitable spectrum analyser, a display will be obtained of the amplitude of the three frequencies, on a horizontal frequency scale as shown in figure 18.1.

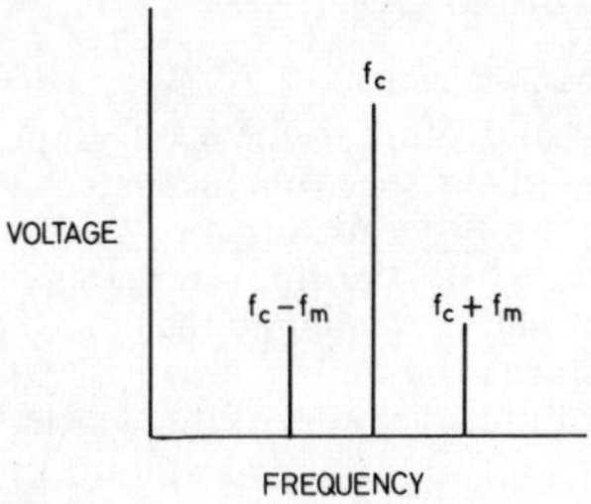

FIG. 18.1. DISPLAY OF AMPLITUDE MODULATED WAVEFORM ON SPECTRUM ANALYSER (TIME DOMAIN DISPLAY)

A spectrum analyser should not be confused with the use of a sweep generator or wobbulator in conjunction with an oscilloscope, as described in Chapter 13. Although the display is voltage against frequency on the oscilloscope screen in the case of the sweep generator, this is displaying the frequency response of a piece of equipment *e.g.* the response of an i.f. amplifier. The input to the amplifier is from the sweep generator, which varies in frequency. Its variation is linked to the horizontal deflection, so that the horizontal scale is in frequency. The output of the amplifier feeds the Y-deflection, so producing a display of the voltage output of the amplifier at various frequencies. If the input voltage is constant at all frequencies, the resultant display is the frequency response of the amplifier.

In a spectrum analyser the input can be from almost any source, commonly of radio frequency, or frequencies. With a normal oscilloscope we look at the way the signal amplitude varies with time and display this using a horizontal time scale, usually called a 'TIME domain display'. With the spectrum analyser we wish to display the input in terms of frequency, using a horizontal frequency scale. This type of display is known as a 'FREQUENCY domain display'. If we consider an amplitude modulated wave, this is commonly displayed in both ways in TIME domain as shown in figure 18.2 or in FREQUENCY domain as in figure 18.1. The input to the spectrum analyser might be from a transmitter, either amplitude or frequency modulated, when the various sidebands will be displayed. The degree of modulation can be determined together with any distortion. For example, if the transmitter produces outputs at harmonic frequencies these will be displayed on a suitable spectrum analyser. One may consider that the spectrum analyser is a radio receiver and the output is fed to produce a Y-deflection proportional to the carrier input to the demodulator (not the a.f. output). In fact, the same display could be obtained by using a receiver and tuning it over the range of frequencies concerned, noting the

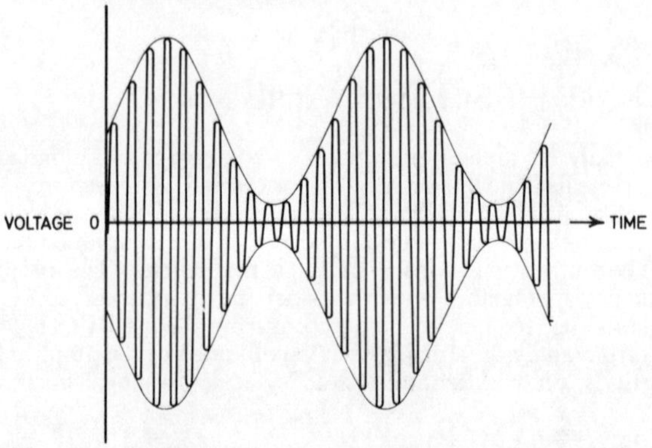

FIG. 18.2. MODULATED CARRIER (FREQUENCY DOMAIN DISPLAY)

output at each frequency and plotting them on graph paper. This is very laborious and the function of the spectrum analyser is to produce this graph automatically and display it on the oscilloscope screen. To do this we need to tune the receiver quickly over the frequency range concerned, feed the output to the Y-plates and link the tuning to the horizontal sweep.

The BASIC principle is shown in figure 18.3, which is really a superhet receiver. Assume that we wish to examine the range of frequencies from 100–200 MHz. This input is fed to a mixer, which is also fed with the output of an oscillator.

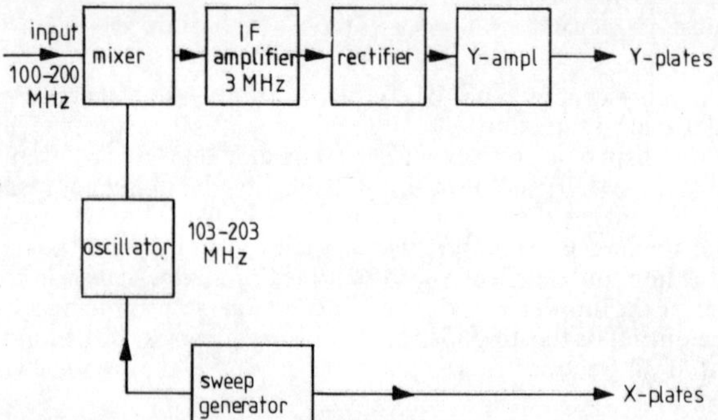

FIG. 18.3. BLOCK DIAGRAM OF BASIC SPECTRUM ANALYSER

The oscillator is varied in frequency by the sweep generator over the range 103 to 203 MHz. The output of the mixer is then fed to an i.f. amplifier having a centre frequency of 3 MHz. Thus, if the oscillator is, say, at 153 MHz, an input frequency of 150 MHz will cause a difference frequency of 3 MHz which is amplified by the i.f. amplifier. The output is rectified, amplified in a Y-amplifier, and fed to the Y-plates. Thus, at this frequency, the amplitude of the Y-deflection will be proportional to the input. As the oscillator frequency changes then the 'receiver' is looking at other input frequencies and producing the corresponding Y-deflections. The sweep generator, as well as varying the frequency of the oscillator, supplies the X-deflection. Thus any point on the horizontal scale will represent the frequency being looked at in the input. This simple arrangement is not practicable because of second channel interference. When the oscillator is at 153 MHz an input of 150 MHz will produce a difference frequency of 3 MHz to the i.f. amplifier, but also a frequency of 156 MHz. Hence, the analyser would look at two frequencies at the same time, and more complex arrangements must be used to resolve this problem.

The operating frequencies of spectrum analysers are very large and variable. The tuning range may be, for example, 1 kHz to 1 GHz, or 5 Hz to 50 kHz, or might be 10 MHz to 40 GHz. These figures are really the frequency range over which the 'receiver' will tune. The range of frequencies displayed on the screen, *i.e.* the horizontal range of frequencies, is also very variable and is known as the 'frequency span' (or dispersion). This is usually variable over a large range on a single spectrum analyser and might be from 200 Hz/div to 100 MHz/div. With a 10-division screen this means a span from one side to the other of the screen of 2000 Hz to 1000 MHz. The span is varied in steps over this range. The vertical display may be linear, *i.e.* the vertical deflection is proportional to the input. This is not very good for displaying inputs over a large range of amplitudes, hence a dB vertical display is commonly provided, which may cover, say, a range of 60 dB, *i.e.* a voltage ratio of 1000/1. In some cases a squared output may be used, *i.e.* an output proportional to the square of the input. This is useful if the magnitudes of the signals at various frequencies are nearly of the same amplitude. To be able to indicate frequencies which are closely spaced, the selectivity of the i.f. amplifier must be high. For example, if we wish to distinguish between two signals 100 kHz apart, the bandwidth of the i.f. amplifier must be less than this. This ability to separate frequencies is known as the 'resolution'. This is commonly made variable and may be linked to the frequency span (a higher resolution is required when the frequency span is small). This might be variable from, say, 30 Hz to 3 MHz. It must be realized that the figures quoted are possible examples; a particular analyser may vary very much from these figures.

To work with a large range of input voltage, suitable attenuators must be placed in the circuit so that overloading does not occur. The actual sweep SPEED (*i.e.* time to go from one side of the screen to the other) is variable and does not DIRECTLY alter the display. However, the maximum sweep speed has to be linked to the resolution required, owing to the response time of the various circuits.

In practice, multiple superhet circuits are used, one example being given in figure 18.4.

The input is fed to mixer M_1 (usually through a calibrated attenuator) and it is assumed that the tuning range is 400–800 MHz. The first i.f. amplifier

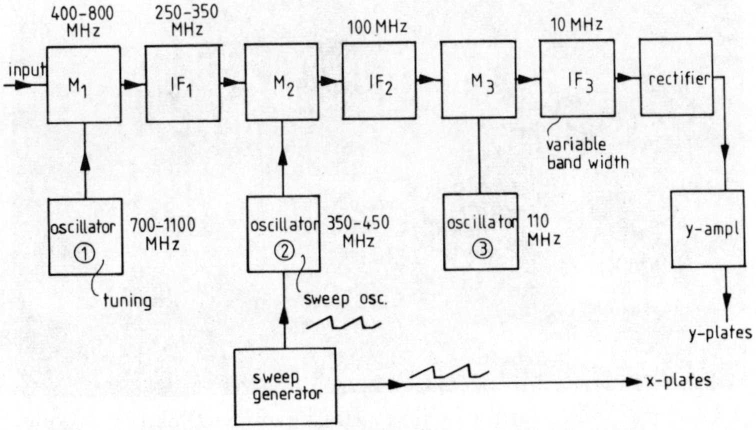

FIG. 18.4. MORE DETAILED BLOCK DIAGRAM OF SPECTRUM ANALYSER

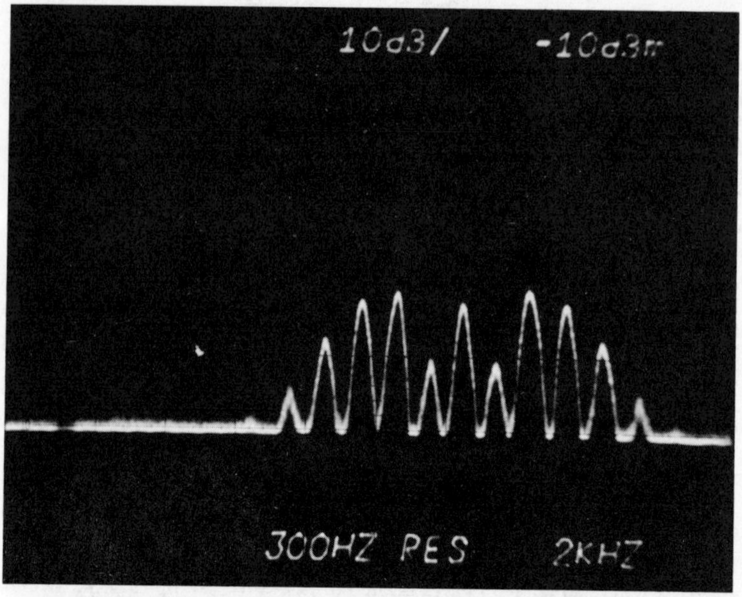

FIG. 18.5(a). DISPLAY WITH 100 MHz F.M. WITH APPROX. 8 kHz DEVIATION AND 1 kHz MODULATING FREQUENCY. 10 dB/DIV VERTICAL SCALE AND 2 kHz/DIV HORIZONTAL SCALE. RESOLUTION 300 Hz

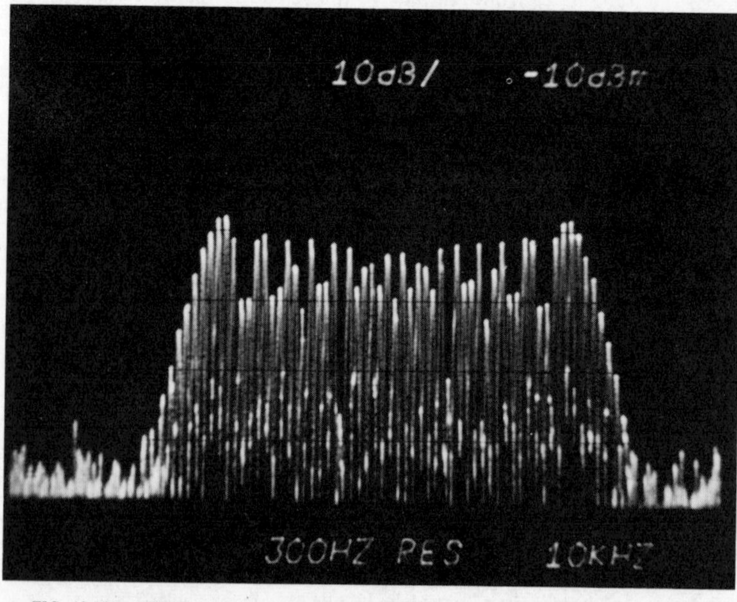

FIG. 18.5(b). DISPLAY WITH 100 MHz F.M. WITH APPROX. 25 kHz DEVIATION AND 1 kHz MODULATING FREQUENCY. 10 dB/DIV VERTICAL SCALE AND 10 kHz/DIV HORIZONTAL SCALE. RESOLUTION 300 Hz

(IF_1) has a centre frequency of 300 MHz, hence the oscillator must tune from 700 to 1100 MHz (*i.e.* 400 + 300 to 800 + 300 MHz). This is the main tuning control and settles the centre frequency of the display. Assuming a maximum frequency span of 100 MHz, the bandwidth of this i.f. amplifier must be 100 MHz so that all frequencies in this band are fed to the mixer M_2. This mixer is fed from an oscillator operating from 350 to 450 MHz to produce a second i.f. of 100 MHz. This oscillator is the sweeped oscillator, the frequency being varied in a linear way by the sweep generator, which also produces the horizontal deflection. The second i.f. amplifier IF_2 feeds a third mixer M_3, which is fed with a fixed oscillator frequency of 110 MHz to produce an i.f. of 10 MHz, which is fed to the third i.f. amplifier IF_3. This last i.f. amplifier is made with a variable bandwidth so as to produce different resolutions. It might have a bandwidth variable from 10 kHz to 1 MHz. The resulting output is then rectified and fed to the Y-amplifier and Y-plates. The rectifier and/or Y-amplifier must produce a linear output, a dB output, and sometimes a squared output.

One reason for using the multiple superhet principle is to get a large percentage frequency variation at the final i.f. amplifier to obtain adequate resolution. For example, consider two input signals at 500 and 501 MHz. They have a percentage difference at this frequency of only

$$\frac{1}{500} \times 100 = 0{\cdot}2\%.$$

With the first oscillator set to 800 MHz they will produce i.f. frequencies of $800 - 500 = 300$ and $800 - 501 = 299$ MHz. If the second oscillator is at 400 MHz, then one second i.f. would be $400 - 300 = 100$ MHz and the other $400 - 299 = 101$ MHz. When fed to the mixer M_3 the outputs would be $110 - 100 = 10$ MHz and $110 - 101 = 9$ MHz. The percentage frequency difference is now

$$\frac{1}{10} \times 100 = 10\%,$$

hence it is much easier to get the required resolution.

The diagram of figure 18.4 must be considered as an example, as different arrangements may be used (some of which are most complicated) in order to get the required performance. It is very important that spurious signals are kept to a minimum in the mixing processes or serious spurious displays appear on the screen.

Figure 18.5 shows displays obtained on a spectrum analyser of a frequency modulated signal with two different modulation indexes.

GLOSSARY

Accelerating Voltage: The voltage used to accelerate the electrons to a high velocity.

A.C. Coupling Position of Control (Trigger circuit)*:* The condition where the source of triggering is a.c. coupled to the trigger unit.

A.C. Fast Position of Control: Same as A.C.–L.F. reject.

A.C. H.F. Reject Position of Control: This is used in the trigger circuit and rejects the high frequency content of the signal fed *to the trigger circuit.* The d.c. component is also rejected.

A.C. L.F. Reject Position of Control: This is used in the trigger circuit to reduce the low frequency content of the signal *to the trigger circuit.* Also rejects d.c.

Add Position of Control: The control position when the inputs of two channels are added together before being displayed on the screen.

A Intensified by B Position of Control (Dual timebases)*:* A position where timebase B is delayed by timebase A, and timebase B brightens or intensifies the sweep produced by timebase A for sweep period of B. In some oscilloscopes the timebases are named the opposite way.

Alternate Display Position of Control: This is where, in a switched beam oscilloscope, one waveform is displayed on one timebase sweep and the other waveform on the next sweep, and so on.

Alternate Trigger Position of Control: The position where triggering is from signal A when A is being displayed, and from signal B when signal B is being displayed. The phase relationship between signals is now incorrect. May be used to display two voltages of unrelated frequencies.

Anode of Cathode-ray Tube: The electrode or electrodes, given a positive voltage, which accelerate the electrons and focuses them. An electron gun usually has three anodes.

Astigmatism: The condition where the spot is not circular. A control is usually provided (may be preset) to reduce any astigmatism.

'A' Timebase (Dual timebases)*:* The first (and often the normal) timebase which produces the delay for timebase B. In some oscilloscopes it is marked the other way round.

Attenuator: A device for reducing the voltage, usually by a known amount.

Automatic Trigger Position of Control: There are a number of variations. In general the timebase runs continuously when in this position without the need for trigger pulses. The display locks when trigger pulses occur.

Bandwith: The difference between the upper and lower frequencies at which the voltage response is 0·707 (-3 dB) of the response at a reference frequency. Usually upper and lower frequencies are stated rather than the difference. The reference frequency should be 20 times the lower frequency limit and $\frac{1}{20}$th of the upper frequency limit. The reference frequency may not be the same for upper and lower frequencies. If the lower limit extends to d.c. then the response should be the same on d.c. as at the reference frequency.

Beam Finder: A button which when pressed reduces the magnitude of the X and Y deflections so that the display is contained within the limits of the screen.

Beam Switch (Dual or multitrace oscilloscopes)*:* The electronic switch used in multitrace oscilloscopes to switch the beam from one input to the other.

Bistable Storage Tube: A storage tube producing only one degree of brightness, *i.e.* a portion of the trace is either displayed or not, there being no half-tones.

Blanking: When part of the trace, *e.g.* flyback, is not required to be seen it is blanked by suitable blanking pulses.

Bootstrap Sweep Generator: A sweep generator using positive feedback of unity value so that the voltage across the charging resistor remains constant. Hence the rate of rise of capacitor voltage in the CR timing circuit is constant with time.

Brightness Control: This control varies the brightness of the trace. It varies the voltage on the grid of the cathode-ray tube and hence varies the beam current.

Bright-up Pulse: The pulse fed to the grid or deflection blanking electrode to produce a trace.

'B' timebase (Dual timebases): The second timebase which produces the sweep after being delayed by the A timebase. In some oscilloscopes it is marked the other way round.

Burn: If a stationary stop of high brightness occurs this may burn the phosphor of the screen, resulting in a dark area when used later. It can also apply to the burning of the storage target in transmission type storage tubes.

Calibrate Position of Control: A position of a variable control when the calibration is correct.

Calibrator: A device which produces known voltages and currents in order to calibrate the oscilloscope. May be on the oscilloscope or a separate piece of equipment.

Chopped Display: In this method of display part of the waveform of one input is traced out and then part of that due to the second input. This is done rapidly at, say, 100 kHz to 1 MHz.

Common Mode Signal (Differential amplifiers): This is a signal which is fed to both inputs in the same phase. It should result in no deflection.

Current Probe: A probe used to display a current waveform on an oscilloscope.

Curve Tracer: A special type of oscilloscope to display characteristics of devices, usually diodes and transistors.

DC/AC Coupling Position of Control: This may be used on the Y-input. On D.C. both a.c. and d.c. are fed to the Y-amplifier and deflecting plates. On A.C. the d.c. is rejected and the deflecting is due to the a.c. component only, down to about 5 Hz.

D.C. Coupling Position of Control (Trigger circuit): The condition where the source of triggering is d.c. coupled to the trigger circuit.

Deflection Blanking: This method of blanking uses a small deflecting plate (usually on A_1) which deflects the beam so that it does not pass through the hole in the anode when the beam is required to be blanked.

Deflection Factor (Cathode-ray tube): The inverse of deflection sensitivity and is the deflection voltage required to produce a deflection of 1 cm for a given anode voltage.

Deflection Plates: The plates that produce the deflection. Two pairs are used: X and Y plates.

Deflection Sensitivity:

TUBE: The ratio of deflection to deflecting voltage in cm/volt for a certain anode voltage.

OSCILLOSCOPE: Often expressed as ratio of the deflection voltage to the deflection expressed in volts/division or mV/division. Should really be 'deflection factor'.

Delayed Gate Trigger Position of Control (Dual timebases): See *Delayed Triggered Sweep.*

Delayed Sweep: The use of two timebases, one to give a delay and the other to produce the sweep. In this way a portion of a waveform can be magnified so that detail can be seen.

Delayed Triggered Sweep Position of Control (Dual timebases): This is a condition where a delay is produced by timebase A, but it does not trigger timebase B. At the end of the delay time a gate is opened allowing timebase B to be triggered.

Delay time (Dual timebases): The delay produced by timebase A before operating (or gating) timebase B.

Delay Time Multiplier Control (Dual timebases): The control which varies the time that timebase A operates before triggering (or gating) timebase B. When timebase A sweep speed is multiplied by the reading of this control the resultant is the time delay.

Diode probe: A probe using a diode or diodes so that an r.f. signal is demodulated before being fed to the oscilloscope.

Dispersion: See *Frequency Span.*

Double-Gun Tube: A cathode-ray tube having two guns. May have separate X-plates or common X-plates.

Dual Trace: An oscilloscope producing two traces, *i.e.* has two Y-inputs. This may be produced by the use of a split-beam or double-gun tube or by the use of beam switching.

Electron Gun: The part of the cathode-ray tube used to produce and focus the electrons into a narrow beam. It may also include the deflecting plates.

Enhancement Control (Bistable storage tube): The use of pulses to bring the written part above the first crossover point of the secondary emission ratio curve. Enables a higher writing speed to be obtained.

Equivalent Time Sampling (Sampling oscilloscope): A method of sampling in which samples are taken on different cycles so that the actual timebase sweep is much slower than the equivalent time.

Erase (Storage oscilloscopes): This is the removal of the trace ready for the writing of another trace.

Fade Negative (Storage oscilloscope): A condition in which a part of the trace of screen begins to dim.

Fade Positive (Storage oscilloscope): This is when the background of a storage oscilloscope gradually fades in a bright direction.

Flood gun: The gun used in a storage oscilloscope to produce a uniform cloud of low velocity electrons on the target.

Fluorescence: The light produced on a phosphor screen when an electron beam falls on it.

Flyback: The return of the beam from the right-hand to the left-hand side of the screen after the completion of the sweep.

Focus Control: This control varies the focal length of the electrostatic lens, and hence the focus of the beam. The control varies the voltage on the second anode.

Frame of Oscilloscope: The case of the oscilloscope into which plug-ins can be inserted. Consists of the tube, power supplies and part of the X and Y amplifiers.

Frequency Span (Spectrum analyser): The frequency range covered by the display.

Gate Out Socket: The output socket which produces a pulse corresponding to the gating of a signal, *e.g.* might be a pulse corresponding to the sweep time.

Geometry Control: The deflections may not be truly recti-linear and the geometry control is used to improve this so that the trace does not show barrel or pincushion distortion.

Gnd (Ground)*:* The earth or usually the case of the oscilloscope, and the common terminal to X and Y inputs. When this is a position of a switch it indicates that the amplifier input is short circuited but not the input socket.

Graticule: This may be separate or internal. The separate graticule consists of a clear plastic sheet with suitable engravings, and is placed in front of the screen. In the case of the internal graticule this is engraved on the inside of the tube and has the advantage that parallax errors do not occur.

Graticule Illumination Control: The control that varies the brightness of the illumination of the graticule.

Grid of Tube: The electrode next to the cathode of an electron gun which controls the number of electrons going to the screen. It operates at a negative voltage with respect to the cathode.

Half-tone Storage Tube: A storage tube which will produce a half-tone display, *i.e.* various brightnesses in comparison with a bistable tube, which only produces one value of brightness (and zero brightness).

Helix PDA Tube: A PDA tube using a high resistance helical coating on the inside of the tube to give the required electrostatic field.

H.F.: Used for two different purposes:

(a) A trigger position used when the input is of high frequency (say above 1 MHz) and assists in triggering.

(b) A condition where only high frequency signals are fed to the trigger circuit, also known as 'l.f. reject'.

Hold-off Control: This varies the time after flyback before the timebase can be triggered again.

Input Impedance: The impedance looking into the input socket of the oscilloscope. The impedance is equivalent to a resistor and capacitor in parallel. For the Y-input may be 1 MΩ and 15–40 pF.

Integrate Control (Bistable storage tube)*:* The use of a number of repetitive traces to build up sufficient charge, the flood guns being switched off.

Intensity Control: A control which varies the intensity of the display. Same as BRIGHTNESS CONTROL.

Intensity Modulation: This is modulation of the intensity of the beam by an external signal (or internal in the case of blanking). This is done by application of a voltage to the grid (or cathode) of the tube.

Internal graticule: See *Graticule*.

Jitter: Variations in the position of the trace when displaying repetitive waveforms, usually on the X-axis. May be due to instability of signal or oscilloscope and causes the waveform to move slightly in horizontal direction.

Level control: The control which varies the amplitude of the voltage at which triggering occurs.

Magnified Sweep Control: The magnitude of the X-sweep is increased so that the corresponding time per division is reduced. May be ×5 or ×10.

Miller integrator: The use of negative feedback to produce a constant rate of rise of voltage across the charging capacitor of the sweep circuit. The charging capacitor is placed between the input and output of the amplifier.

Mixed A and B position of control (Dual timebases)*:* See *Mixed Sweep*.

Mixed Sweep (Dual timebases): This uses both timebases: the timebase A producing both the first portion of the trace, and the delay for timebase B. The second part of the trace is that due to timebase B and will be at a faster sweep speed.
Mode position of control: The method of operating the oscilloscope, *e.g.* a switch may be labelled MODE, the switch changing from 'Channel 1' to 'Channel 2' to Channel 1 + Channel 2', etc.
Monoaccelerator Tube: A tube where all the acceleration of the electrons is done before deflection, *i.e.* before the electrons enter the deflecting plates.
Multitrace Oscilloscope: An oscilloscope displaying more than one waveform on the screen, commonly two waveforms but may be up to four.

Non-real time sampling (Sampling oscilloscope): See *Equivalent Time Sampling*.

PDA Tube: A tube where post deflection acceleration is used, *i.e.* the electrons are accelerated after passing through the deflecting plates.
Persistence control (Storage oscilloscope): The time for the trace to disappear. This is usually variable by varying the erase pulses fed to the target.
Phosphor: The material used on the screen to produce the visible display.
Photographic Writing Time: The speed of the spot in, say, cm/μs, which will give a trace on the photographic film. This depends on the oscilloscope, the camera, and the film.
Plug-in: A unit, *e.g.* amplifier or timebase, that fits into a frame so making a complete oscilloscope. Various plug-ins can be used with the same frame.
Power Switch: The switch which switches the oscilloscope ON, *i.e.* supplies power to it from the mains or batteries.
Probe: A device used to connect the oscilloscope to the circuit under test. May be a voltage or current probe, and may be passive or active.

Ramp Generator: See *Sweep Generator*.
Random Mode Sampling: Where samples are taken at random.
Readout: The use of the beam to write alphanumeric displays on the screen as well as the normal trace.
Real Time Sampling (Sampling oscilloscope): Sampling which takes place on one or more cycles and the actual sweep speed is the real sweep speed.
Reset or ready position of control (Single sweep): The condition where the single trace sweep generator can be triggered.
Retrace: See *Flyback*.
Rise Time: The time for the voltage to rise from 10% to 90% of its final value.
Roll off: The manner in which the frequency response decreases, usually at high frequencies. For example, it may be 6 dB per octave.

Sampling Oscilloscope: An oscilloscope that takes samples of the waveform commonly one sample only from a particular cycle, and known as a 'non-real time oscilloscope'.
Sawtooth Out Socket: The output of the sweep generator.
Scan: See *Sweep*.
Sensitivity: Often used to express the voltage required to produce a deflection of 1 div or 1 cm, but to be correct should be the deflection produced by 1 volt. See also *Deflection Sensitivity*.
Sequential Sampling: Where samples are taken in sequence.
Set Speed Control: See *Speed, Set Control*.
Signal delay: The use of a delay line in the Y-channel longer than the triggering delay so that the first portion of the waveform can be seen.

Single Sweep: A method of operation of the sweep generator so that only one sweep is produced when triggered. It will not trigger again until reset.
Slope Control: The control which determines whether triggering takes place on a waveform with a positive slope or a negative slope.
Spectrum Analyser: A device for displaying the magnitude of the voltage input against a frequency scale on the oscilloscope.
Speed, Set Control: A preset control used to set the sweep speed to its correct value.
Split or Dual-beam Tube: A tube using a single gun, but the beam is split into two to produce two traces.
Stability Control: A control which varies the timebase between self-running, able to be triggered, and impossible to trigger. It must be set correctly for satisfactory operation of the timebase.
Storage Writing Speed: The maximum speed of the spot in cm/μs that will give a visible stored display on a single transient.
Store Position of Control (Storage oscilloscope): This is the condition where the charge pattern is stored (and there is no display) and the flood guns are off.
Sweep: The movement of the beam from the left-hand side to the right-hand
Sweep Generator:
(a) The device that produces the horizontal linear motion of the spot across the screen.
(b) A separate oscillator which is frequency modulated so that a display of voltage against frequency is obtained, *e.g.* response of amplifier.

Timebase See *Sweep Generator*.
Time/div or *Time/cm:* The speed of the sweep produced by the sweep generator.
Time Domain Reflectometer: A method, generally using a sampling oscilloscope, of measuring discontinuities in cables, etc.
Time Expander (Sampling oscilloscope): Expands a portion of the waveform, but maintains the same samples/division.
Trace Rotation Control: A control which varies the current in a coil surrounding the tube and causes the trace to rotate through a small angle so that the trace can be correctly positioned relative to the graticule.
Transmission Storage Tube: A storage tube using a target through which electrons can pass to produce a trace on the screen.
Trigger Coupling: The method used to couple the trigger source to the trigger circuit, *e.g.* a.c., d.c. and l.f. reject.
Trigger-in or External Trigger Socket: The socket by which external triggering signals can be fed into the oscilloscope.
Trigger Level Control: See *Level Control.*
Trigger Pulse: The pulse used to trigger the oscilloscope. This may refer to an internal pulse or the pulse applied to the trigger-in socket.
Trigger Source: The source from which the trigger pulses are obtained for triggering the sweep generator, *e.g.* it might be internal or external.
T.V. (Television) Position of Control: A position on the trigger input control that places a synchronizing separator in the trigger circuit so that the synchronizing pulses are extracted from the composite T.V. signal.
T.V. Field (Frame) Position of Control: A trigger position which uses a synchronizing separator and feeds the timebase with a pulse corresponding to the field synchronizing pulses.
T.V. Line Position of Control: A trigger position which uses a synchronizing separator and feeds the timebase with line synchronizing pulses.

Unblanking: The beam is normally cut off, *i.e.* blanked, and is unblanked for those periods when the beam is required to be seen, *i.e.* during the scan or sweep period.

Vectorscope: A special oscilloscope used to display U and V signals of a colour television signal.

View or Read Position of Control (Storage oscilloscope)*:* This is the condition where the storage trace is displayed. In some cases the display time is limited.

Volts/div. or *volts/cm Control:* The control that varies the sensitivity or deflection factor of the oscilloscope, *i.e.* the voltage required to deflect the spot one division or 1 cm.

Wobbulator: See *Sweep Generator* (b).

Write Position of Control (Storage oscilloscope)*:* The writing of the waveform by the writing gun using high velocity electrons.

X-position or Shift Control: The control that varies the horizontal position of the trace.

Y-axis Alignment Control: A control which varies the current in a coil so that the Y-deflection is exactly at right angles to the X-deflection.

Y-position or Shift Control: The control which varies the vertical position of the trace.

Z-axis Input: The input that produces intensity modulation of the trace, *i.e.* varies its brightness.

INDEX